Nano Materials Induced Removal of Textile Dyes from Waste Water

Authored by

Diptonil Banerjee

Department of Physics
Faculty of Engineering and Computing Sciences
Teerthanker Mahaveer University
Moradabad, UP
India

Amit Kumar Sharma

Department of Physics
Faculty of Engineering and Computing Sciences
Teerthanker Mahaveer University
Moradabad, UP
India

&

Nirmalya Sankar Das

Department of Physics (BSH)
Techno International - Batanagar
Maheshtala, WB
India

Nano Materials Induced Removal of Textile Dyes from Waste Water

Authors: Diptonil Banerjee, Amit Kumar Sharma and Nirmalya Sankar Das

ISBN (Online): 978-981-5050-29-5

ISBN (Print): 978-981-5050-30-1

ISBN (Paperback): 978-981-5050-31-8

need for a court order if at any point you breach any terms of this License Agreement. In no event will any delay or failure by Bentham Science Publishers in enforcing your compliance with this License Agreement constitute a waiver of any of its rights.

3. You acknowledge that you have read this License Agreement, and agree to be bound by its terms and conditions. To the extent that any other terms and conditions presented on any website of Bentham Science Publishers conflict with, or are inconsistent with, the terms and conditions set out in this License Agreement, you acknowledge that the terms and conditions set out in this License Agreement shall prevail.

Bentham Science Publishers Pte. Ltd.
80 Robinson Road #02-00
Singapore 068898
Singapore
Email: subscriptions@benthamscience.net

BENTHAM SCIENCE

CONTENTS

PREFACE

It has been a long time since Feynman's pioneering lecture unlocked the world of Nanomaterials. A large number of extensive research works have already enlightened the properties and key applications of Nanoscience. Nano-science has been strongly correlated with quantum physics and chemistry. Most of the basic features of Nanoscience are basically extended from those two classic subjects. Thus, proper expertise in basic Nano-science is still not independent; it often requires a strong basic understanding of traditional subjects like advanced physics and chemistry. On the other hand, Nanoscience has already been applied from the smallest electronic chip to the large displays of modern smartphones. To extend the applications of Nanoscience in a further variety of sectors, the correlation between Nanoscience and Nanotechnology must be opened up in front of the general engineering community in addition to classic science scholars.

Another key aspect of nano-science is its mimicking nature. Different nanostructures have strong similarities with the natural pattern, and spontaneously open up their possible applications in different fields. One of such fields which is a matter of most acute concern nowadays is the environment. Actually, in science and technology, whatever be the fields of research, ultimately has a connection with either energy or the environment. It is an irony that from the very advent of scientific and technological development, we always harnessed energy issues with the cost of the environment. Now it is nature's turn to repay the debt. As we know that today we are standing beneath a tremendous environmental crisis in terms of air, soil, water, noise and radioactive pollution, thus, a potential field of science and technology such as nanotechnology should be used to address the problem.

Keeping this in mind, this book brings a small effort of the authors to give the reader an idea of how nanomaterials may be used to address the environmental issue. As the entire topic is too vast, we have taken only a single part, *i.e.*, water pollution, and even after that, we have only concentrated the textile dye-induced water pollution and their negative impact and have shown how nanomaterial, can help us in getting rid of this problem through various processes.

This book is properly designed to solve basic queries of common academicians and technologists about the fundamentals of Nanoscience and nanomaterial-induced removal of textile dyes. Its basic concepts, chronological development and applications have been thoroughly discussed with appropriate examples and comparisons. We strongly believe that this effort shall be very important and useful for the budding engineers and scientists who are interested in the environmental aspect of Nanoscience.

The efforts have been made so that the style of the writing can be kept simple and easy to understand, and the essence of the subject can be fed even to a school student. Also, we have tried to keep the volume of the book reasonable so that the journey into this subject from the introduction to the advanced application can be finished within a couple of hours, say within a certain four-hour journey from Kolkata to Mumbai. In our previous book (by DB and NSD), "Nano Science - Concepts and Fundamentals" published by NOLEGEIN (an imprint of Consortium e-Learning ISBN: 978-93-87376-77-9), we put our effort into making the beginner interested in the subject for further study and offer deep learning about a specific topic. Now in this venture, we have taken advancement one step forward to deal with a particular topic likenanomaterial-induced removal of textile dyes from water.

The entire book has been divided into eight chapters, of which the first two chapters are mainly related to history and basic science behind the technology and resemblances between nature and nano-science. Chapter 3 has been dedicated to discussing the ways and means of seeing nanomaterials, *i.e.*, the basic principle of microscopies, mainly electron and scanning probe microscopes. Chapter 4 is all about the basics of textile dyes and their impact on the environment. In Chapter 5, we have given ideas about the basic structures and properties of a few nanomaterials having potential as dye remover. The next two chapters have detailed theories and experiments regarding ways and means of removing dyes through photocatalysis and adsorption. The last chapter is basically a concluding Chapter 5 discussing the efficiencies of different nanomaterials as dye remover.

We really enjoyed a lot writing this book and sincerely hope that readers will enjoy reading the book as well.

Last, it is a humble request of the author to the reader to kindly provide feedback, suggestions and unbiased, critical comment for further improvement of the book. Though we have tried our best to make the content of the book error-free still, if some mistakes are found, kindly let us know and thus help us to make the project perfect and error-free.

CONSENT FOR PUBLICATION

None.

CONFLICT OF INTEREST

The authors declare no conflict of interest.

Diptonil Banerjee
Department of Physics
Faculty of Engineering and Computing Sciences
Teerthanker Mahaveer University
Moradabad, UP
India

Amit Kumar Sharma
Department of Physics
Faculty of Engineering and Computing Sciences
Teerthanker Mahaveer University
Moradabad, UP
India

&

Nirmalya Sankar Das
Department of Physics (BSH)
Techno International – Batanagar
Maheshtala, WB
India

iv

ACKNOWLEDGEMENTS

It is really difficult to acknowledge the contribution of everyone behind a huge project like this. There are so many well-wishers and active contributors who have helped our project to take its final shape. The contribution may be actively or through moral support throughout the tenure. Still, few names deserve special acknowledgement in this section.

The authors first want to thank **Dr. K. K. Chattopadhyay**, Prof. Department of Physics, Jadavpur University, Kolkata, for his tireless effort to make the book information rich by providing various materials and data. The authors want to thank all the colleagues of the current institution for their kind help and support, and in this consequence, DB wishes to mention the name of **Dr. Ajay Kumar Upadhyay** (Associate Professor, Dept. of Physics, TMU) especially.

DB wishes to mention the name of a few students like **Debaleena, Dheeraj Kumar, Unmesha, Pratyusha, Dipsikha, Dimitra, Arun Kumar, Robin Chopra, Alok Kumar, Sudarshan Sarkar, Gulshan, Pramod** and many others who have done their work under the supervision of DB and generate data, many of which have been used here.

The authors want to thank all the **publishers** who have kindly provided permission to reuse different content already published in their different journals.

We are very thankful to our **family members** for providing constant moral support.

And last but notleast, it is the **GOD** who motivated us to start such a novel effort of spreading knowledge to others.

Diptonil Banerjee
Department of Physics
Faculty of Engineering and Computing Sciences
Teerthanker Mahaveer University
Moradabad, UP
India

Amit Kumar Sharma
Department of Physics
Faculty of Engineering and Computing Sciences
Teerthanker Mahaveer University
Moradabad, UP
India

&

Nirmalya Sankar Das
Department of Physics (BSH)
Techno International – Batanagar
Maheshtala, WB
India

CHAPTER 1

Introduction to Nanomaterials: Interaction with Water

Abstract: The first chapter helps the reader to get acquainted with the basic features of the nanomaterials. Here, the basic properties of nanomaterials, basic synthesis processes, and different applications of nanomaterials are discussed. The novel features of the material are also mentioned. How the suppression of degrees of freedom of electrons that affects the electrical, optical, and other properties, has been discussed in detail. Special emphasis has been given to the resemblance of nano-systems in mother nature, and as a result, few examples have been mentioned. As the objective of the book is to discuss the nanomaterial-induced removal of dye materials from water, thus the interaction between water and the nanomaterials plays one of the major roles. Keeping this in mind, we have discussed the concept of surface tension, surface energy, surface energy components, hydrophobicity, and lotus effect in much depth with their consequences to different particular applications. The discussion has also been done on the concept of contact angle, hysteresis, porosity, and related topic. Thus, this chapter familiarizes the reader acquainted with the basic characteristics, properties, and applications of nanomaterials and provide useful information regarding the interaction of nanomaterials with water, which is the central theme of this book.

Keywords: Contact Angle, Density of the State, Hydrophobicity, Hysteresis, Lotus Effect, Nanomaterial, Quantum Confinement.

INTRODUCTION TO NANOMATERIALS

The term "nano" does not refer to a substance, science, or technology. It is simply a dimension of 10^{-9} meters. Thus, any material having this dimension along any direction is called a nanomaterial. The science that deals with the dynamics of the material in nanoregime is called nanoscience, and the technology established manipulating nanomaterials is called nanotechnology. Nanotechnology is an interdisciplinary science in which material scientists, mechanical and electronic engineers, biologists, chemists & physicists work together to extend the nanoscale boundaries. Nano is a Greek word which means micro or small. Every particle whose size is 100 nm or smaller is considered a nanoparticle. Nanoscience is the study of the fundamental principles of a molecule and structure with at least one dimension, roughly between 1 & 100 nm.

Diptonil Banerjee, Amit Kumar Sharma and Nirmalya Sankar Das

In 1959, Richard Feynman, an American physicist at the California Institute of Technology, said in one of his lectures, *"There's plenty of room at the bottom."* These lines laid the foundation for nanotechnology, which is why Richard Feynman is known as the "Father of Nanotechnology." To indicate how small an object is, we compare it to human hair. We know that the average diameter of human hair is about fifty thousand nanometres. In comparison, the smallest features etched on a commercial microchip are typically less than 100 nm. The human eye can resolve objects with a diameter up to 10,000 nm. Fig. **(1.1)** shows the size dependence of different objects in the universe.

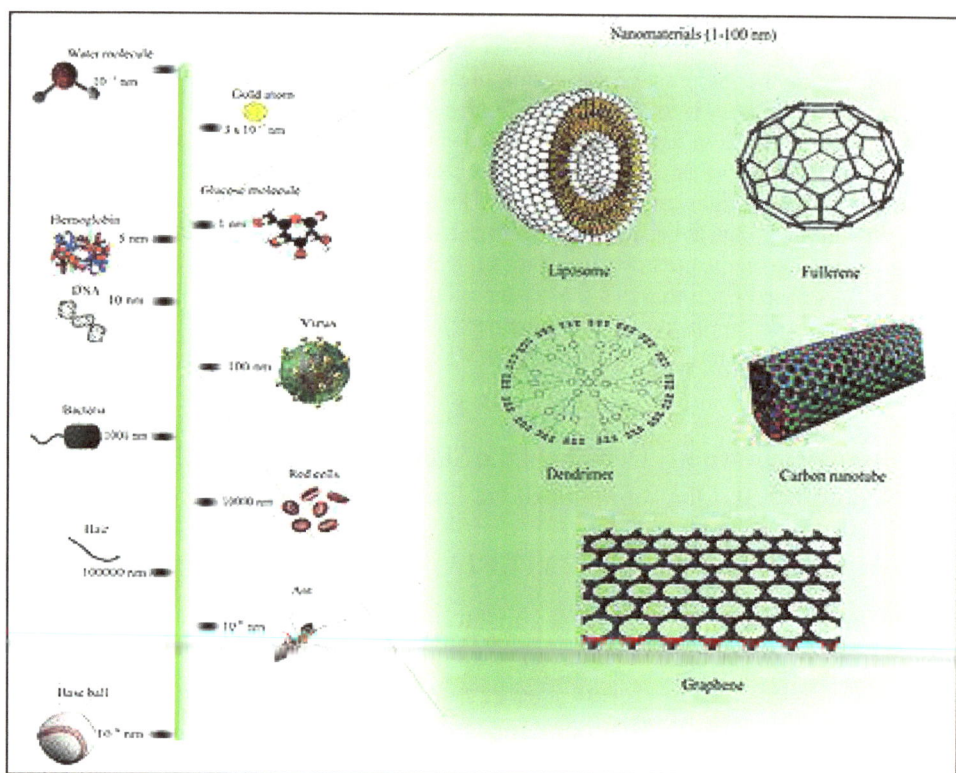

Fig. (1.1). Size dependence of different objects in the universe.

Nanoscience is a study that deals with small things and the use of nanotechnology in various places. The chemical and physical characteristics of nanoparticles change due to their extremely small sizes. For example, the cubes of sugar are less sweet than the castor sugar because the surface area of the castor sugar is larger than the cubes of sugar. With the help of nanotechnology, we have the ability to control atoms and molecules. The materials which we use in nanotechnology are called nanomaterials.

Following are the reasons for unique properties seen in nanomaterials compared to their bulk form:

The Higher Ratio between Surface and Volume

The very high surface-to-volume ratio gives rise to higher numbers of dangling bonds, which, in turn, develop very high chemical activities within the nanomaterial, making them different from its bulk form (detailed mathematical treatment has been given in the next chapters).

Quantum Effect

From the very basic particle in a box problem one faces in quantum mechanics, it is seen that the energy gap between the two successive energy levels or rather any two energy levels, depends inversely on the square of the box dimension, thus enhancing the optical gap of the material. This optical gap governs the optical properties of the nanomaterial (detailed mathematical treatment has been given in the next chapters).

Density of the States

Density of the state, *i.e.*, the numbers of states per unit energy interval per unit volume, are highly dimension dependent and thus shows marked differences for 0, 1, 2, and 3-dimensional structure of the same material (detailed mathematical treatment has been given in the next chapters).

It is to be noted that there are number of quality books related to the basic properties of nano, even by the authors themselves. Thus, here the same particular topics are not being focussed in much depth. However, the reader may focus on the following reference [1-4].

Synthesis of Nanomaterial

In order to explore the unique physical properties & phenomena and also to realize the useful applications of nanostructures and nanomaterial, the ability to fabricate and process nanomaterial and nanostructures is the first hurdle in nanotechnology.

The following schematic diagram in Fig. **(1.2)** shows the two significant approaches in synthesizing nanomaterial detail, which has been discussed in the the next chapter.

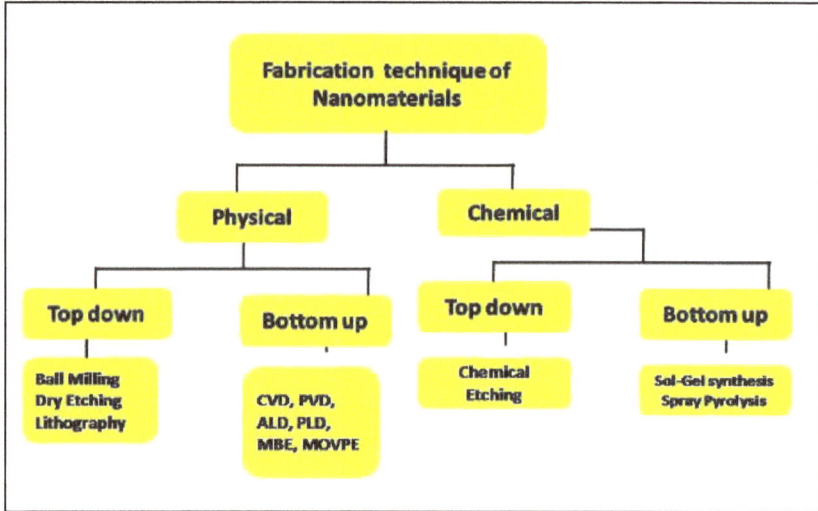

Fig. (1.2). Schematic diagram of different synthesis approaches of nanomaterials.

For synthesizing and processing nanomaterial and nanostructures, the following challenges must be faced carefully:

1. Overcome the huge surface energy, a result of enormous surface area or large surface to volume ratio.
2. Ensure all nanomaterials with the desired size, uniform size distribution, morphology, crystallinity, chemical composition, and microstructure which results in desired physical properties.
3. Prevention of nanomaterial and nanostructures from coarsening through either Ostwald ripening or agglomeration.

Many technologies have been explored to fabricate nanostructures and nanomaterials. These technical approaches can be grouped in several ways, like growth media or the form of products.

According to the Growth Media

This is subdivided into three parts as per the three states of matter.

a. LASER induced pyrolysis that used for nanoparticle synthesis or atomic layer deposition; excellent means for ultra-smooth, thin film deposition is the example of vapour phase growth.

b. Self-assembled monolayers or colloid-assisted development of nanoparticles may be achieved through the Liquid phase growth.

c. The last remained part is, of course, no doubt Solid phase formation that includes phase segregation that facilitates dispersion of metal nanoparticle into the matrix of amorphous glass.

According to the Form of Products

a. Nanoparticles by means of colloidal processing, flame combustion and phase segregation.

b. Nano rods or nanowires by template-based electroplating, solution liquid- solid growth (SLS), and spontaneous anisotropic growth.

c. Thin films by molecular beam epitaxy (MBE) and atomic layer deposition (ALD).

Top-Down and Bottom-Up Approach

The top-down approach uses traditional methods to guide the synthesis of Nanoscale materials [5]. The paradigm proper of its definition generally dictates that in the top-down approach, it all begins with a bulk piece of material, which is then gradually or step by step removed to form objects in the nanometre-size regime. Well-known techniques, such as photolithography and electron beam lithography, anodization, and ion and plasma etching, all belong to this type of approach. The bottom-up approach is exactly the opposite of top-down approach. In this case, instead of starting with large materials and chipping them away to reveal small bits, it all begins with atoms and molecules that get rearranged and assembled for the formation of the large nanostructures. It is the new paradigm for synthesis in the nanotechnology world as the bottom-up approach allows the creation of diverse types of nanomaterial, and it is likely to revolutionize the way of material fabrication. A schematic representation of Bottom-up and Top-down approaches is shown in Fig. (**1.3**).

Nanoparticle Synthesis

There are two approaches for synthesis of nanomaterials and the fabrication of nano structures.

Top-Down approach   **Bottom-Up approach**
                                          (or self-assembly approach)

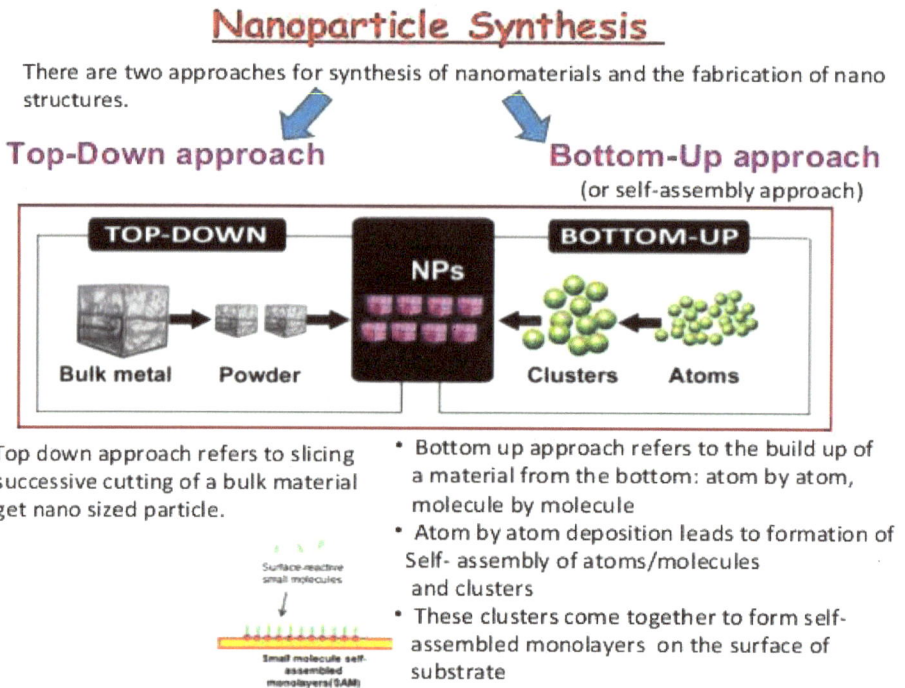

- Top down approach refers to slicing or successive cutting of a bulk material to get nano sized particle.

- Bottom up approach refers to the build up of a material from the bottom: atom by atom, molecule by molecule
- Atom by atom deposition leads to formation of Self- assembly of atoms/molecules and clusters
- These clusters come together to form self-assembled monolayers on the surface of substrate

Fig. (1.3). Schematic diagram of Top Down and Bottom Up Approach [5].

Attrition or Milling is a top-down method in making nanoparticles, whereas colloidal dispersion is a good example of bottom-up approach in the synthesis of nanoparticles.

Self Assembly

One of the bottom-up methods is nature's way; —*self-assembly*. Self-organizing processes are common throughout nature and involve components from the molecular (*e.g.*, protein folding) to the planetary scale (*e.g.*, weather systems) and even beyond (*e.g.*, galaxies). The key to using self-assembly as a controlled and directed fabrication process lies in designing the components required to self-assemble into desired patterns and functions. Self-assembly reflects information coded - as shape, surface properties, charge, polarizability, magnetic dipole, mass, *etc.* in individual components; these characteristics determine the interactions among them [6].

Nanomaterials can be classified dimension-wise into the following categories shown in Table **1.1**.

Table 1.1. Classification of Nanomaterials.

Classification	Examples
Zero dimension< 10nm	Particles, quantum dots, hollow Spheres, *etc.*
One dimension < 100nm any two dimensions	Nanorods, nanowires, nanotubes, *etc.*
Two dimensions < 100nm in any one dimension	Flakes, sheets, *etc.*

History of Nanomaterials

There are many evidences present which it is proved that nanotechnology was used even in ancient time, of course, without having any knowledge about it. The Lycurgus Cup, shown in Fig. **(1.4)** is the first nanomaterial used unknowingly by Roman in the 4th century.

Fig. (1.4). Lycurgus Cup [7].

Table **1.2**, along with Fig. **(1.5)**, summarizes some important points from the history of nanotechnology according to their timeline:

There are other examples of the ancient use of nanomaterials, like a Damascus sword that used steels of Indian origin containing carbon nanotubes. These CNTs give the sword exceptional flexibility and sharpness. The picture of the sword is shown in Fig. **(1.6)**.

Table 1.2. Some milestones in the history of nanotechnology, according to their timeline.

2000 years ago	Nanocrystal of Sulphide used by Greeks and Romans to dye hair.
1000 years ago	Gold nanoparticles of various sizes used to fabricate many colours in stained glass windows.
1959	Physicists Richard Feynman gives the sentence, "There's plenty of room at the bottom."
1974	The word "Nanotechnology" was first time used by professor Norio Taniguchi.
1981	IBM develops Scanning Tunnelling Microscope.
1986	The first book on nanotechnology was named "Engines of creation" which was written by K.E. Drexler.
1991	S. Ijima discovered Carbon nanotubes.
1999	R. Freitas wrote the first book on nanomedicine named "Nano Medicine".
2000 The The The The The The The	The National Nanotechnology Initiative was launched by the US

Applications of Nanotechnology

Nanotechnology has several applications in science and technology, as shown in Fig. **(1.7)**.

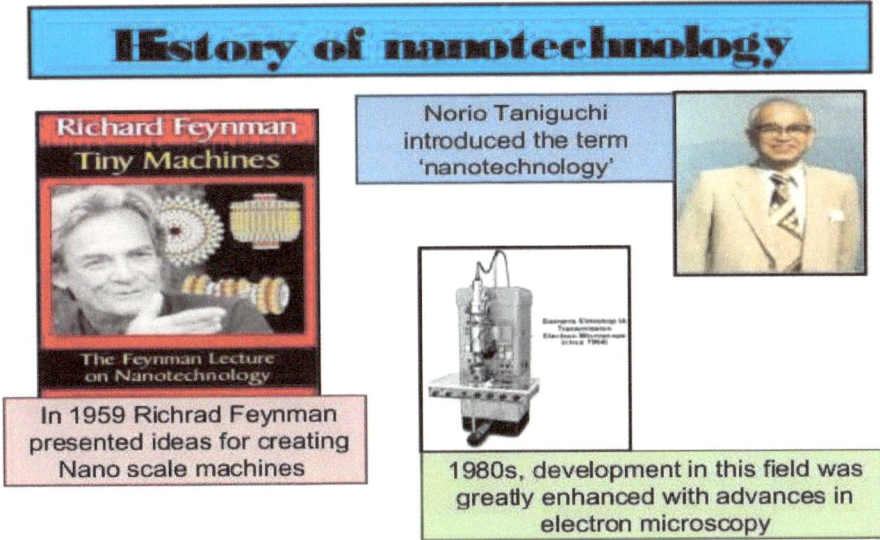

Fig. (1.5). Some key facts about the history of nanotechnology.

Fig. (1.6). Photograph of Damascus Sword [8].

Few specific fields are:

a) Energy

Nanotechnology has brought a revolution in the energy sector, which is one of the most serious concerns in the present day. For instance, nanomaterial-based illumination technology can reduce the power significantly while using LED compared to the ordinary bulb. The latter can use only 5% of the electricity fed to it. In another example, if we talk about present solar-driven cars, the best one can covert 40 % of the solar energy, whereas the ordinary commercial car, it cannot exceed 15–20 %. The use of Nanomaterials with proper band gap can help conversion of the energy in many high extend.

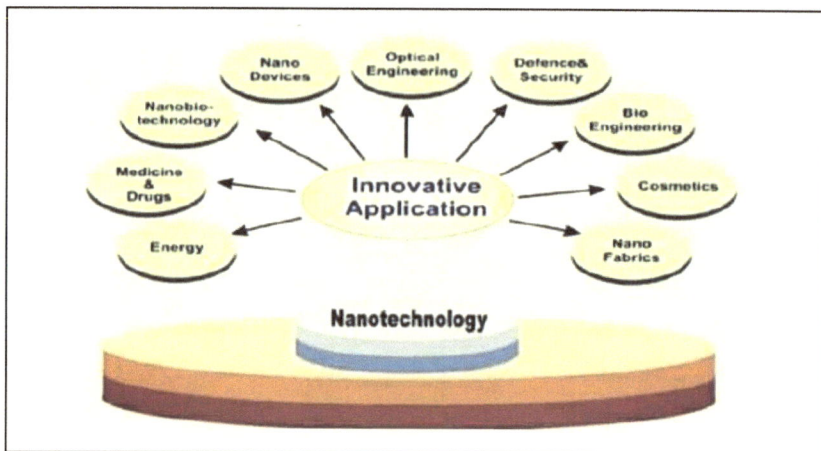

Fig. (1.7). Applications of Nano technology.

b)   Defence and Security

The application of nanotechnology has enough prospects to redesign the present defense and security system. This is possible by miniaturizing the deadly weapons or developing unmanned vehicles for combat. Also, there is plenty of scopes to develop different sensors, display, satellite component and others, all of which may prove their importance in a different part of defense and security system.

c)   Information & Communication

Conventional silicon-based electronics are now somewhat obsolete due to its very high cost and other related issues. The nanomaterial-based development of transistor, capacitors and other components are so far the hope for the new era of electronics. Also, researchers are now being able to develop molecular transistors, which shows hope in this sector in the coming years.

d)   Medicine

The effectiveness of nanomaterials in the field of medicine and drug delivery came to the mind of the researchers from the fact that the size of nanomaterials is almost similar to the most of the biological molecules. It thus readily suggests that nanomaterial-based technology can effectively be used *in-vivo* and *in-vitro* medical applications. There is a remarkable success of nanomaterial-based treatment that offers targeted drug delivery to a particular position of the body where cancer cells evolve. The other application is the development of contrast agents, or different analytical tools, tools that are needed for physical therapy and many others.

Nature and Nanotechnology

Although with the discoveries of different sophisticated electron microscopes, the concepts of nanomaterials have emerged in the last few decades, and the associated nanotechnologies have replaced the conventional concepts of microelectronics in large aspects, but surprisingly from the very beginning of the creation of the universe nature has marked resembles with few concepts of the nanomaterials and related science and technology.

For example, we are quite familiar with the term an ecological pyramid as shown in Fig. (**1.8**).

Here it is seen that (in general) from primary to tertiary consumers, bigger is the size of the consumer lesser is its amount of activity. Just in analogy in material science, it is also seen that the more "nano" the material is, more is its activity.

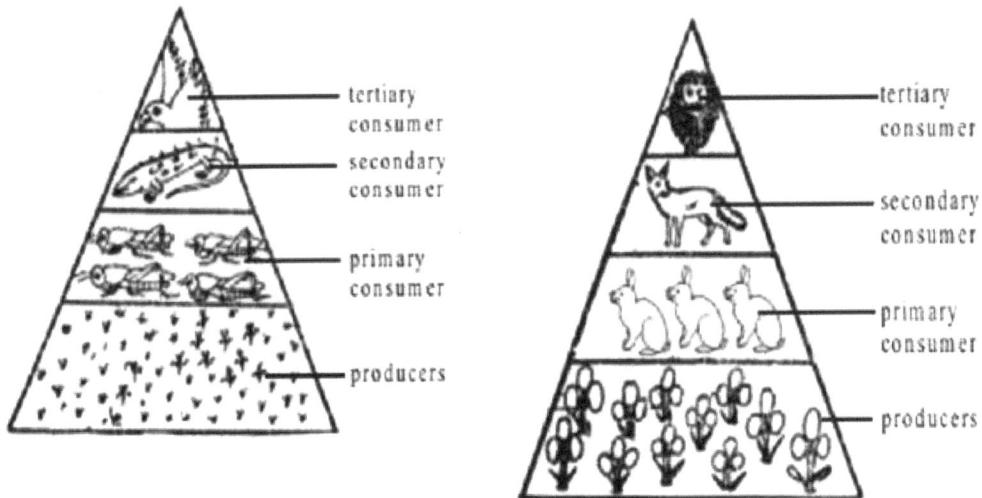

Fig. (1.8). Ecological food pyramid .

Also, when one works with simple Nasturtium leaf, a small component of nature, and studies it under high-resolution microscopes, one would see that as magnification is increased higher and higher, more and more fine structures of the crystals get unveiled, as shown in Fig. (**1.9**) [9].

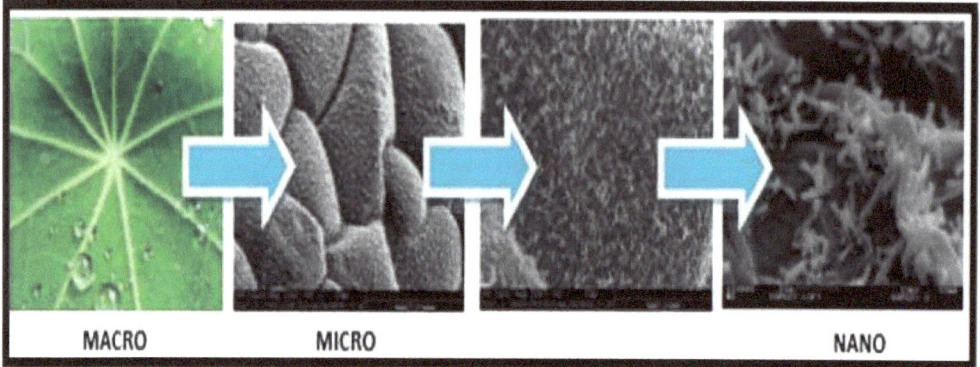

MACRO MICRO NANO

Fig. (1.9). Close-up views at progressive magnification of a Nasturtium leaf revealing the presence of surface nanocrystals [9].

Another example of the natural "Nano" effect is seen in Gecko, who can walk on almost any substrate in any direction, vertically upward or at a certain angle. It can hang upside down on a smooth surface like cleaning glass. Walking on a dirty or wet surface is also not a problem for them. They, in principle, do not have any suction-like features and do not produce any sticky substance. Here it is to be noted that Gecko has a series of structures called sensors containing numerous projections called septa. Each site has a dimension of 5 (m in diameter and 100 (m in length and consists of number of tiny projections called spatulas of dimension 200 nm. This helps the surface area of the Gecko's feet increase enormously. Also, the spatula is very flexible; thus, they can be molded into the molecular structure into any substrate, increasing the adhesion, which is completely Van der Waals force. Also, as the surface energy of such structures are such, the particles, dirt or mud will prefer to stay on the substrate rather than stick on the feet. Thus the feet of the Gecko remained clean. The structure can clearly be seen in the below figure (Fig. **1.10**).

What Distinguishes Nanomaterial from Bulk?

It is to be noted that though the micro-form of the material shows almost the same properties as that of the bulk material, its nano-form shows significantly different properties and this is mainly because of the following reasons:

1. In nano-dimension the numbers of atoms at the surface increase significantly.

2. It has a surface area much higher compared to its bulk form.

3. It shows confinement effect and thus very different optical properties.

4. It has a different density of state in 0, 1, 2, 3 dimensional material and thus fascinating electrical properties.

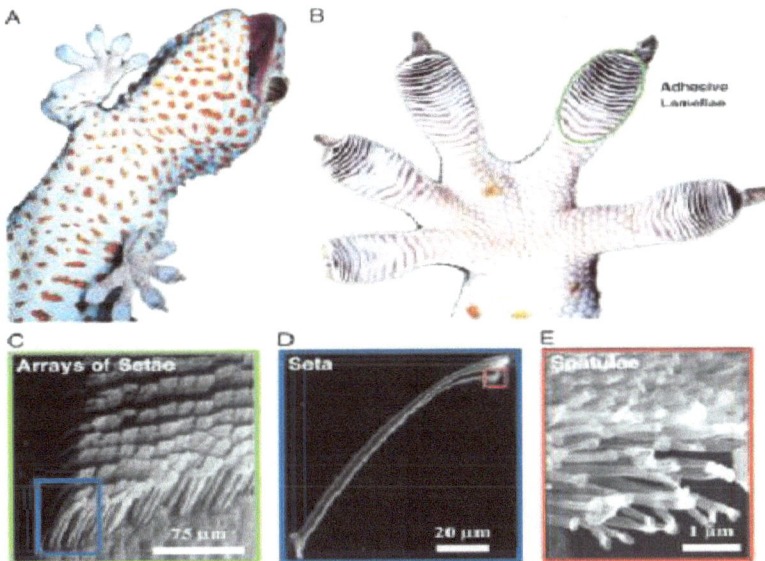

Fig. (1.10). (A-E) The macroscopic and nanoscopic structures of Gecko's feet with different magnifications.

The advantages of using nanomaterials firstly involve the miniaturization of the system. In this present era when we all want the entire world within our arm. This supports the Moor's law speculating numbers of transistor gets increased exponentially obviously forcing the dimension of the electronic component to be reduced.

The second advantages or rather the second characteristics of nanomaterial, are the enhanced surface area and thus many more numbers of dangling bonds present in the surfaces exposed to the environment. These enhanced dangling bonds are basically chemically unstable and are always in need of interaction with other materials in order to make the bond and thus get the minimum energy configuration. This helps nanomaterial to be exceptionally chemically active and performance in any surface-induced properties like catalysis, adsorption, *etc.*, is exceptional. As mentioned before, as a result of the changes that occur in particles with reduced particle size, nanomaterials can have extremely high biological and chemical reactivity. For example, catalytically active nanomaterials enhance either chemical or biochemical reactions by tens of thousands, and even a million times. This attribute explains, even 1 g of nanomaterial, can be more effective than 1 ton of a similar but macro substance.

Another aspect we must take into account is that a free surface is a place of accumulation (sink) of crystallographic defects. At small particle sizes, the surface concentration of such defects increases considerably. Hardeveld and Hartog, in 1969, calculated classically and showed that the largest changes of proportions between facets, edges, corners, and micro defects at the surface occur between 1 and 5 nm. As a result, strong lattice distortion and even a change of lattice-type can take place on the surface layer. In fact, due to the accumulation of structural defects and chemical impurities on the surface, we can observe the purification of the bulk area of the nanoparticles.

An important specific characteristic of nanomaterial properties (we mean here polycrystalline materials with grain size less than 40 nm) is an increase in the role of interfaces with the decrease of the size of grains or crystallites in nanomaterials. Experimental research has shown that the state of grain boundaries has a non-equilibrium character, conditioned by the presence of a high concentration of grain boundary defects. This non-equilibrium is characterized by extra energy of the grain boundaries and by the presence of long-range elastic stress. At the same time, the grains have ordered crystallographic structure, while the grain boundary defects act as a source of elastic strains. Non-equilibrium of the grain boundaries initiates the occurrence of the lattice distortion, the change of interatomic distances, and the appearance of sufficient displacement of atoms, right up to the loss of an ordered state. Another important factor peculiar to nanoparticles is their tendency to aggregate. The possibility of migration (diffusion) of either atoms or groups of atoms along the surface and the boundaries, as well as the presence of attractive forces between them, often leads to processes of self-organization into various cluster structures. This effect has already been used for the creation of ordering nanostructures in optics and electronics.

One more important aspect of nanomaterial properties is connected with the fact that, during transport processes (diffusion, electro- and thermal conductivity, *etc.*), there are certain effective lengths of free path of a carrier of this transport (Le), such as photon and electron mean free paths, the Debye length, and the exciton diffusion length for certain polymers. While proceeding to sizes smaller than Le, transport speed starts to depend on both the size and the shape of the nanomaterial; in general, the transport speed increases sharply.

The principal characteristics of nanomaterials are conditioned by not only by their small the size, but also by the appearance of new quantum mechanical effects in a dominating role at the interface (Esaki 1991; Serena and Garcia 1997. Those quantum size effects occur at a critical size, which is proportionate with the so-called correlative radius of one or another physical phenomenon, for example, with the length of the free path of electrons or photons, the length of coherence in a superconductor, sizes of magnetic domains, and so on. As a rule, quantum size effects appear in materials with crystalline sizes in the nano range D < 10 nm. As a result, in nanomaterials with characteristic size, one can expect the appearance of effects that cannot be observed in bulk materials.

Interaction between Solids and Liquids

Surface Energy

Surface energy quantifies the disruption of intermolecular bonds that occur when a surface is created. It can also be defined as the excess energy at the surface of a material compared to the bulk. The surface energy across an interface or the surface tension at the interface is a measure of the energy required to form a unit area of new surface at the interface. The intermolecular bonds or cohesive forces between the molecules of a liquid cause surface tension. When the liquid encounters another substance, there is usually an attraction between the two materials. The adhesive forces between the liquid and the second substance will compete against the cohesive forces of the liquid. Liquids with weak cohesive bonds and a strong attraction to another material will tend to spread over the material. Liquids with strong, cohesive bonds and weaker adhesive forces will tend to form a droplet when in contact with another material. The surface free energy of materials is a characteristic factor, which affects the surface properties and interfacial interactions such as adsorption, wetting, adhesion, *etc.*

Wetting

The property of a liquid to keep in touch with a solid surface arising from the interaction between molecules of both solids and liquids at their actual points of contact is known as wetting. The degree of wetting is measured by the energy balance between the grip and the bonding force. The bonding force between the liquid and the solid determines the shape that the droplet would take and the corresponding angle of contact. Cohesive forces within the liquid cause the drop to ball up and avoid contact with the surface.

Surface energy comes into play in wetting phenomena to check the wet condition, take a drop of water on a solid specimen. When the surface strength of a specimen differs due to the addition of a drop, the specimen is known as wetting. The distribution variable can be used to measure this statistically:

$$S = Y_s - Y_l - Y_{s-l} \qquad\qquad (1.1)$$

where S is the spreading parameter,

Y_s is the surface energy of the substrate,

Y_l is the surface energy of the liquid and

Y_{s-l} is the interfacial energy between the substrate and the liquid.

If, $S < 0$ the liquid partially wets the substrate and if $S > 0$ the liquid completely wets the substrate.

Contact Angle

The contact angle (θ) is defined as the angle made by the intersection of the liquid/solid interface and the liquid/air interface. It can be alternately described as the angle between solid sample's surface and the tangent of the droplet ovate shape at the edge of the droplet.

A high contact angle indicates low solid surface energy, and there is a low degree of wetting. A low contact angle indicates high solid surface energy, and a high or sometimes complete degree of wetting. Some of the factors affecting the contact angle are roughness, functional group present on the surface, impurities, porosity and surface energy.

One way to quantify a liquid's surface wetting characteristics is to measure the contact angle of a drop of liquid placed on the surface of an object. Liquids wet surfaces when the contact angle is less than 90 degrees. For a penetrant material to be effective, the contact angle should be as small as possible. In fact, the contact angle for most liquid is very close to zero degrees.

In liquid penetrant testing, there are usually three surface interfaces involved, the solid-gas interface, the liquid-gas interface, and the solid-liquid interface. For a liquid to spread over the surface of a part, two conditions must be met. First, the surface energy of the solid-gas interface must be greater than the combined surface energies of the liquid-gas and the solid-liquid interfaces. Second, the surface energy of the solid-gas interface must exceed the surface energy of the solid-liquid interface. A penetrant's wetting characteristics are also largely responsible for its ability to fill a void. Penetrant materials are often pulled into surface-breaking defects by capillary action. The capillary force driving the penetration into the crack is a function of the surface tension of the liquid-gas interface, the contact angle, and the size of the defect opening. Table **1.3** summarizes the relation between contact angle and solid-liquid interaction.

Table 1.3. Description of contact angle, degree of wetting and strength between different interactions.

| Contact Angle | Degree of Wetting | Strength of | | Pictorial Representation |
		Solid/liquid Interaction	Liquid/liquid Interaction	
$\theta = 0°$	Perfect wetting	Strong	Weak	
$0 < \theta < 90°$	high wettability	Intermediate	Intermediate	

(Table 1.3) cont.....

$90° \leq \theta < 180°$	low wettability	Weak	Strong	
$\theta = 180°$	Perfectly non-wetting	Weak	Strong	

Solid surfaces have been divided into high-energy solids and low-energy types. Solids such as metals, glasses, and ceramics are known as 'hard solids' because the chemical bonds that hold them together (*e.g.*, covalent, ionic, or metallic) are very strong. Thus, it takes a large input of energy to break these solids, so they are termed "high energy". Most molecular liquids achieve complete wetting with high-energy surfaces. The other type of solids is weak molecular crystals (*e.g.*, fluorocarbons, hydrocarbons, *etc.*), where the molecules are held together essentially by physical forces (*e.g.*, van der Waals and hydrogen bonds). Since these solids are held together by weak forces, it would take a very low input of energy to break them, and thus, they are termed "low energy".

Hydrophobicity and Hydrophilicity

A hydrophilic molecule is one that has a tendency to interact with or be dissolved by water and other polar substances. This property is known as hydrophilicity. In other words water gets attracted by hydrophilic materials and the associated substance readily form a bond with water. This bond is found to be hydrogen bond due to the higher electro-negativity of the –OH group. It is observed that (will be discussed later in detail) a hydrophilic material has overall high surface energy and contact gets varied just reverse way as the polar part of the surface energy. One of the most common examples of hydrophilic substances is sugar that we use in daily cooking. One can see that if sugar is readily kept open in the atmosphere, it absorbs water and becomes sticky. Hydrophilic materials found their application in various fields and industries that, include medical, coating, mechanical and many others. There are also hydrophilic materials that don't get completely dissolved into water, initiating the concept the colloid.

Hydrophobicity ("having a horror of water") is the physical property of a molecule that is seemingly repelled from water. A hydrophilic molecule or portion of a molecule is one that has a tendency to interact with or be dissolved by water and other polar substances. Hydrophobic molecules have a tendency to repel water and tend to be non-polar and, thus, prefer other neutral molecules and non-polar solvents. Super-hydrophobic surfaces have contact angles greater than 150°, showing almost no contact between the liquid drop and the surface. This is sometimes referred to as the "Lotus effect".

There arc also substances like soap molecules that simultaneously have hydrophobic and hydrophilic characters due to the presence of both polar and non-polar parts. Thus, they can simultaneously attract water as well as soap. These molecules are called amphiphilic.

Lotus and Petal Effect

The term lotus effect comes from the non-wetting behaviour of the lotus leaves that do not get dunked in water even after they spend their whole life in the muddy water. It is seen that the lotus leaves are super hydrophobic in nature and offer a contact angle as high as 170 °. This lotus leaf is one of the most efficient materials to offer self-cleaning (to be discussed in detail in the coming sections.) Where the dirt particle sitting on the surface gets collected by the water droplets during rolling over the surface due to the extreme resilience of the surface towards the water. Thus the surface remained clean. Few other plants like Tropaeolum (nasturtium), Opuntia or Alchemilla, can also show superhydrophobicity as well as self-cleaning properties. Not only plants, but different living animals, especially insects, also show such low adhesion properties to water. Though the lotus effect had been a much-known phenomenon from the last couple of hundreds of years, it was not until 1977, when Barthlott and Ehler studied the phenomena of water adhesion in detail, the people of the scientific community called the self-cleaning effect and superhydrophobicity "lotus effect" (Fig. **1.11a**).

On the contrary, there are instances seen in nature, most prominently in rose petals, where the water shows a very low contact area with the surface, but also shows a very high rolling angle suggesting self-cleaning is not the property that one can expect from such surfaces. In this kind of structure, the rolling angle can be so high that water droplets can exist on the surface, even if its orientation is made changed to upside down. This is called "Petal effect".

It has been seen that the rose petals have a particular hierarchy of micro- and nanostructures that helps the surface to achieve certain roughness values and equilibrium states adequate to offer a very high adhesive force to the water droplet. One can now also develop surfaces that resemble the rose petal structures consisting of a series of micropapillae and that, in turn, are divided into numbers manifolds. In such a surface, as described just before, water simultaneously shows quite a large contact angle and a high rolling angle (Fig. **1.11b**).

Fig. (1.11). Shows the **(a)** Lotus effect **(b)** Petal effect.

When comparing the "petal effect" to the "lotus effect", there are some prominent differences. The lotus petal has a randomly rough surface and low contact angle hysteresis, which means the water droplet, is not able to wet the microstructure spaces between the spikes. This allows air to remain inside the texture, causing a heterogeneous surface composed of both air and solid. As a result, the adhesive force between the water and the solid surface is extremely low, allowing the water to roll off easily (*i.e.* "Self-cleaning" phenomenon).

The main reason behind this different behaviour shown in the Louis effect and the petal effects are the corresponding equilibrium states that are to be discussed in the coming section. In the case of the lotus effect, as the nanostructures are much smaller in dimensions, air cannot penetrate within the nano-asperities below the water droplet (Cassie state). Whereas in the case of petal effects, the structure dimensions are relatively big, and thus water can enter into the nano-asperities below the water droplet (Wenzel state). Since the liquid can wet the larger-scale grooves, the adhesive force between the water and solid is very high. This explains why the water droplet will not fall off even if the petal is tilted at an angle or turned upside down. However, this effect will fail if the droplet has a volume larger because the balance between weight and surface tension is surpassed. The water contact angle with the substrate is basically the measure of hydrophobicity. As stated before, higher is the water contact angle, lesser is the contact between the water droplet and the substrate, and the higher is the hydrophobicity. The 90° value of the contact angle is a kind of critical angle in case of solid, liquid interaction that separates the liquid phobic and public region. Following the same trend here also, if a water droplet makes a contact angle greater than 90° the interaction is called hydrophobic, and on the contrary, if it is lesser than $90 s^\circ$ it is called hydrophilic. Obviously, for extreme values, when the contact angle reaches less than 5°, the surface is called super hydrophilic, and if it exceeds 150°, it is called superhydrophobic. In the latter case, only $< 5^\circ$ of the total area of the droplet remained in contact with the surface. Surfaces with superhydrophobic properties are extremely important when it comes to self-cleaning applications. The self-cleaning property protects any substrate from the growth of different fungi or algae. Not even that it has an important contribution to plant's daily life cycle by protecting plant leaves from external contamination that hampers photosynthesis.

Application of Hydrophobic Surfaces

The properties of hydrophobicity or hydrophilicity have their own versatile applications. When the self-cleaning property was observed in nature, it readily proved its efficiency in different fields of applications associated with this property.

The main application of hydrophobic surfaces is related to the coating industry. There are always significant efforts by the researchers to develop hydrophobic surfaces by mimicking the same available in nature that shows the lotus effect. This effect adds efficiencies in the working of coatings or paints, roof tiles, fabrics and others. Also, it found its application in the automobile industry, especially in a coating on glass used as a viewfinder or window. When used on a microwave antenna hydrophobic surface can reduce the rain fade effect as well as the formation of ice or snow by its self-cleaning properties and thus enhancing signal receiving efficiency. These surfaces also show their efficiency in dew harvesting or water funnelling for further use in irrigation. If the hydrophobic surfaces are properly patterned, they may be a useful tool for surface-based bio-analysis done in "lab-on-a-chip" or in microfluidic devices.

One of the most common ways to develop a hydrophobic surface is the fluorochemical or silicone treatments done on the structured surface. Also, some composites, inherently having micro/nano scale structures or clusters show such hydrophobic effect. One of the main fields of application of such hydrophobic surfaces is the medical field, where Teflon micro-particles aid the surfaces with enhanced hydrophobicity that may alternatively be achieved by a combination of polyethylene glycol (PEG) in a combination with different hydrocarbons like glucose or sucrose. Different nanostructures adequate for giving lotus effect may also be produced by Femtosecond pulsed LASER treatment.

a) Self-Cleaning

The self-cleaning process is one of the most important properties of hydrophobic surfaces that makes them resistant to external pollution from dust, soil or other agents. Self-cleaning is nothing but rolling off of water/other liquids over dirt deposited on the surface and thus taking the particle away with it. Thus, generally, the substrate is kept tilted at a small angle (< 20 °), and the dirt particle is placed externally. The whole process is schematically shown in Fig. **(1.12).**

Here one of such experimental results has been shown in Figs. **(13a-d)** for stearic acid, modified ZnO thin film, grown on cotton fabric. It is seen that the inherent hydrophobicity of the sample makes it an extremely good candidate as a self-cleaning agent. The dust particles were completely removed as the water drop was released and got rolled off without wetting.

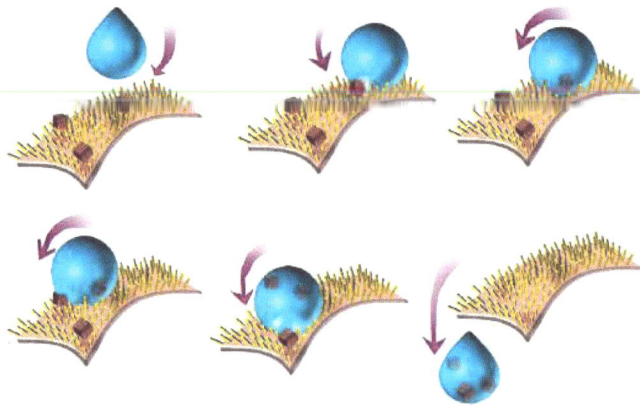

Fig. (1.12). Schematics of self-cleaning process.

The macro-scale size of the particle is much larger compared to the pore dimension, and the roughness of the sample hinders the penetration of the dust particles deep into the surface. This, along with two other effects, *i.e.,* The Low contact area between dirt and surface and between water and surface help increase adhesion between dirt and water, leading to enhanced self-cleaning performance.

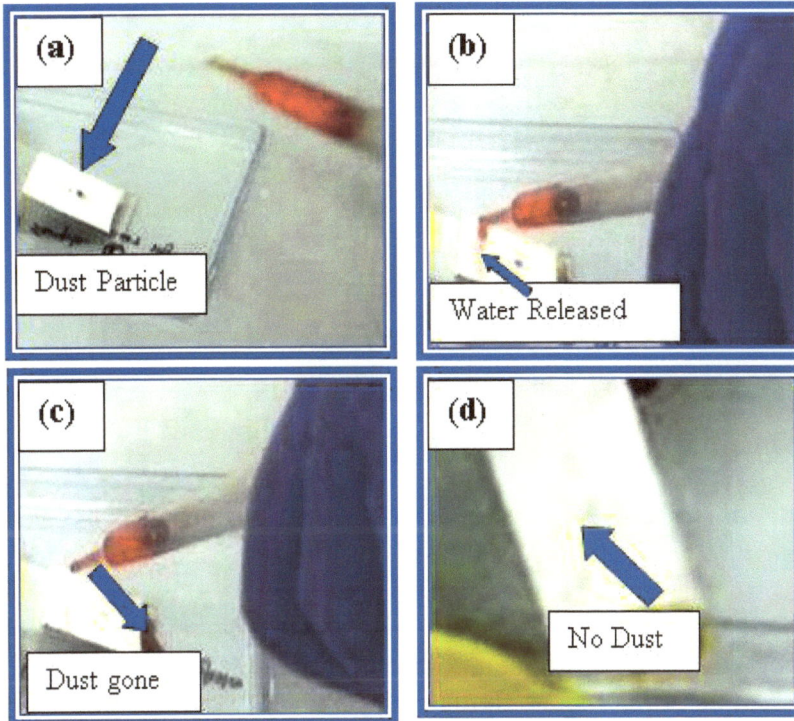

Fig. (1.13). Steps for self-cleaning process (**a**) dirt on the surface, (**b**) interaction of water with dirt and (**c, d**) self-cleaning.

b) Selective Filtering of one Liquid Over Another (Oil, Water Separation)

The sample mentioned in the case of testing self-cleaning properties has been used for partial filtration of two liquids, oil and water. This would help reduce water pollution created, from any crack in the oil container in the ocean. Here commercially available petrol with water was used in order to perform the test. Also, to achieve better visual effect, methyl orange was added to the water. The partial separation of the oil, and water has been depicted in Figs. (**14a-f**).

Fig. (1.14). (a-f) Steps of oil water separation.

It is seen that the water stays over the ZnO-coated cloth (S1), whereas oil readily comes out through it. Also, when oil and water (orange coloured) poured simultaneously, the separation between oil and water is readily done. This efficient partial admittance of water may be due to the hydrophobic nature of the sample, capillary action and abundant nanopores, as suggested by Chen *et al.* [10]. This makes the sample efficient for solving the problem of water pollution that may be occurred from oil spillage and industrial discharge of organic pollutants.

For a hydrophobic or super-hydrophobic surface, the passage of the water is blocked due to the negative capillary pressure *i.e.*, Due to the absence of irritability gradient. The negative capillary pressure is given by the Young-Laplace equation:

$$P_{cp} = \frac{2S}{r} \cos \theta \tag{1.2}$$

Where θ is the contact angle between the liquid and capillary wall, r is the pore radius, and S is the surface tension of the liquid.

There may be two cases; first when θ< 90 °, *i.e.*, Wettable pore surface liquid will be allowed to pass through the surface due to the positive capillary action or, on the contrary, when > 90 °, the liquid stays on the top of the surface without being passed through it. Thus, if there is a mixture of two liquids where θ> 90 ° for one liquid (like water in present cases) and for other liquid < 90 ° (like oil in present cases) the partial separation of oil-water (or any two liquids in general) can be possible as shown above. However, for a non-wetting surface, the transmission of liquid is possible if an additional pressure called $P_{breakthrough}$, given by:

$$P_{breakthrough} = -P_{cp} = -\frac{2S}{r}\cos\theta \tag{1.3}$$

The last equation is out of the scope of this work and hence has not been discussed here. The phenomena discussed above are the basis of the directional fluid gating mechanism. All the situations are schematically shown in Fig. **(1.15)**.

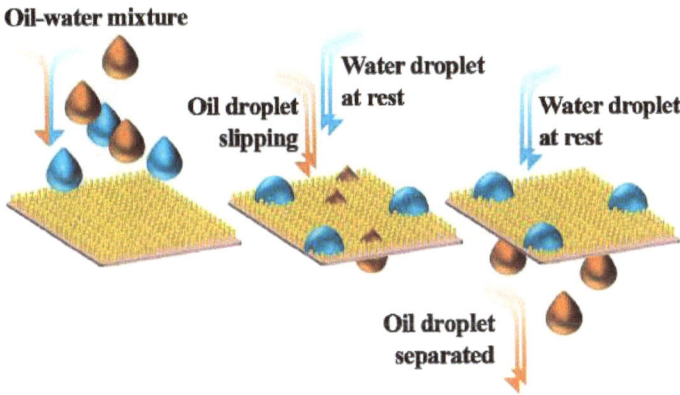

Fig. (1.15). Schematics of oil-water separation.

Fig. **(1.16)** (a-h) shows the partial absorbance of both petrol and diesel by the sample mentioned before from their mixture with water. It is seen that the absorption is almost instant, suggesting it to be a material of potential for partial separation of oil from water. To quantify the performance of the sample, the absorption capacity (Φ) has been defined as:

$$\Phi = \frac{W_2 - W_1}{W_1} \tag{1.4}$$

Where W_1 is the weight of fabric before immersion and W_2 is the weight of fabric after absorption.

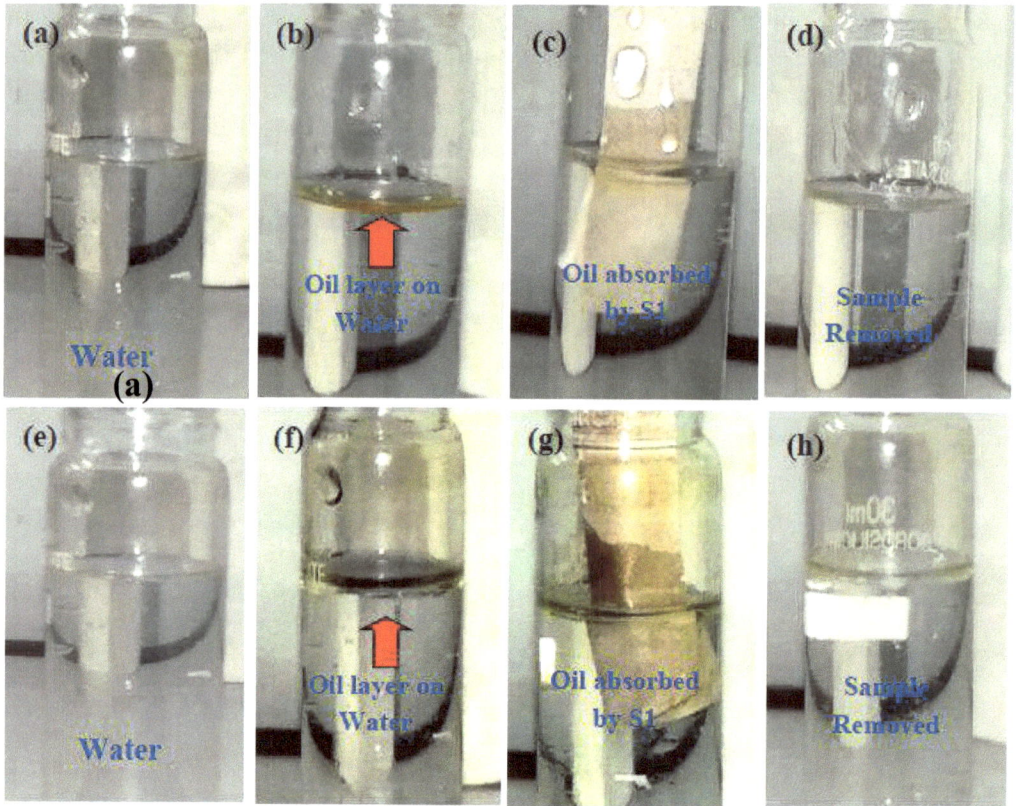

Fig. (1.16). Different steps of oil being absorbed by sample S1 from the oil-water mixture: (**a-d**) petrol and (**e-h**) diesel.

Equilibrium State and Calculation of Surface Energy

It is to be noted that for a hydrophobic surface, water can stay on it with two different equilibrium states [11]. In the first case called Cassie state the water sits on the asperities of the surfaces, and air may be entrapped in the pores. For the second one, called Wenzel state, there is no question of air trapping within the pore of the surface as the water gets penetrated within the asperities. Both the states are schematically shown in Fig. (**1.17**). In both cases, though the water can have a very large contact angle, hysteresis takes a high value for Wenzel states. Generally, most of the natural phenomena are associated with the Cassie states.

Cassie State   Wenzel state

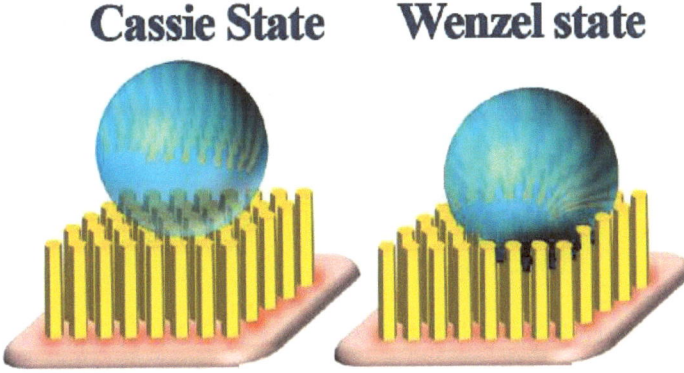

Fig. (1.17). Cassie and Wenzel equilibrium states.

The surface energy of the as-prepared samples was calculated using a Young equation that runs as follows:

$$S_{lv}\cos\theta = S_{sv} - S_{SL} - \pi_e \tag{1.5}$$

Where, θ is the contact angle between solid-liquid interfaces, S is surface energy, subscript lv, sv signifies the liquid-vapour and solid-vapour interface, respectively.

π_e is the equilibrium pressure of the adsorbed vapour of the liquid on the solid and, in most cases, can be taken to be zero [12].

Considering the polar (S_p) and a dispersive (S_d) component of the surface energy, the dispersive component of the interfacial tension between the solid and liquid is given by the Good–Girifalco–Fowkes combining rule [13, 14]:

$$S^d_{sl} = S^d_{sv} + S^d_{lv} - 2\times[S^d_{sv}\times S^d_{lv}]^{1/2} \tag{1.6}$$

One can express the polar component of the interfacial tension in a similar way and by rearranging Young's equation as shown by Oss *et al.* [15] Young's equation takes the final form as:

$$[1+\cos\theta] = 2/\ S_{lv} \times [\{S^d_{sv} \times S^d_{lv}\}^{1/2} + \{S^p_{sv} \times S^p_{lv}\}^{1/2}] \tag{1.7}$$

Thus equation 1.7 can be used to determine the surface energy of the as-synthesized material by using two different liquids of known liquid-vapour surface energy. We can take as example two liquids like water ($S^p_{lv} = 51.0$ mJ/m^2, $S^d_{lv} = 21.8$ mJ/m^2) and glycerol ($S^p_{lv} = 30.0$ mJ/m^2, $S^d_{lv} = 34.0$ mJ/m^2)   for this purpose. The approach mentioned above can also give an indirect method to predict the porosity of the surface. If we consider the surface chemistry remained the same for all the samples, then the observed contact angle data can give an estimation regarding the porosity of samples if one follows the Cassie's equation given:

$$\cos\theta = \Gamma\cos\theta_1 + (1-\Gamma)\cos\theta_2 \tag{1.8}$$

Where, is the contact angle (experimentally measured), θ_1 is the contact angle of water on a flat surface of the material under investigation and $\theta_2 = 180\,°$ is the contact angle of water on air. Γ is the solid fraction of the surface area at the top of the asperities, and hence $f_p = (1 - \Gamma)$ would represent the surface porosity. Wenzel equation is also available for relating the contact angle with the porosity and runs as:

$$\cos\theta = r\cos\theta_1 \tag{1.9}$$

Where r is the ratio of the real surface area of the projected surface area.

The question that comes very obviously to our mind if we are to use equation 1.5 or equation 1.6 for a complete unknown system, which we don't know is supposed to stay under Cassie state or Wenzel state. There are two ways to determine. Firstly, one can see the difference between the advancing contact angle and receding angle *i.e.* the contact angle hysteresis. If it has a significant difference, then the system is in the Wenzel state, and if the hysteresis is negligible, then the Cassie equilibrium state is expected. Alternatively, one can synthesize a thin film of the same system with varying roughness. If the contact angle reduces with reduced roughness then it is under Cassie state, or it is a Wenzel state if the contrary is observation.

The reason for the above discussion is that for removing dyes from water, by either photocatalysis or adsorption or by any other processes one should think that the remover should have a very strong interaction with the dyes, *i.e.*, With the water so a high hydrophobicity of the remover is preferable to be avoided.

From Nanoscience to Nanotechnology

It is to be noted that while talking about a historical overview of this newly emerging topic "Nanoscience and Technology," the first name that comes to our mind is Richard Feynman, but there is another man behind the success of this topic. If Richard Feynman was the philosopher of this new field, then Eric Drexler should be considered as the guide to steer this new technology predicting the unlimited scope of nanotechnology for developing molecular nanodevices. This person has a vision that strongly supports the possibility of placing "atoms and molecules as required for any precisely defined reaction in almost any environment."

Actually, this emergence of new technology was obvious as predicted by Gordon Moor in his famous empirical Moor's law which, without any experimental basis, predicts so accurately that the number of transistors on an affordable CPU would double about every 18 month but more transistors is more accurate according to the prediction of Moore.

Since the 1970s, the power of computers has doubled every year or and a half, yielding computers, which are millions of times more powerful than their ancestors of a half-century ago- The law states that processor speeds, or overall processing power for computers will get doubled about in every 18 months and thus predicting steady rate miniaturization of technology. The variation of CPU speed with time has been shown in Fig. (**1.18**). It is to be noted that Moor's law is not only limited to the miniaturization of the chip component now a day; something that follows an exponential variation is said to be following Moor's law.

Coming back to the statement of Eric Drexler, he wrote a book entitled "Engines of Creation: The coming era of nanotechnology," published in the year of 1986, where he strongly mentioned the possibilities of a civilization that would be governed by the different molecular assemblers where anything can be built with absolute precision and no pollution.

Fig. (**1.19**) shows three pioneer workers in the field of Nanoscience and technology. Apart from the versatile applications of Nanoscience and technology, it has some obvious negative impacts as well specially in the field of health and environment. The negativity of the field was predicted long before by Eric Drexler at the time of the birth of this new era. In the same book Engines of Creation: The coming era of nanotechnology" he has expressed his anxiousness in predicting a hypothetical situation called "Grey goo" where the molecular assembler would be out of control, and would be capable of replicating themselves and thus consumes all the living matter in the earth. A few of the negative impacts may be recapitulated here following the review article written by K. Syamala Devi and co-workers [16].

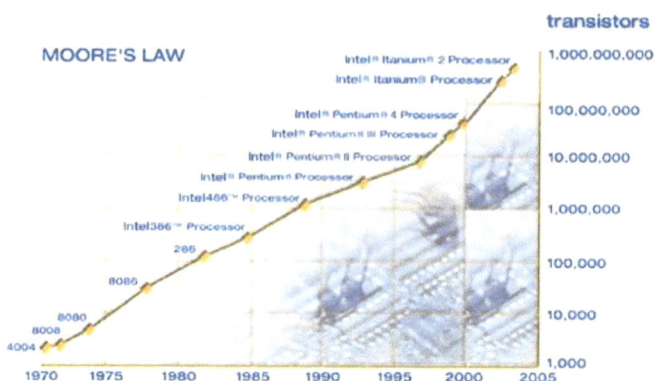

Fig. (1.18). Variation of CPU speed w.r.t. time as predicted through Moor's law.

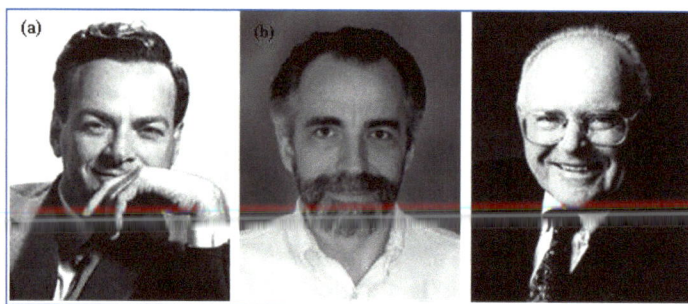

Fig. (1.19). Three pioneer workers in the field of nanoscience and technology (**a**) Richard Feynman (**b**) Eric Drexler, (**c**) Gordon Moore.

The enhanced surfaces of nanomaterials and thus the increased amount of dangling bonds make them extremely reactive, and sometimes the reaction takes place in an uncontrolled way. This may harm the environment and, in turn, the living cells. As we are yet to get finer control on the shape, size and composition of the nanomaterial, may it happen that we indulge in toxicity unknowingly. For example,

CNT being one of the most studied materials, has shown a toxic effect when it come into contact with lung epithelia, but it is still not very clear how toxic it would be or if it would be toxic at all. The shape and size and thus the activity of the nanomaterials are the key parameters that define the toxicity. Though in current days there are huge improvements in the fields of neuroscience and technology still, the lack of adequate information regarding the proper chemical structures as well as proper characterization techniques is the main reason that researchers do not have adequate information regarding the toxicity of the material. Also, a slight deviation from the chemical structure has a huge effect on the properties and drastically changes its toxicity. Thus, one should be more careful regarding the toxicity of each material in their nanoregime for different shapes. The risk assessment should not only be limited to the exposure analysis, but also care should be taken to the probability of the exposure. The life cycle assessment technique is another tool to predict the toxicity of the nanomaterial. With the decrease in size of the nanomaterials, it is becoming more difficult to predict their existence into the environment (Fig. **1.20**). Through primary emission, nanoparticles are directly exposed to the environment, whereas homogeneous nucleations from sulphuric acid or ammonia are the source of secondary particle.

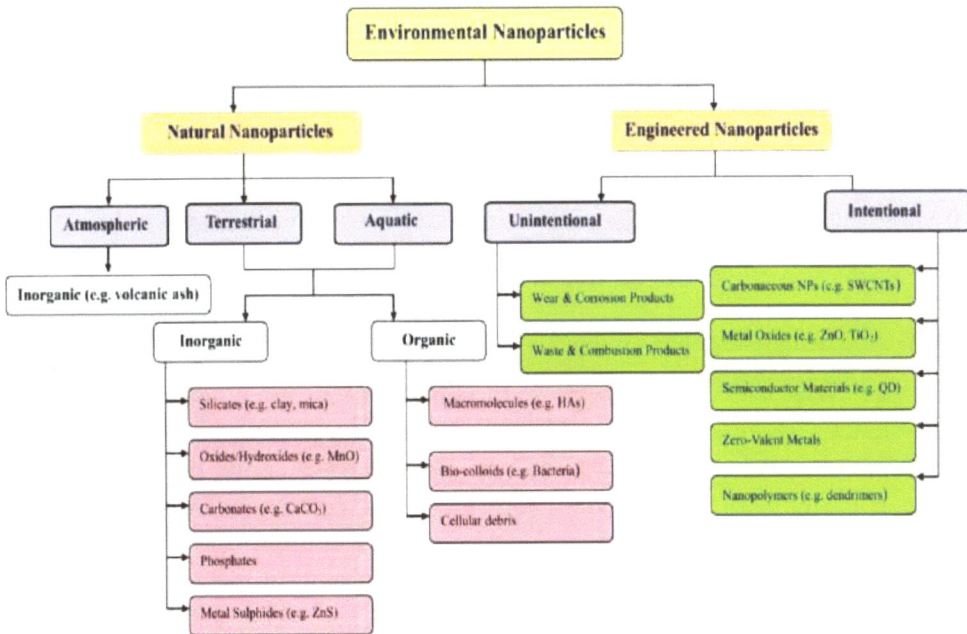

Fig. (1.20). A detail Classification of environmental nanoparticle.

Environmental Analysis

Several nanoscale inclusions have been used for various applications. Among these nanoscale inclusions, graphene has the highest priority for various reasons. Graphene is one of the most advanced materials for structural improvement, the substitution of silicon for electronic devices, as well as thermal transferring, and fire retardant.

Environmental Effects & Risks

The risks associated with current consumer and industrial uses remain unknown and therefore un-quantified. CBNI has begun research regarding plants, animals, and micro-organisms in order to understand the potential impact of nanoparticles upon ecosystems.

As mentioned before, nanotoxicity needs much more systematic research to say anything conclusively. However, based on the existing knowledge, the toxic effect associated with the nano world may be summarized as follows:

➢ Most of the nanomaterial synthesis process needs high energy, thus demanding the same in proportion.

➢ Exposure and propagation of toxic nanoparticles into nature cause serious. environmental harm of toxic, persistent nano substances originating environmental harm.

➢ It shows lower recovery; also, recycling rates are not very high.

➢ During the life cycle, interaction with the environment in each step is unclear.

➢ Huge shortage of trained human resources in this field.

Let us take Graphene, another well-studied material, as an example. It has proved its efficiency in number of fields of applications, however graphene-based nanocomposite has been proven to be toxic, but unfortunately, the degree, as well as type of toxicity, is not known. There are many instances where graphene has been seen to be interacting with other materials or living systems in unexpected ways. It is even predicted by researchers in spite of having fire retardant property, as well as good thermal stability graphene, may offer fire risk if thus contaminated.

CONCLUSION

The first chapter deals with the basic idea of Nanoscience and technology and its historical approach. In this context, it is discussed how the dimension of material takes importance in determining the properties of the material and how the nanomaterial is called novel material when compared with its bulk form. This chapter humbly remembered the contribution of pioneers like Richard Feynman, Eric Drexler and Gordon Moor in the field of basic Nanoscience and technology. This chapter also deals with the basic synthesis processes and applications of Nanoscience/technology. Considering the central theme of the book, special emphasis has been given on resemblance between nature and nanotechnology as well as the interaction between surfaces and the surface energy. In this context, surface energy calculation, hydrophobicity, contact angle, hysteresis, equilibrium state, and porosity determination were discussed qualitatively and quantitatively.

Few applications of hydrophobic surfaces like partial filtration and self-cleaning have also been the central theme of discussion in this chapter. At the end of the chapter, we have tried to discuss the impact of nanotechnology, especially the negative impact. In this consequence, we have mostly focussed on the environmental and health effect of Nanoscience and technology.   .

REFERENCES

[1]   Sengupta, A.; Sarkar, C.K., Eds.; *Introduction to nano: basics to Nanoscience and nanotechnology*; Springer, 2015.
       http://dx.doi.org/10.1007/978-3-662-47314-6

[2]   Hornyak, G.L.; Moore, J.J.; Tibbals, H.F.; Dutta, J. *Fundamentals of nanotechnology*; CRC press, 2018.
       http://dx.doi.org/10.1201/9781315222561

[3]   Karkare, M. *Nanotechnology: fundamentals and applications*; IK International Pvt Ltd., 2010.

[4]   Banerjee, D.; Das, N.S. *Nanoscience Concepts and Fundamentals: A Student's Textbook*; Consortium eLearning Network Pvt. Ltd., 2020.

[5]   Khan, F.A. Synthesis of Nanomaterials: Methods & Technology.*Applications of Nanomaterials in Human Health*; Springer: Singapore, 2020, pp. 15-21.
       http://dx.doi.org/10.1007/978-981-15-4802-4_2

[6]   Yamamoto, T.; Fukushima, T.; Aida, T.; Shimizu, T. Self-Assembled nanomaterials II: nanotubes. **2008**.

[7]   https://interestingengineering.com/7-scintillating-facts-about-the-earliest-known-use-of-nanotechnology-the-lycurgus-cup [Accession date 12.07.21]

[8]   https://www.nationalgeographic.com/science/article/carbon-nanotechnology-in-an-17th-century-damascus-sword [Accession date 12.07.21]

[9]   Filipponi, L.; Sutherland, D.; Center, I.N. Introduction to Nanoscience and nanotechnologies. In: *NANOYOU Teachers Training Kit in Nanoscience and Nanotechnologies*; , 2010; pp. 1-29.

[10]   Chen, F.F.; Zhu, Y.J.; Xiong, Z.C.; Sun, T.W.; Shen, Y.Q. Highly flexible superhydrophobic and fire-resistant layered inorganic paper. *ACS Appl. Mater. Interfaces,* **2016**, *8*(50), 34715-34724.
http://dx.doi.org/10.1021/acsami.6b12838 PMID: 27998140

[11]   Pozzato, A.; Zilio, S.D.; Fois, G.; Vendramin, D.; Mistura, G.; Belotti, M.; Chen, Y.; Natali, M. Superhydrophobic surfaces fabricated by nanoimprint lithography. *Microelectron. Eng.,* **2006**, *83*(4-9), 884-888.
http://dx.doi.org/10.1016/j.mee.2006.01.012

[12]   Roy, R.K.; Choi, H.W.; Park, S.J.; Lee, K.R. Surface energy of the plasma treated Si incorporated diamond-like carbon films. *Diamond Related Materials,* **2007**, *16*(9), 1732-1738.
http://dx.doi.org/10.1016/j.diamond.2007.06.002

[13]   Fowkes, F.M. Additivity of intermolecular forces at interfaces. i. Determination of the contribution to surface and interfacial tensions of dispersion forces in various liquids1. *J. Phys. Chem.,* **1963**, *67*(12), 2538-2541.
http://dx.doi.org/10.1021/j100806a008

[14]   Good, R.J.; Girifalco, L.A. A theory for estimation of surface and interfacial energies. III. Estimation of surface energies of solids from contact angle data. *J. Phys. Chem.,* **1960**, *64*(5), 561-565.
http://dx.doi.org/10.1021/j100834a012

[15]   Van Oss, C.J.; Chaudhury, M.K.; Good, R.J. Interfacial Lifshitz-van der Waals and polar interactions in macroscopic systems. *Chem. Rev.,* **1988**, *88*(6), 927-941.
http://dx.doi.org/10.1021/cr00088a006

[16]   Devi, K.S.; Alakanandana, A.; Lakshmi, V.V. Impacts of Nano Technology on Environment-a Review. *Asia Pacific Journal of Research,* **2018**, 45-51.

CHAPTER 2

Uniqueness of Nanomaterials and Associated Science

Abstract: This chapter deals with the concept of nanomaterials, especially basic quantum mechanics and solid-state physics. It is now a well-established fact that the dynamics of materials in nano-regime cannot be described by simple Newtonian mechanics, and, therefore, one should opt for quantum mechanics. Thus, the focus of this chapter will be on basic quantum mechanics and solid-state physics in order to gain an understanding of the system's transport properties. A few characteristic phenomena of nanomaterials like the density of state, quantum confinement, excitonic radius, *etc.*, will also be discussed in detail. To acquaint the reader with the Fermi energy and related properties, efforts will be made to provide a fundamental concept of statistical physics, specifically the Fermi-Dirac statistics. The authors have focused their efforts exclusively on dealing with different length scales, namely Ballistic transport and the associated Landauer-Buttiker formula of quantum transport in nanomaterials. Also, the relationship between exciton radius and quantum confinement and optical properties of nanomaterial has also been explained in detail. The concept of tunnelling, which is the foundation of quantum transport, has been explained, and an effort has been made to acquaint the reader with quantum conductance and Coulomb blockade.

Keywords: Ballistic transport, Crystallography, Density of state, Exciton, Quantum confinement.

A BRIEF REVIEW OF CRYSTAL STRUCTURES

Depending upon the extent of periodicity of atoms in a solid (mainly), the material has been classified into amorphous material and crystalline material. It is a well-known fact, as described in Fig. **(2.1),** that when a material has periodicity over a substantial distance, it is called crystalline material, and when there exists no periodicity or periodicity over a very short range, it is called amorphous material. Both materials have distinct properties that differ greatly from one another, and the properties of the same material in amorphous form differ significantly from those of the crystalline form. As a result, it is to be mentioned that in a crystalline material, when the same periodicity exists throughout the material, it is called a single crystal, and when periodicity changes from one crystal grain to another separated by a region called grain boundary, it is called a poly-crystalline material.

Diptonil Banerjee, Amit Kumar Sharma and Nirmalya Sankar Das

Table **2.1** summarizes the basic differences in the properties of amorphous and crystalline material.

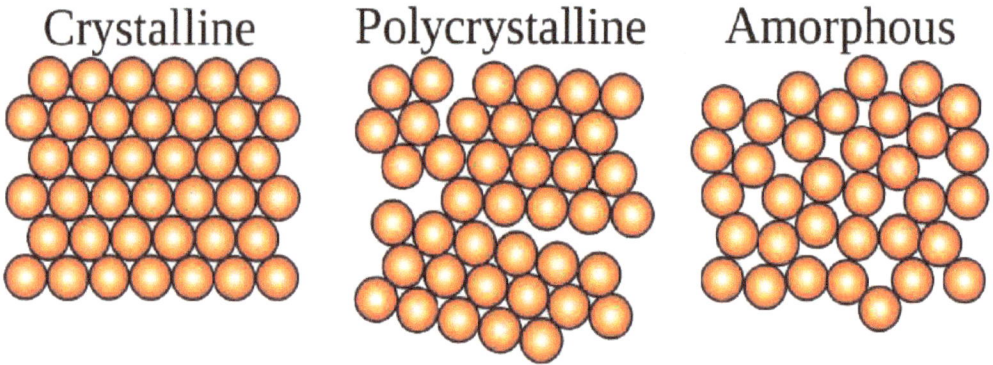

Fig. (2.1). Basic schematic representation of crystalline and amorphous structures.

Table 2.1. Differences in the properties of crystalline and amorphous material [1].

Crystalline Material	Amorphous Material
Atoms are arranged in regular 3 dimensions	They do not have a regular arrangement
Sharp melting point	No particular melting point
Anisotropic	Isotropic
True solid	Pseudo solid
Symmetrical	Unsymmetrical
More rigid	Less rigid
Long-range order	Short-range order
Example: Potassium nitrate, copper	Example: Cellophane, polyvinyl chloride

A crystalline material has two units, depending on which the entire structure of a crystal gets developed, *i.e.*, which determines the periodicity of the crystal. The first part is a set of imaginary points arranged in proper periodic ways within the crystal that determines the periodicity of the material. These are called lattice points. The second component, called basis, is the presence of an actual atom or collection of atoms at each lattice site. Thus, one can say, as shown in Fig. (**2.2**), that a unit cell of a crystal is nothing but the building block of the crystal system whose repetitive arrangement along all three directions develops the entire crystalline material. If a unit cell effectively contains only one single atom, it is called a primitive unit cell as in the case of a simple cubic system, or on the contrary, if the effective numbers of atoms in the unit cell are more than one, it is called a non-primitive unit cell as in the case of body-centered or face-centered cubic system. Both systems are shown in Fig. (**2.3**).

Lattice + Basis = Crystal,

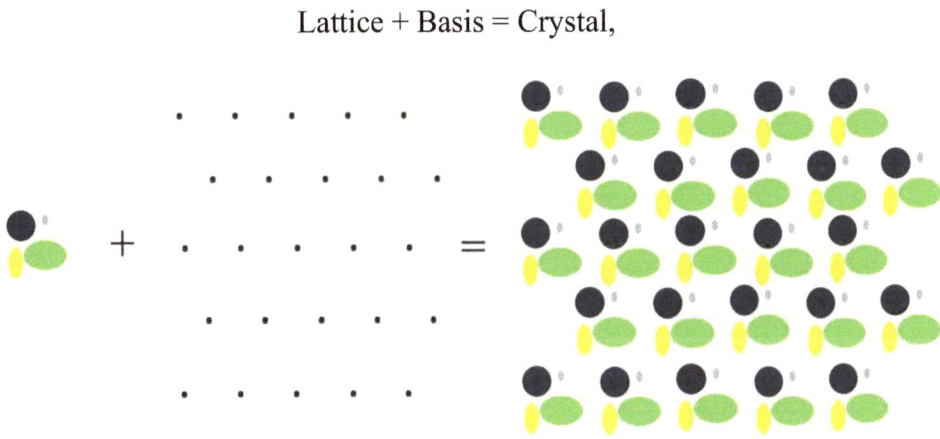

Fig. (2.2). Basic schematic of lattice and basis.

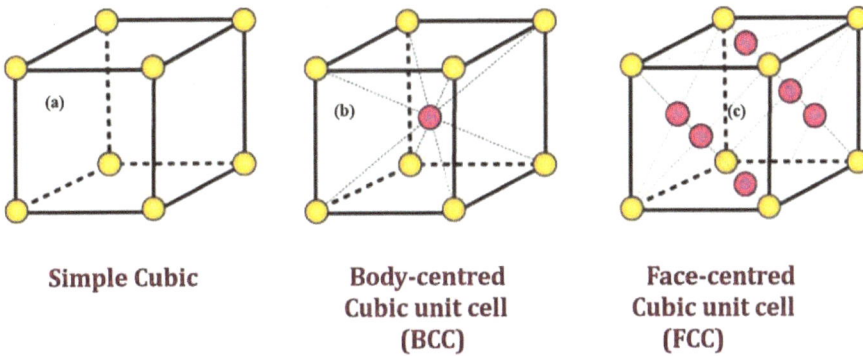

Simple Cubic   **Body-centred Cubic unit cell (BCC)**   **Face-centred Cubic unit cell (FCC)**

Fig. (2.3). Type of unit cells: Primitive (**a**) non-primitive (**b, c**) unit cell.

It is to be noted that mainly we have six parameters to determine a unit cell uniquely those are:

1. Lattice parameters **a, b, and c** along three directions X, Y and Z, respectively.
2. The angles α, β and γ where conventionally the angle between b and c is represented by α, between a and c by β and between a and b by γ.
3. It is to be noted that the edges of the unit cell may or may not be mutually perpendicular.

Concepts of Bravais Lattice

One of the alternative definitions of crystalline material is that except for the surface basis, all other basis points within the crystal should see an identical environment and thus, if one point gets shifted from a particular position to another effectively, there would be no change in the surroundings of that particular basis in the new position. The way by which points change its position within crystalline materials is called symmetry operation. There are mainly four types of symmetry operations, *i.e.*, translation, rotation, reflection and inversion and obviously, all possible combinations of such symmetry operations.

It has been shown that there are only limited ways by which it is possible to change the position of a basis within the crystal, which will provide the basis identical environment in its new position both in 2 D as well as 3 D. It is 5 and 14 numbers respectively in 2 D and 3D space by which such symmetry operations are possible. These are called Bravais lattices.

In 3 D, these fourteen lattices are further classified as shown in Table **2.2** below, where a, b and c are the magnitudes of the unit vectors and α, β and γ are the angles between the unit vectors.

Table 2.2. Classification of crystal structure as per Bravais structure.

Crystal System	Cell Shape	Cell Axes	Classes
Triclinic	General parallelepiped	$a \neq b \neq c$ $\alpha \neq \beta \neq \gamma \neq 90°$	2

(Table 2.2) cont.....

Monoclinic (Simple, Base centered)	Right prism with parallelogram as base	$a \neq b \neq c$ $\alpha = \gamma = 90° \neq \beta$	3
Orthorhombic (Simple Body centered, Face Centered, Base centered)	Rectangular parallelepiped	$a \neq b \neq c$ $\alpha = \beta = \gamma = 90°$	3
Tetragonal (Simple, Body centered)	Square prism	$a = b \neq c$ $\alpha = \beta = \gamma = 90°$	7
Trigonal (Rhomohedral)	Cube deformed along one diagonal	$a = b = c$ $\alpha = \beta = \gamma \neq 90°$	5
Hexagonal	Prism on a 60° paralleogram	$a = b \neq c$ $\alpha = \beta = 90°, \gamma = 120°$	5
Cubic (Simple, Body centered, Face Centered)	Cube	$a = b = c$ $\alpha = \beta = \gamma = 90°$	7

It is to be noted that the crystalline system has energy, less compared to the amorphous phase of the same system and thus the material has an inherent tendency to be formed in the crystalline phase. The formation energy of the system should determine the symmetry of the system and thus the shape of the as synthesized structured material.

We will end the discussion on basic crystal structure here. However, this is a vast topic, and interested readers are advised to go through the following references [2-5].

Basics of Free Electron Theory and Transport Phenomena

The most fundamental atomic structure tells that an atom is composed of a positively charged nucleus (where positive charged protons and neutral neutrons are staying together) at the center surrounding which the negatively charged electrons revolve in certain circular or elliptical orbits. The numbers of electrons in certain orbits are governed by the famous $2n^2$ law, where n is the number of orbits. Now we can imagine that when there are negatively charged electrons, and positively protons coming within the closed vicinity of each other, they will interact and that attractive interaction will be governed by famous Coulomb law of electrostatic interaction. Coulomb's interaction, being governed by inverse square law, becomes lesser and lesser effective as the number of orbits increases, *i.e.*, the distance between electron and nucleus increases. Thus the inner electrons we call core electrons are under the strong influence of the nucleus, and only the outermost electron we call valence electrons, are very loosely bound and thus can be considered free electrons which, when an electric field is applied, move as per the field direction thus conducts electricity and responsible for transport phenomena.

Thus, under the influence of an external electric field, an atom can be considered to be composed of negatively charged valence electron and positively charged ion cores. All the ion cores equally share these free electrons, and there are a large number of energy overlaps (or can be said splitting) that occurs and thus, a discrete energy level now gets transformed to an energy band and thus the concept of band spectra, valence band and conduction band comes. The gap between the valence and conduction band determines the nature of the materials. When the two bands are overlapping, we call it metal, for materials having gaps between these two bands between 0.6 1.2 eV are called, and materials having gaps greater than 3 eV are called insulators.

We all know that metals are good conductors due to the instantaneous availability of a large number of free electrons, whereas insulators are non-conducting in nature. The semiconductor can conduct electrons only at temperatures greater than the absolute temperature. The conductivity increases with the increased temperature due to more and more free elections available in the conduction band.

Now the points are to be noted that in the case of metal, where there are numbers of free electrons already available in ambient conditions, we call this set of electrons "free electron gas". The catch here is that, though, when so many charged particles are remained and moved so close to each other (talking about free/valence electrons) it is very expected that they will interact with each other suggesting there should exist some potential energy which however completely ignored in case of the free electron theory. It is noteworthy that though the valence electrons in metals are free to move on the surface of the material, they are not allowed to come out of the surface due to an infinite potential barrier the electron observed in the metal-vacuum interface. Thus, it requires external energy applied to the material in terms of light, heat, electric field or impact. The energy should be higher than the work function of the material. The postulates we have made so far when used in mathematical derivation can explain very satisfactorily Ohm's law and also the famous Widmann-Frentz law can be obtained from this. But still, this concept has serious drawbacks in explaining certain facts, like heat capacity and paramagnetic susceptibility of conduction electron. This was because here, electrons are considered distinguishable particles; thus, Maxwell-Boltzmann statistics were applied. But soon, it was realized that the electron should be considered as spin half fermions, and thus quantum statistics should be used to calculate the possible energy states keeping Pauli's exclusion principle valid. The transport in metals has first successfully explained by Drude and Lorentz with their free electron theory model, which is solely based on classical or Newtonian Mechanics. Here they have considered that electrons behave like classical gas having only kinetic energy, as mentioned before. Under the application external electric field, the electrons move along a perfectly straight line unless they suffer a collision with other electrons or ion cores. The time between two successive collisions is called Relaxation time/mean free time, and the path traversed within this relaxation time is called mean free path, one of the characteristic lengths in the transport phenomena to be discussed in the coming section of this chapter. After suffering each collision, the velocities get changed randomly, with all the previous memory completely erased.

Coming to the mathematical derivation of Dtude Lorentz theory, let us assume τ to be relaxation time or mean free time and v_d to be the average velocity where the average has been taken over all the collisions thus, the mean free path (λ) is simple.

$$\lambda = v_d \times \tau \tag{2.1}$$

If an external electric field of strength ξ is applied to the electron of charge q and mass m, then the Lorentz force (F) experience b ay charge particle is F = q which according to Newton's law of motion is:

$$F = q\xi = m\frac{dv}{dt} \tag{2.2}$$

Thus, after rearranging and integration:

$$\frac{dv}{dt} = \frac{q\xi}{m} \quad \text{and} \quad v = \frac{q\xi}{m}t + c \tag{2.3}$$

at t = 0, v = 0 leaving integration constant c = 0:

Thus we get $v = \frac{q\xi}{m}t$ \hfill (2.4)

Equation 2.4 gives:

$$\frac{dx}{dt} = \frac{q\xi t}{m} \tag{2.5}$$

which in turn gives $x = \int_0^\tau \frac{q\xi t}{m} dt$ \hfill (2.6)

or $x = \frac{q\xi \tau^2}{2m}$ \hfill (2.7)

However, under this x is nothing but mean free path λ and thus the drift velocity is given by:

$$v_d = \frac{\lambda}{\tau} = \frac{q\xi \tau}{2m} \tag{2.8}$$

Considering both direction of movement, the above expression should be multiplied by factor 2, and the expression for drift velocity becomes now $v_d = \frac{q\xi \tau}{m}$

If now it is assumed that carrier density is n in a conductor of cross-section area A, which is again subjected to the same electric field strength ξ for a time duration of τ. If the carrier moves a distance l under the electric field applied, then the total amount of charge is:

Q = nqAl \hfill (2.9)

Thus, corresponding current is given by I = Q/t = nqAl/τ = nqAv$_d$ (As t = τ thus l = λ).

Thus, current density:

$$J = \frac{I}{A} = \frac{nq^2\tau}{m}\xi \qquad\qquad (2.10)$$

which is analogous to the expression obtained from modified Ohm's law $J = \sigma\xi$

$$\sigma = \frac{nq^2\tau}{m} \qquad\qquad (2.11)$$

There are following assumptions on which the entire relation theory stands:

1. The energy of the electron is completely kinetic, and the interaction energy *i.e.*, the potential energy, is completely ignored.

2. Mean free time is independent of the position and electron of velocity, which is a good approximation and thus, the probability of collision is $1/\tau$.

3. The relation between thermal conductivity (K) to electrical conductivity at a particular temperature is constant and given by the following relation:

$$\frac{K}{\sigma T} = L = \frac{\pi^2 k^2}{3q^2} \qquad\qquad (2.12)$$

where k is Boltzmann constant and L is called Lorentz number with a value approximately equal to 2.45×10^{-8} WΩ K^{-2}

This is known as Widemann-Franz law.

This result gives moderately accurate results when it is the electrical conductivity of metal at normal temperature is concerned. However, it shows a substantial discrepancy to tally with the experimental values of heat capacity, which is rather obvious since it completely overlooks the mutual interaction between the other electrons. The theory was modified further by Sommerfeld, who introduced quantum mechanics into the theory and thus, a door towards "world of nano" got opened at that very time.

As per Sommerfeld's theory, the Drude – Lorentz assumption was modified in the following points:

1. Electrons are not classical particles, they are Fermions.

2. Being Fermions, electrons are identical and indistinguishable.

3. Being Fermions, density of state of electrons is governed by Pauli's exclusion principle.

Thus the modifications that were made in Sommerfeld theory are the following:

1. Fermi-Dirac statistics was introduced in dealing with the problem mathematically.

2. Quantization of energy level has been introduced.

3. In calculating the density of state, Pauli's exclusion principle has been incorporated.

The detail theory may be found in reference [3, 5, 6].

As per the theory, the expression for the Fermi energy (E_F) at temperature T can be related to that at absolute zero (E_{F0}) is given by:

$$E_F \approx E_{F0}\left[1 - \frac{\pi^2}{12}\left(\frac{kT}{E_{F0}}\right)^2\right] \qquad (2.13)$$

Which, when employed to calculate the specific heat of metals, hold a good agreement with the experiments.

Basic Quantum Mechanics Describing Nanoworld

The reason that nanomaterial shows unique properties compared to its bulk is because of the fact that here size is too small and comparable to the atomic range. In this size regime, whole Newtonian mechanics falls down and the system is entirely governed by the quantum mechanics. The existence of a certain system/particle is no longer described by its actual existence, but the probability density of the associate wave function describing the system. The transport phenomena have seen a completely new aspect as conventional Ohm's law completely failed here because of the fact that the system size here gets itself smaller than the corresponding characteristics length *i.e.* mean free path. Here the transport phenomenon is governed by the probabilistic ballistic transport. Thus, it would be a good effort if we can recapitulate few basic of the quantum mechanics. In doing so, we will start from the Bohr's atomic model and put our main emphasis on the few applications of Schrodinger equation (especially the particle in box problem). The detailed discussion of quantum mechanics is no doubt not the focus of this book and thus, we will take only a few concepts of quantum mechanics, the detail of which is available for the reader in the following references [7-9].

Bohr's Model of Hydrogen Atom

This is a semi-classical approach proposed by Neils Bohr to explain the line spectra of hydrogen atom, but the first effort is to use the quantization concept [10]. The models kept the idea of Rutherford's planetary motion concept of electrons with the modifications that electrons are allowed to move only in few stationary orbits where there is no continuous radiation of energy. Thus the Coulomb attraction between an electron of charge −e with the nucleus containing z number of protons is given by:

$$F = \frac{1}{4\pi\varepsilon_o} \frac{Ze^2}{r^2} \tag{2.14}$$

Where r is the radius of the orbit and ε_o is the permittivity of free space (= 9×10^9 Nm^2C^{-2}). The force is equal and opposite in direction to the force generated from centripetal acceleration $m_ev^2/r = Ze^2/r^2$, with m_e being the electron mass. This gives the total energy of the electron is the sum total of the potential energy and kinetic energy ½ m_ev^2.

$$E = \frac{m_e v^2}{2} - \frac{1}{4\pi\varepsilon_o} \frac{Ze^2}{r} = -\frac{1}{4\pi\varepsilon_o} \frac{Ze^2}{2r} \tag{2.15}$$

(where we have used the relation $m_ev^2/r = Ze^2/r^2$ or $r = Ze^2/mv^2$).

Up to this still, one expects that during motion, the electron continuously would emit energy and ultimately fall into the nucleus following a spiral motion, thus collapsing the entire atomic system. Bohr resolved the ambiguities by proposing his quantum constrain on the angular momentum (L) of the electron giving:

$$L = m_e vr = \frac{nh}{2\pi} \tag{2.16}$$

where h is the Plank's constant and n is the arbitrary quantum number (n = 1, 2, 3….*etc.*). As stated before, when an electron exists in that stationary orbit, continuous energy emission would be forbidden. A little simplification of equations 2.15 and 2.16 would give:

$$E_n = -\frac{1}{4\pi\varepsilon_o} \frac{Ze^2}{2r_n} \tag{2.17}$$

and

$$r_n = \frac{n^2 a_0}{Z} \tag{2.18}$$

where $a_0 = \dfrac{h^2 \varepsilon_o}{\pi m_e e^2} = 0.053\,nm$ is the Bohr radius of hydrogen atom and the energy of the electron in the first Bohr orbit (n = 1) of the hydrogen atom (Z = 1) being given by:

$$E_1 = -\frac{1}{4\pi\varepsilon_o} \frac{e^2}{2r_1} = -\frac{1}{4\pi\varepsilon_o} \frac{e^2}{2a_0} = -\frac{1}{8\varepsilon_o^2} \frac{e^4 m_e}{h^2} = -13.6\,eV \tag{2.19}$$

Thus when energy of the electron in nth orbit (E_n) is expressed in terms of E_1 it would be given as:

$$E_n = E_1 \left(\frac{Z^2}{n^2}\right) \tag{2.20}$$

The above equation is adequate to explain the discrete line spectra of the hydrogen atom that occurs due to changes in the orbit of an electron in a quantized way under the application of external energy. This change is given by:

$$\Delta E = E_f - E_i = \frac{hc}{\lambda} = E_1 Z^2 \left(\frac{1}{n_f^2} - \frac{1}{n_i^2} \right)$$ (2.21)

Here ΔE is the energy change of the one-electron atom that goes from an initial quantum number n_i to a final quantum number n_f.

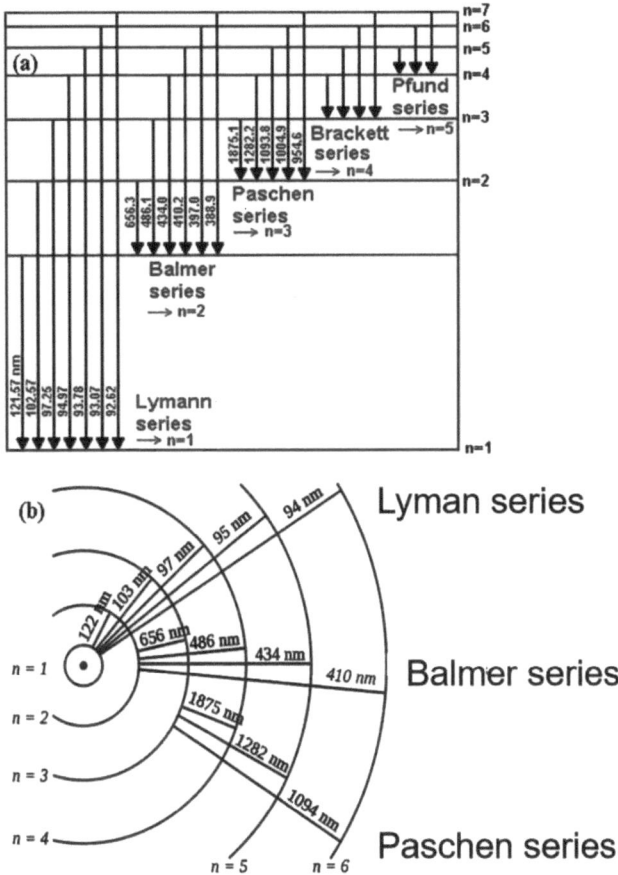

Fig. (2.4). (a) Hydrogen line spectra **(b)** description of the transitions between various Bohr orbits.

Fig. (**2.4**) shows the different spectral lines that get created due to transition between different orbits. This model showed a huge success in explaining the size effect in the optical properties of quantum dots.

Schrodinger Equation

Following up the Bohr's successful development of the atomic model using quantized energy level, quantum mechanics have been developed slowly based on the following landmark work like De-Broglie's wave-particle duality, Heisenberg's uncertainty principle, Davisson-Germerexperiment. Following the wave-particle duality every system is no longer described by the system itself, but as a wave function ($\psi(\mathbf{r}, t)$) which is the representation of the space-time behavior of the system. Here ψ is a complex quantity, and in this way, the existence of any particle (say an electron) is measured in terms of the probability density ($P(\mathbf{r},t)$) given by:

$$\psi^{*}(\vec{r},t)\psi(\vec{r},t) = P(\vec{r},t) \tag{2.22}$$

With the fact that the particle should exist somewhere in the universe, we have the normalization condition.

$$\iiint\limits_{x,y,z} P(\vec{r},t)dxdydz = \iiint\limits_{x,y,z} \psi^{*}(\vec{r},t)\psi(\vec{r},t)dxdydz = 1 \tag{2.23}$$

In this branch of science, Newton's equation of motion is replaced by a new kind of equation proposed by Erwin Schrödinger's equation describing the dynamics of a quantum mechanical system. The equation while got developed remembered the basic facts that [11]:

1. The wave equation must satisfy de Broglie relation,

$$\lambda = \frac{h}{p} \ (i.e. \ E = \hbar\omega, p = \hbar k)$$

2. The solution should reproduce the Davisson - Germer experiment of electron diffraction and should give a travelling wave solution for a free particle.

Without going through the derivation of Schrodinger equation which is out of the scope of this book, we take the final form of the time-dependent Schrodinger equation describing the dynamics of a particle of mass m and the equation runs as:

$$i\hbar \frac{\partial \psi}{\partial t} = \left[-\frac{\hbar^2}{2m}\nabla^2 + V\right]\psi \tag{2.24}$$

with ∇ being the Laplacian operator.

The space and time part of the total wave function $\Psi(x,t)$ can be separated by expressing:

$$\psi(\mathbf{r},t) \text{ as } \psi(\mathbf{r}, t) = \phi(\mathbf{r})\eta(t) \tag{2.25}$$

Solution of **time dependent** part takes the form:

$$\eta(t) = e^{\frac{-i}{\hbar}Et} \tag{2.26}$$

and the spatial part, *i.e.* **time independent** part as:

$$\left[-\frac{\hbar^2}{2m}\nabla^2 + V\right]\psi = E\psi \tag{2.27}$$

In dealing with the wave function one should remember the following facts:

1. ψ must be continuous,

2. ψ should have a continuous derivative (unless the potential in infinite),

3. ψ should be single-valued and smooth as the probability of finding the particle is $|\psi(x)|^2$.

Applications of Schrödinger Equation

Schrödinger equation has huge applications, and to be precise, the entire non-relativistic quantum mechanics is nothing but the application of the Schrodinger equation with or without perturbation. Thus we will not endeavor to discuss all the applications of this equation which is a matter of several big fat books. We will only include the part that would be of interest from the point of view of the focus of this book.

Particle Confined in a 1D Box

Here we assume that a particle of mass m gets confined within a potential box of dimension L between x = 0 to x = L. The potential U has been given below and the whole state of affairs has been shown in Fig. (**2.5**).

U = 0 for 0 < x < L

$U \to \infty$ for x ≥ L and x ≤ 0.

The situation is schematically shown in Fig. (**2.8**).

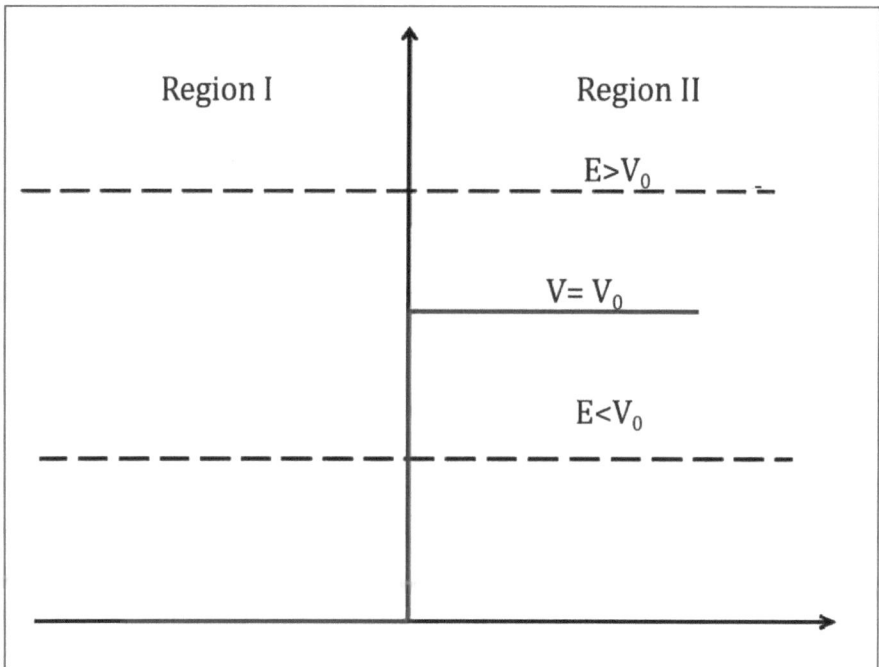

Region I   Region II

E>V_0

V= V_0

E<V_0

Fig. (2.5). Particle in a infinite quantum well.

Considering the potential pattern, it is confirmed that the particle must be found somewhere within the box between x = 0 to x = L. The Schrodinger equation for such a system runs as:

$$\frac{d^2}{dx^2}\psi(x) + \frac{2mE}{\hbar^2}\psi(x) = 0 \tag{2.28}$$

with the assumption $k^2 = \frac{2mE}{\hbar^2}$ and $\psi(0) = \psi(L) = 0$ and considering the normalization condition we have the final expression of the wave functions.

$$\psi_n(x) = \left(\frac{2}{L}\right)^{1/2} \sin\frac{n\pi x}{L} \tag{2.29}$$

with quantized energy level $E_n = \frac{n^2 h^2}{8mL^2}$, $\quad n = 1, 2, 3, \dots$

Here, we have made use of the boundary conditions.

And $\psi(0) = \psi(L) = 0$ $\quad \frac{d}{dx}\psi(x)\Big|_{x=0} = \frac{d}{dx}\psi(x)\Big|_{x=L} = 0$

When the following system has been extended to 3 dimensions, we have:

$$\nabla^2 \psi = \frac{2mE}{\hbar^2}\psi(\mathbf{r}) = 0 \tag{2.30}$$

and the corresponding wave function takes the form:

$$\psi_n(x, y, z) = \left(\frac{2}{L}\right)^{3/2} \sin\left(\frac{n_x \pi x}{L}\right) \sin\left(\frac{n_y \pi y}{L}\right) \sin\left(\frac{n_z \pi z}{L}\right) \tag{2.31}$$

$$E_n = \left[\frac{h^2}{8mL^2}\right]\left(n_x^2 + n_y^2 + n_z^2\right) \tag{2.32}$$

These results can also be used for boxes of unequal dimensions L_x, L_y and L_z as:

$$E_{n_x n_y n_z} = \left[\frac{h^2}{8m}\right]\left[\frac{n_x^2}{L_x^2} + \frac{n_y^2}{L_y^2} + \frac{n_z^2}{L_z^2}\right] \tag{2.33}$$

It is to be noted that the energy of a particular quantum state or, more precisely, the energy difference between any two energy levels depends on $1/L^2$, which is the basis of one of the characteristic phenomena of nano world, *i.e.*, Quantum confinement.

Potential Step: Reflection and Tunneling: Quantum Leak

In such case the potential distribution has been summarized in Fig. (**2.6**).

$V = 0$ for $x < 0$ [**Region I**]

$V = V_0$ for $x \geq 0$.   [**Region II**]

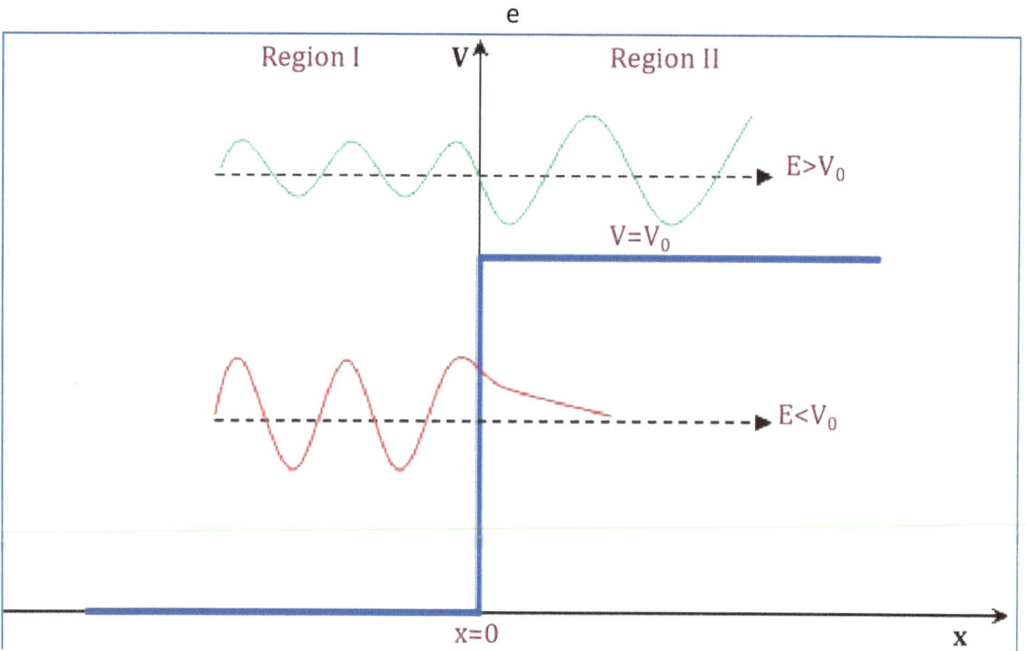

Fig. (2.6). Step potential at x = 0.

Here of course, we have to solve the Schrödinger equation in two regions separately.

Case I: $E > V_0$

In this region energy of the particle is greater than the potential energy and thus Schrödinger equation may be written as:

$$\frac{d^2\psi(x)}{dx^2} + k_1^2\psi(x) = 0 \text{ [in Region I: } x < 0]$$

(2.34)

and

$$\frac{d^2\psi(x)}{dx^2} + k_2^2\psi(x) = 0 \text{ [in Region II: } x \geq 0]$$

(2.35)

Where $k_1^2 = \frac{2mE}{\hbar^2}$ and $k_2^2 = \frac{2m(E-V_0)}{\hbar^2}$; k_1, k_2 being real quantities.

The solutions in the two regions have the form:

$$\psi(x) = Ae^{ik_1x} + Be^{-ik_1x} \qquad \text{for } x < 0 \text{ [Region I]}$$

(2.36)

$$\psi(x) = Ce^{ik_2x} + De^{-ik_2x} \quad \text{for } x \geq 0 \text{ [Region II]}$$

(2.37)

We have as obvious A, B, C, D are arbitrary constants that we have to determine from the boundary conditions. However, D may be taken to zero as there is no wave coming from right in the medium II and A, B, C are the representative of the amplitude of the incident, reflected and transmitted wave, respectively. Considering the boundary condition that the waves should be continuous at the interface, we finally have:

$$\frac{B}{A} = \frac{k_1 - k_2}{k_1 + k_2}$$

(2.38)

and

$$\frac{C}{A} = \frac{2k_1}{k_1 + k_2}$$

(2.39)

Thus the transmission (T) and reflection (R) coefficients are obtained as:

$$T = \frac{k_2}{k_1}\left|\frac{C}{A}\right|^2 = \frac{4k_1 k_2}{(k_1 + k_2)^2} \tag{2.40}$$

and

$$R = \left|\frac{B}{A}\right|^2 = \left(\frac{k_1 - k_2}{k_1 + k_2}\right)^2 \tag{2.41}$$

It is clearly seen that we have T + R = 1 *i.e.*, the sum of the transmitted and reflected wave should reproduce the incident wave. The above derivation says that in spite of particle energy being higher than the energy of the barrier, there is a finite probability some fraction of the wave gets reflected, unlike one expect in classical physics.

Case II: E < V₀

In this case Schrödinger equation will be different in region II only and given by:

$$\frac{d^2\psi(x)}{dx^2} - k_3^2\psi(x) = 0 \text{ (Region II)} \tag{2.42}$$

where $k_3^2 = \frac{2m(V_0 - E)}{\hbar^2} = -k_2^2$; k₃ being a real decay constant. Here the solution would look like:

$$\psi(x) = A_1 e^{ik_1 x} + B_1 e^{-ik_1 x} \text{ for x < 0 [Region I]} \tag{2.43}$$

$$\psi(x) = C_1 e^{-k_3 x} + D_1 e^{k_3 x} \text{ for x ≥ 0 [Region II]} \tag{2.44}$$

Here also, all the A₁, B₁, C₁ and D₁ are constants with similar meaning. As $x \to \infty$, $e^{k_3 x} \to \infty$. Therefore D₁ should be zero to avoid unrealistic interpretation. Again applying the boundary condition as before and approaching the same steps, ultimately, the expression for R would take the form as:

$$R = \left|\frac{B_1}{A_1}\right|^2 = \left(\frac{k_1 - ik_3}{k_1 + ik_3}\right)^2 = 1 \tag{2.45}$$

For the region II $(x > 0)$, the transmission is not zero, but an exponentially decaying function and the square of the amplitude is given by:

$$|C|^2 = |A|^2 \frac{4k_1^2}{\left(k_1^2 + k_3^2\right)^2} = |A|^2 \frac{4E}{E + (V_0 - E)} = |A|^2 \frac{4E}{V_0} \tag{2.46}$$

Thus, contrary to the previous case here, we can see that here we have a finite transmission coefficient of the particle even when the energy of the particle has energy less than that of the potential barrier again, unlike the result one gets in classical mechanics.

Eq. (**2.43**) suggests that the amplitude of the transmitted wave gets decreases exponentially.

Fig. (**2.7**) shows the nature of the wave function in a different region.

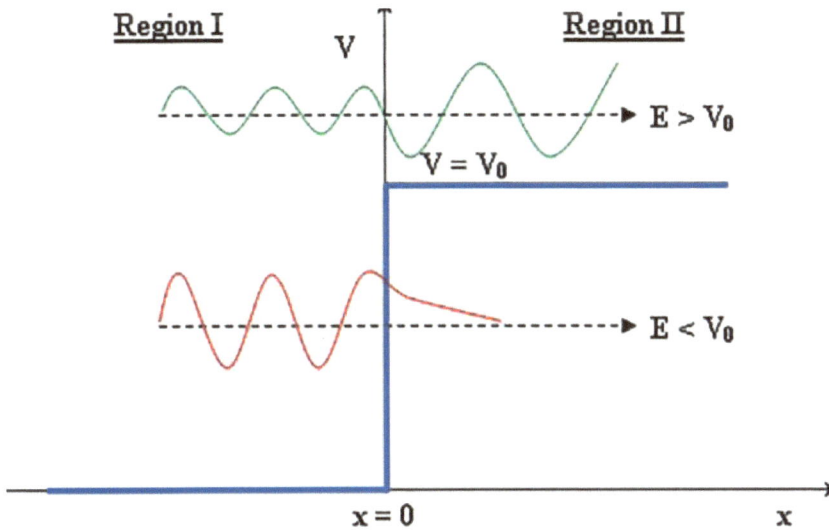

Fig. (2.7). Wave functions for a step potential for both the two cases $E < V_0$ and $E > V_0$.

Referring Fig. (**2.8**), we give the step potential finite width, say d and thus, there would be three regions I, II, III, as shown. One can see that there is a finite amplitude of the wave in region III, which is nothing but what we call as tunneling effect or quantum leak, which is the basis of several scientific phenomena like electron field emission or alpha particle decay and many other.

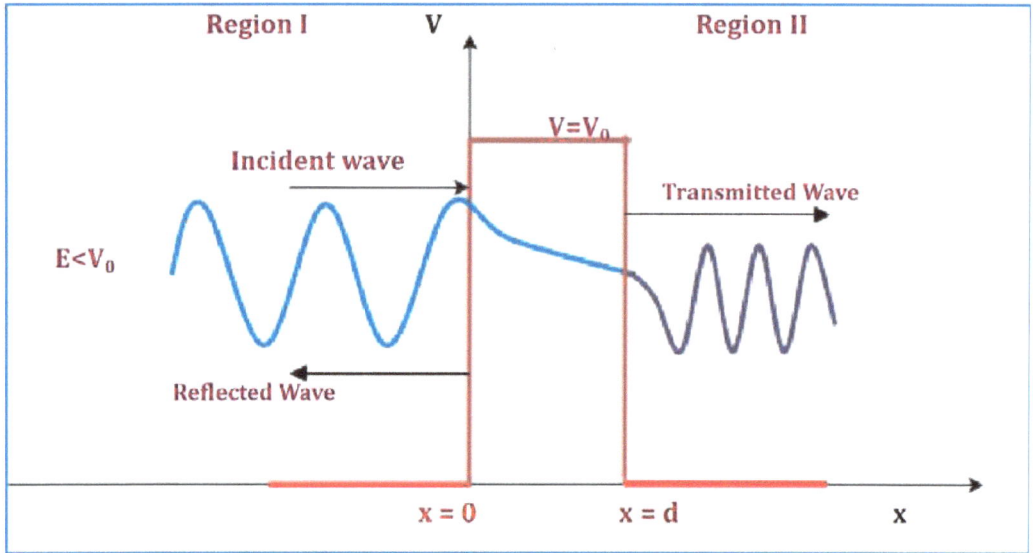

Fig. (2.8). Wave functions in a finite potential barrier.

One can denote the incident and transmitted wave as ψ_{in} and ψ_{tr} respectively given by:

$$\psi_{in} = A o^{ik_1 x} \qquad \psi_{tr} = F o^{ik_3 x}$$

Any standard quantum mechanics book will show that after suitable mathematical simplification, the transmission coefficient may be re-written as:

$$T = \left[1 + \frac{V_0^2}{4E(V_0 - E)} \sinh^2(k_3 d) \right]^{-1} \tag{2.47}$$

where k_3 is obvious.

In the special case when $k_3d \gg 1$, equation 2.47 may be reduced to:

$$T = \frac{16E(V_0 - E)}{V_0^2}(e^{-2k_3d}) \tag{2.48}$$

It is to be remembered that as $E < V_0$ thus, due to the presence of the factor $E(V_0 - E)$ T will be monotonically decreasing.

Quantum Confinement Effect in Nanomaterials

Let us recapitulate Eq. (**2.33**) we derive while dealing with the particle in a box problem in 3D:

$$E_{n_x n_y n_z} = \left[\frac{h^2}{8m}\right]\left(\frac{n_x^2}{L_x^2} + \frac{n_y^2}{L_y^2} + \frac{n_z^2}{L_z^2}\right)$$

For simplicity, we assume a cubic box and thus Eq. (**2.32**) should be used:

$$E_n = \left[\frac{h^2}{8mL^2}\right]\left(n_x^2 + n_y^2 + n_z^2\right)$$

Now let us take two energy levels and compute the energy gap between them. Thus we have:

$$\Delta E = E_{1,0,0} - E_{1,1,0} = \frac{h^2}{8mL^2} \tag{2.49}$$

It is thus clearly seen that the lesser the size of the particle greater the gap between two energy levels and the more confined the electron is within a particular potential box. This is the well-known quantum confinement effect.

It clearly suggests that when the system goes into a nano - regime the optical gap of the material gets widened and thus there is always a blue shift seen in the UV-Vis transmission spectroscopy.

Also, Bohr radius (r_B) plays a key role in the quantum confinement effect and the system with radius r will show a strong and weak quantum confinement effect subjected to the condition.

$r < r_B$ or $r > r_B$ respectively.

Under the quantum confinement effect, the blue shift in the band gap can be formulated as:

$$\Delta E = E_{g(nano)} - E_{g(bulk)} = \frac{h^2 n^2}{8\mu^* r^2} - \frac{1.8e^2}{\varepsilon r} - 0.248 E_R^* \qquad (2.50)$$

Where r is the radius of the nanoparticle, n is an integer, $\mu^* = \left[\left(m_h^*\right)^{-1} + \left(m_e^*\right)^{-1}\right]^{-1}$ is the reduced mass of the electron and hole effective asses, respectively, ε is the dielectric consort of the semiconducting material, E_R^* is the effective Rydberg energy.

The first term on the RHS of equation 2.50 is associated with quantum localization energy as has been derived from a particle in a box problem. The second term is nothing but a coulomb interaction term having 1/r dependence, and the last term is the representation of spatial correlation. When system size is large, *i.e.*, r is large, one has $\Delta E = 0$ giving tends to zero and $E_{g(nano)} \approx E_{g(bulk)}$.

Exciton

Broadly speaking, excitons are nothing but bound electron and hole pairs that are under the influence of each other, and the influence is nothing but the simple Coulomb attractive interaction. Thus, when semiconductors get excited, electrons may make a transition from the valence to the conduction band, making a vacancy in the valence band that is a creation of a hole is taken place. Now, if the electron and holes are under the influence of each other, they are called an exciton. If the electron makes, a move in the conduction band hole makes a movement in the valence band. Thus the importance of the exciton is that it can carry energy without effectively carrying the charge. The exciton resembles hydrogen atoms except for the fact that binding energy is much smaller and the size of the exciton is much bigger compared to the hydrogen atom. Also, in hydrogen atoms, the nucleus is assumed to be stationary, whereas, in the case of excitons, electron-hole pairs circle around each other with respect to their center of mass, as shown in Fig. **(2.9)**.

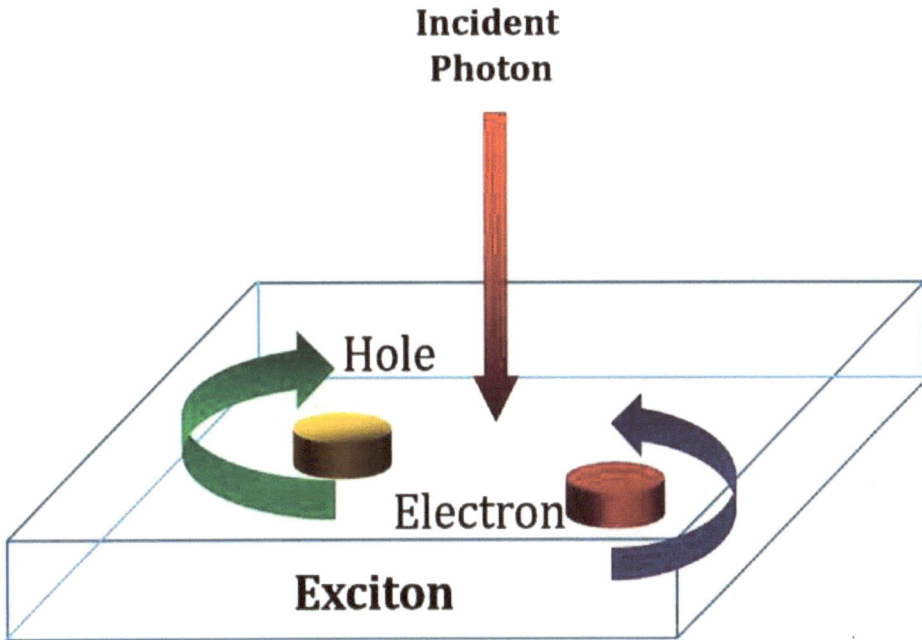

Fig. (2.9). Schematic diagram of the formation of an exciton inside a material.

Such a system can also be described Schrodinger equation as follows:

$$-\left(\frac{\hbar^2}{2m_h^*}\nabla_h^2 + \frac{\hbar^2}{2m_e^*}\nabla_e^2\right)\psi - k\frac{e^2}{|\vec{r}_e - \vec{r}_h|}\psi = E\psi \tag{2.51}$$

where "**m**" s are the masses and "**r**" s are the position vector of electron (e) and (h), respectively. The excitons can be divided into **Frenkel excitons** and **Mott-Wannier exciton.**

In **Frenkel Excitons,** one can see the following features:

a) For strong electron-hole attraction, as in ionic crystals, the electron and the hole are tightly bound to each other within the same or nearest-neighbor unit cells.

In materials with a relatively small dielectric constant, the Coulomb interaction between an electron and a hole may be strong, and the excitons, thus, tend to be smaller, of the same order as the size of the unit cell.

a)   Molecular excitons may even be entirely located on the same molecule, as in fullerenes.
b) This Frenkel exciton, named after Yakov Frenkel, has typical binding energy on the order of 0.1 to 1 eV.

Whereas **Wannier-Mott excitons** are characterized by the following properties:

a) In most semiconductors, the Coulomb interaction is strongly screened by the valence. Electrons *via* the large dielectric constant, *i.e.*, electrons and holes are only weakly bound.
b) Wannier-Mott excitons are typically found in semiconductor crystals with small energy gaps and high dielectric constants, but have also been identified in liquids, such as liquid xenon. They are also known as large excitons.
c) In semiconductors, the dielectric constant is generally large. Consequently, this tends to reduce the Coulomb interaction between electrons and holes, which has a radius larger than the lattice spacing.
d) Small effective mass of electrons that is typical of semiconductors also favors large exciton radii.
e) Within this approximation, the electron and the hole are considered two particles moving with the effective masses of the conduction and the valence bands, respectively.

Few More Points about Exciton, Bohr Radius and Quantum Confinement

The distance between the electron and hole in the exciton is known as the existential Bohr radius.

The electron and hole and exciton Bohr radius are given by:

$$a_e \; \frac{0.0529\varepsilon}{m_e^*/m_0} \; \text{nm (for electron)} \qquad a_h \; \frac{0.0529\varepsilon}{m_h^*/m_0} \; \text{nm (for hole)} \qquad (2.52a)$$

$$a_B \; \frac{0.0529\varepsilon}{\mu/m_0} \; (\text{Exciton Bohr radius}) \qquad (2.52b)$$

Let us now recall the de-Broglie hypothesis, *i.e.*, $p \approx hk$

and K.E. $= p^2/2m = k_B T$

Thus, considering the above fact, we have,

$$\lambda_e = \frac{h}{p_e} = \frac{h}{\sqrt{2m_e^* K_B T}} \text{ and } \lambda_h = \frac{h}{p_h} = \frac{h}{\sqrt{2m_h^* K_B T}} \qquad (2.53)$$

If the diameter of the particle is d, then as per Heisenber's uncertainty principle, gives:

$\Delta p \approx h/d.$

The confinement is associated with a kinetic energy of a magnitude.

$$E_{confinement} \frac{\Delta p^2}{2m^*} \approx \frac{h^2}{2m^* d^2} \geq k_B T \qquad (2.54)$$

The last term in Eq. **(2.54)** comes due to the fact that confinement energy will be significant when it is comparable to or greater than kinetic energy due to thermal motion.

Thus quantum confinement will be effective when:

$$d \approx \frac{h}{\sqrt{2m_{e,h}^* K_B T}} \qquad (2.55)$$

Thus quantum confinement will be effective only when the particle diameter will be comparable to the de-Broglie wavelength (or more precisely, it should be $1/2\,\pi$ times de-Broglie when reduced Planck constant is used in the derivation).

The relation between the band gap in nano and bulk form are related to each other by the relation.

$$E_{gn} = E_g + \frac{\pi^2 h^2}{2\mu^* R^2} - 1.8\,\frac{e}{\varepsilon R} \qquad (2.56)$$

Here the second term is the single-particle ground state energy and the last term is Coulomb interaction energy. As $E_{gn} - E_g \geq 0$ it implies:

$$R \le \pi^2 h^2 \varepsilon / 3.6 \mu^* e \tag{2.57}$$

Which is $\pi^2 e / 3.6$ times the exciton Bohr Radius (a_B).

Fig. (**2.10**) shows the effect of quantum confinement on the band gap of a material.

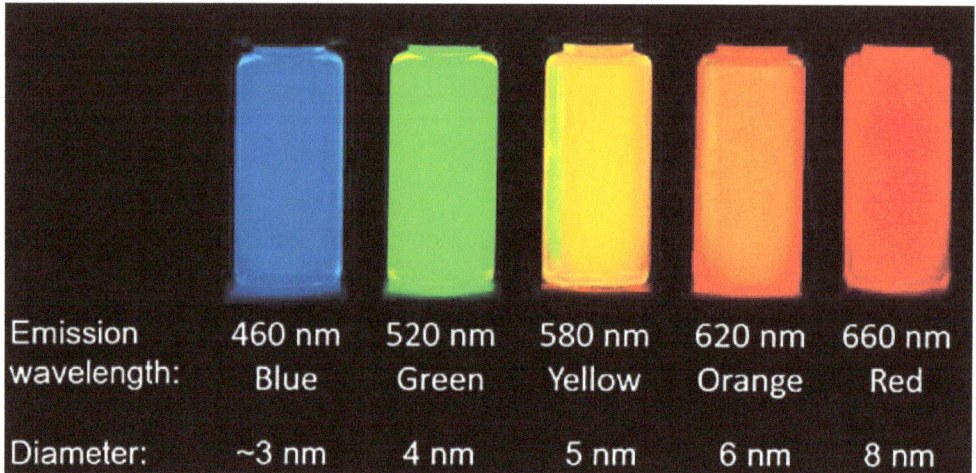

Emission wavelength:	460 nm Blue	520 nm Green	580 nm Yellow	620 nm Orange	660 nm Red
Diameter:	~3 nm	4 nm	5 nm	6 nm	8 nm

Fig. (2.10). Effect of quantum confinement on band gap.

Coulomb Gap

One of the major applications of Nanoscience and technology is the single-electron transistor (SET) which is a modified and miniature version of the MOSFET (metal oxide semiconductor field-effect transistor). Here the channel has been replaced by an island, which is, in principle, a quantum dot. The electron gets transferred by way of tunneling, and in this way, the Fermi level of the system gets tuned. Here the conductivity got quantized, and the characteristics look like a staircase.

SET works for a system of high resistivity, and it has been seen that the system works in a high resistive medium having resistance around 25.8 kΩ. If we analyze the reason for such a high value of resistance, we may have an idea about the Coulomb gap.

Fig. (**2.11**) shows the basic block diagram of a SET where which is the most important is the formation of GATE capacitance.

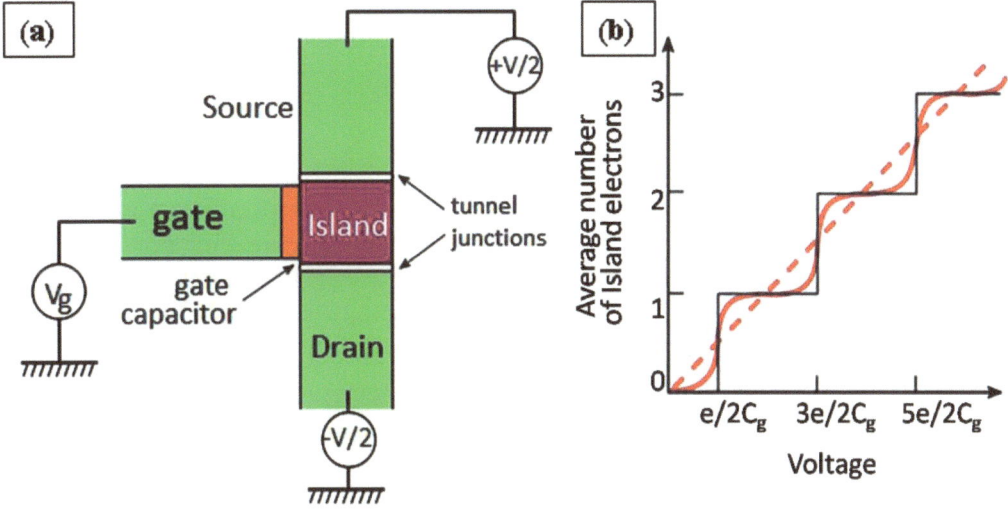

Fig. (2.11). Schematic of a SET (**a**) and the corresponding staircase like current characteristics (**b**).

We have charging energy of the capacitance $E_C = e^2/2C_g$ where e is the electronic charge, C_g is the gate capacitance.

Now we have uncertainty in the charging energy (E_c) given by:

$\Delta E_C \approx h/\Delta t$ where Δt is related to time constant $\Delta t = R_t C_g$

Thus $\Delta E_C \approx h/ R_t C_g$

For maintaining electron isolation in the quantum dot, we must have $\Delta E_C \ll E_C$ *i.e.*

$h/ R_t C_g \ll e^2/2C_g$ giving:

$R_t \gg 2h/e^2$

Putting the value of h and e we have $R_t \approx 25.8$ KΩ

In the entire process addition of an extra electron into the quantum dot by tunneling, requires extra energy of $e^2/2C_g$ due to mutual Coulomb repulsion between like charges.

Thus a Coulomb gap $E_{gap} = \dfrac{e^2}{C}$ appears in the density of electron states.

Density of States

The term density of states (DOS) is defined as the number of states per unit volume per unit energy intervals. This is highly dimensional dependent and depending upon the degrees of freedom of the charge carrier DOS gets changed and that in turn changes the transport properties of the system, thus bulk conductors may behave as nano insulators or metal can behave as semiconductors or *vice-e-versa*. To have an idea about the expression of DOS in 3D let us again start from Eq. **(2.32)**.

$$E_n = \left\lfloor \frac{h^2}{8mL^2} \right\rfloor \left(n_x^2 + n_y^2 + n_z^2 \right)$$

Remembering de-Broglie hypothesis p = $\hbar k$.

we have:

$$p_x = \hbar k_x = \frac{2n_x \pi \hbar}{L}, p_y = \hbar k_y = \frac{2n_y \pi \hbar}{L}, p_z = \hbar k_z = \frac{2n_z \pi \hbar}{L} \tag{2.58}$$

So the number Δn_x for which the momentum component will be within p_x and $p_x + dp_x$.

$$\Delta n_x = \frac{L}{2\pi \hbar} dp_x \tag{2.59}$$

Considering all three components, we have a total number:

$$\Delta n_x \Delta n_y \Delta n_z = \frac{V}{8\pi^3 \hbar^3} dp_x dp_y dp_z \tag{2.60}$$

where V= L^3 volume of the cube.

The quantity $dp_x dp_y dp_z$ is nothing but the volume of a spherical shell in the momentum space having diameter p and thickness dp and is $4\pi p^2 dp$. Now if we think of the contribution coming from both the up and down spin particles then 3.16 can be re-written as:

$$\Delta n_x \Delta n_y \Delta n_z = \frac{2V}{h^3} 4\pi p^2 dp \tag{2.61}$$

If we use the relation

$$p^2 = 2mE \tag{2.62}$$

We finally get the expression for the density of states as:

$$D(E) = \frac{V}{2\pi^2} \left(\frac{2m}{\hbar^3}\right)^{\frac{3}{2}} E^{\frac{1}{2}} \tag{2.63}$$

So the density of states is a parabolic function of energy.

The density of states can be expressed in the general form of d dimensional:

$$N(E)d(E) \propto E^{\frac{d}{2}-1} dE \qquad d = 1,2,3 \tag{2.64}$$

Here we have used the concept of phase space, and in 3D, it is spherical in nature. Also, we have taken only 1/8th part of the momentum sphere since we need all ns to be simultaneously positive and being fermion, the electron has 2 spins.

For a 2-D system, the momentum space would be circular, and because of the same reason, we will take only the first quadrant of the circle and thus, we have to divide the area by 4. For 1 D the multiplication factor is ½. The variation of density of states in 3, 2, 1 and 0-D is very interesting. Every time when we go to a lower dimensionality system, the dependence of the density of states on energy changes by $E^{-\frac{1}{2}}$. The variations of density of states for different dimensions with energy are shown in Fig. (**2.12**). In 0 dimension, the treatment is rather complicated and is beyond the scope of the book, however, the variation would follow δ function.

Transport Phenomena in Nanomaterials (Landauer-Buttiker Formula)

This is one of the simplest treatments regarding the **Landauer-Buttiker (LB)** formula and has mostly been taken from ref. 12 and 13 [12, 13]. To start with, let us consider an Ohmic conductor having cross-sectional area A, length L and the associated material resistivity ρand conductivity σ. Thus, as per Ohm's law, we have the resistance (R) of the material given by:

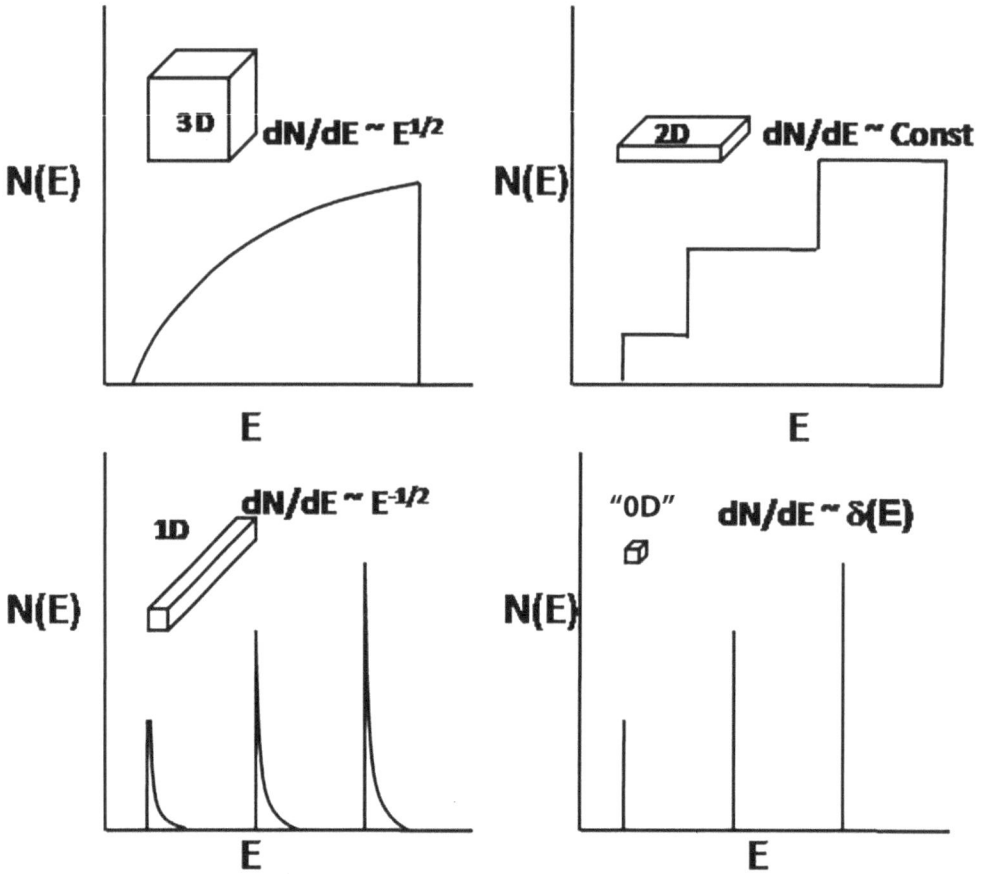

Fig. (2.12). Density of states N(E) for charge carriers as a function of the dimensionality of the semiconductor: (3D) three-dimensional semiconductor, (2D) quantum well, (1D) quantum wire, (0D) quantum dot

$$R = \frac{\rho L}{A} = \frac{L}{\sigma A}$$

Now the question arises what happens when the dimension of the conductor gets very small *i.e.* L ≈ A ≈ 0.

This will lead us to a wrong conclusion which is never possible experimentally that:

$$\lim_{A,L \to 0} R = 0$$

This ambiguity comes when a typically characteristic dimension of the device is smaller than one or more of the following length scales:

- The de Broglie wavelength of the electrons (given by their kinetic energy).
- Their mean free path, (distance between collisions).
- Their phase coherence length (the distance over which an electron can interfere with itself).

Such a sample, or device, is then described as being *mesoscopic* — one that is bigger than atoms, or 'microscopic', dimensions, but yet small enough not to exhibit the Ohmic properties of bulk, or 'macroscopic' materials.

Before we proceed further, we should get acquainted with a few of the characteristic length scale that is extremely important to characterize the transport phenomena within a material.

Mean Free Path, λ_m

The average distance an electron travels before it experiences *elastic scattering,* which destroys its initial momentum. If the size of the system L is smaller than λ_m, carriers can cross the device without scattering **(Ballistic transport).**

Phase-Relaxation Length, l_φ

The average distance an electron travels before information about its initial phase is lost. *Inelastic* scattering, such as in electron-phonon interactions, is responsible for this dephasing, since, in such collisions, the electron's energy is changed, and its quantum-mechanical phase is randomised. Impurity scattering may also contribute to phase relaxation if the impurity has internal degrees of freedom. For the system sizes smaller than the phase relaxation length l_φ, the quantum-mechanical wave function of the carriers has a well-defined phase throughout the system. Quantum interference effects may be observed in transport **(Coherent transport).**

de Broglie Wavelength, λ

An electron with wavenumber k has a de Broglie wavelength of $\lambda = 2\pi/k$. In three dimensions, this can be expressed in terms of the electron energy E as $\lambda = (h^2/2m^*E)^{0.5}$ with m^* the effective electron mass. The de Broglie length defines the scale on which quantum-mechanical effects become important. If the de Broglie wavelength is greater than one or more of the system dimensions, size quantisation of the carrier wave functions will occur. Propagation in those directions is not possible, and the density of states of the system is modified accordingly **(Quantum size effects).**

Magnetic Length, l_B

In the presence of a magnetic inductance B, electron energy is quantised in integer multiples of $\hbar\omega_c$, where ω_c is the cyclotron frequency. The magnetic length $l_B = (\hbar/eB)^{0.5}$ characterize the extent of the electron cyclotron orbit.

For macroscopic dimensions $L \gg \lambda_m$, l_φ, carriers experience frequent elastic and inelastic collisions such their momentum and phase are relaxed and what one gets is **classical diffusive transport.**

Concept

As shown in Fig. **(2.13)**, we assume a semi 1 D conductor (*i.e.*, length L is sufficiently higher compared to its width A) got sandwiched between two contacts.

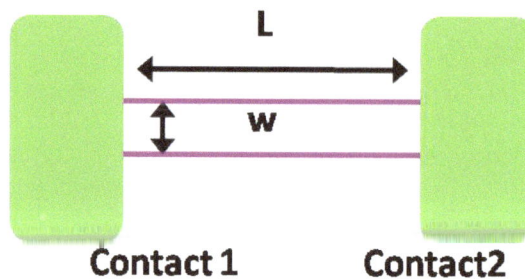

Fig. (2.13). Schematics of a conductor between two contacts.

Here we will not count the numbers of electrons that will pass through the conductors, unlike in the classical case, but here the transport properties will be described as transmission probability (T) with which an electron passes the conductor. Obviously, (1-T) will give the probability of reflection. When transmission probability is 100 % or the transmittance is 1, it is called ballistic transport, and the associated material is a ballistic conductor.

The mathematical expression of the Landauer formula giving the conductance runs as:

$$G_{tot} = \frac{2e^2}{h} MT \tag{2.65}$$

Here M is the most important parameter and is known as the number of transverse mode. The concept has come from the fact that electronic transport happens in discrete channels through a narrow conductor, which we call transverse modes.

The resistance comes mainly from the contacts as well as scattering as divided below:

$$G^{-1} = \frac{h}{2e^2 MT} + \frac{h}{2e^2 M} \frac{1-T}{T} \tag{2.66}$$

The first term comes from the contact and the second term comes from the scattering for perfect ballistic conductor T = 1 and thus, the second term becomes zero.

Buttiker extended the formula to a multi-terminal device where more than two terminals exist. Let us consider (Fig. **2.14**).

Buttiker introduces the concept that there is no principal difference between voltage probes and current probes, so we can simply extend the two-terminal Landauer formulas by summing over all probes:

Thus Buttiker formula runs as:

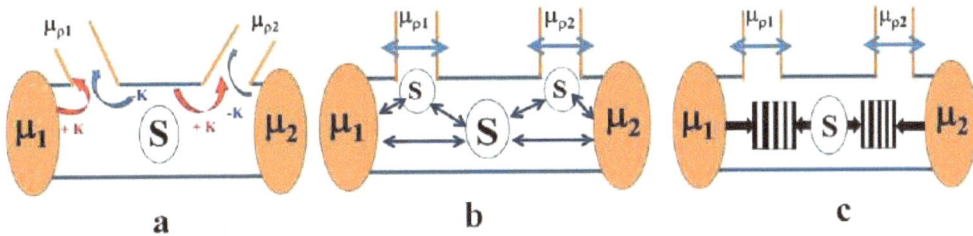

Fig. (**2.14**). Different problems with multiterminal devices arise: (**a**) The terminals may couple differently to different species of states (*e.g.* $^{+-}$ k states). (**b**) Since the terminals are invasive by themselves, they may produce additional sources of scattering. (**c**) A propagating wave may interfere with its own from a scattered reflected part. This is a pure quantum-mechanical effect and the results of a measurement may depend on the exact location of the terminals.

$$I_p = \frac{2e}{h} \sum_q \bar{T}_{q\leftarrow p}\mu_p - \bar{T}_{p\leftarrow q}\mu_q \qquad (2.67)$$

where I stands for current and $\bar{T}_{q\leftarrow p} := M_{q\leftarrow p}T_{q\leftarrow p}$ is the product of transmission probability T from contact p to contact q and the number of transverse modes M between them, and is called transmission function. We can modify this a little and re-write as:

$$G_{pq} = \frac{2e^2}{h} \bar{T}_{pq} \, ,$$

$$V_q = \frac{\mu_q}{e} \, , \qquad\qquad\qquad (2.68)$$

$$\sum_q G_{qp} = \sum_q G_{pg}$$

That finally gives

$$I_p = \sum_q G_{pq} (V_p - V_q) \qquad (2.69)$$

Lementary Fermi-Dirac Statistics

Statistical mechanics deal with particles of a very large number, and it always gives a result either from time average or ensemble averages, where the latter is defined as the set of systems that are microscopically same but microscopically different.

We will not discuss much about every point on the subject, but only touch on the basic points of Fermi-Dirac statistics as electrons belong to the group of Fermions. Broadly, there are two kinds of statistical approaches to discuss any system classical approach and quantum approach. The classical approach belongs, namely to Maxwell-Boltzmann statistics (M-B statistics), whereas quantum statistics further gets divided into Bose-Einstein (B-E) statistics and Fermi-Dirac (F-D) statistics. Table **2.3** summarizes the basic assumptions of all the three statistics.

Table 2.3. Comparison between all the three statistics regarding basic assumptions.

Properties	Statistics		
	M-B	B-E	F-D
Particle	Distinguishable	Indistinguishable	Indistinguishable
Heisenberg's uncertainty principle	Not applicable	Applicable	Applicable
Pauli's principle	Not obeyed	Not obeyed	Obeyed
Particle spin	Spin less	0, 1, 2....	½, 3/2, 5/2....
Wave function	Not applicable	Symmetric	Antisymmetric
Number of particles per energy state	No upper limit	No upper limit	Only one
Distribution function f(E)	$\dfrac{1}{e^{\alpha}e^{\frac{E_i}{kT}}}$	$\dfrac{1}{e^{\alpha}e^{\frac{E_i}{kT}} - 1}$	$\dfrac{1}{1 + e^{(E_i - E_F)/K_B T}}$

F-D statistic was developed on the following assumption.

1) The particles are identical and hence indistinguishable.

2) Pauli's exclusion principle is valid here,

3) The sum total of all the particles belonging to different states is constant. *i.e.*

$$N = \sum_{i=0}^{n} N_i \tag{2.70}$$

4) The total energy of the system is simply the sum of the energy of all the individual particles. *i.e.,*

$$E = \sum_{i=0}^{n} N_i \, U_i \tag{2.71}$$

where U_i is the energy of N_i numbers particles belonging to the ith state.

We have N number of indistinguishable particles in the system and assume that i^{th} state having degeneracy g_i can accommodate N_i number of particles having energy E_i. Now Fermion obeying Pauli's principle, can accommodate only one particle in a single state and thus, there will be only one way (X) by which N_i particles can be arranged into g_i quantum states, and that would be given by:

$$X = \frac{g_i!}{N_i!(g_i - N_i)!} \qquad (2.72)$$

Considering all the states and all the particles, the total number of ways will be given by:

$$W = \prod_{i=1}^{N} \frac{g_i!}{N_i!(g_i - N_i)!} \qquad (2.73)$$

We use the Boltzmann relation* between the entropy S and probability function W, *i.e.*

$$S = K_B \ln W \qquad (2.74)$$

Where K_B is Boltzmann constant.

Using Sirling's theorem* *i.e.*

$$\ln X! = X \ln X - X \qquad (2.75)$$

one finally gets from 2.73

$$K_B \ln W = K_B [\sum_{i=1}^{n} g_i [\ln g_i - \ln[g_i - N_i]] - N_i [\ln N_1 - \ln[g_i - N_1]]] \qquad (2.76)$$

For maximum probability we must have,

$$\delta[\ln(W)] = 0 \qquad (2.77)$$

$$\ln \sum_{i=1}^{n} [\ln(g_i - N_i) - \ln N_i + \alpha + \beta E] \delta N_i = 0 \qquad (2.78)$$

Multiplying 2.71 by α and 2.72 by β (where α, β are constant) and using Lagrange's method of undetermined multipliers, one can obtain from Eqs. (**2.71**, **2.72**, **2.76** and **2.78**).

$$\ln(g_i - N_i) - \ln N_i + \alpha + \beta E = 0 \tag{2.79}$$

Hence the Fermi distribution function $f(E_i)$ comes out to be:

$$f(E_i) = \frac{N_i}{g_i} = \frac{1}{1 + e^{-(\alpha + \beta E_i)}} \tag{2.80}$$

The value of α and β can be shown to be:

$$\alpha = -\frac{E_F}{K_B T} \quad \beta = -\frac{1}{K_B T} \tag{2.81}$$

With T and E_F be the absolute temperature and Fermi energy, respectively. Eq. (**2.80**) so has the form:

$$f(E_i) = \frac{1}{1 + e^{(E_i - E_F)/K_B T}} \tag{2.82}$$

Eq. (**2.82**) gives the average number of particles per quantum state of the system.

Some Features of Fermi-Dirac Distribution Function

Let us take Eq. (**2.82**) and consider two cases $T = 0$ and $T >$

In first case, *i.e.* $T = 0$ we have two cases further depending upon the value of energy.

$f(E_i) = 1$ for $E_i < E_F$

$f(E_i) = 0$ for $E_i > E_F$

Thus the function would be a step function, as has been shown in Fig. (**2.15**). It further suggests that at absolute zero temperature, the probability of occupation of all states having energy, lower than E_F is unity and that of the states having energy higher than E_F = zero.

At temperature $T > 0$, $f(E_i) = 0.5$ when $E_i = E_F$ suggests that in this case, the Fermi level is that energy level for which the occupation probability is half.

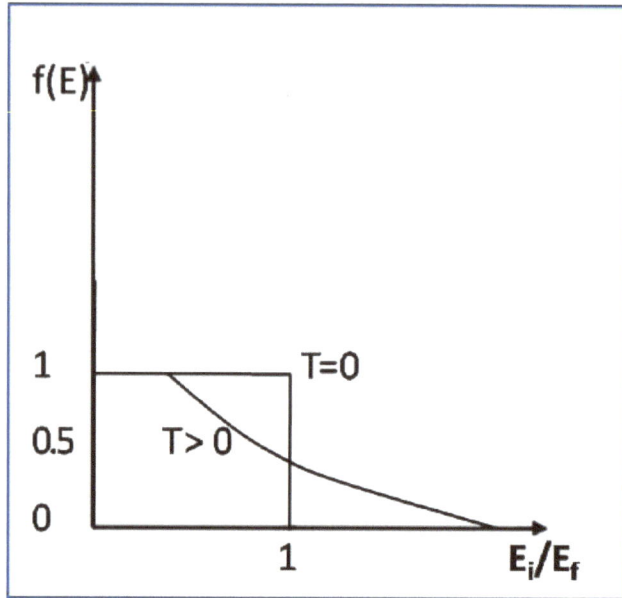

Fig. (2.15). Plot of Fermi distribution function against normalized energy at two different temperatures.

CONCLUSION

This chapter deals with some basic concepts of nanomaterials, basics of free electron theory, and a few applications of the Schrodinger equation. It is shown here the Drude Lorentz theory is pretty good to predict the electrical conductivity of the few metals at moderate temperature however it is not trustworthy to assume the correct value of specific heat because of complete ignorance of mutual interaction Quantum mechanics, which is the key tool for handling nanoworld, rightly predicts the possibilities of reflection or transmission, which is classically not possible. The particle in a box problem suggests that the lesser the system size, the greater the possibility of confinement of an electron into a particular potential well, suggesting a widening of the band gap. This is a direct consequence of quantum confinement. It has also been shown that in nanoregime, the density of state gets markedly changed as the movement of the charge carrier gets more and more constrained, changing a metal into an insulator or a semiconductor into a metal or *vice-e-versa*. The relation between excitonic radius and quantum confinement has been defined and the idea of coulomb gap, and quantum conductance has also been discussed. The last part showshow the conventional

Ohmic approach of electrical gets failed as the dimension of the system gets lesser than the mean free path. In this consequence basic of Ballistic transport and Landauer-Buttiker theory has been discussed.

REFERENCES

[1]　　https://byjus.com/chemistry/difference-between-crystalline-and-amorphous/

[2]　　Cullity, B.D. *Elements of X-ray Diffraction*; Addison-Wesley Publishing, 1956.

[3]　　Kittel, C. Introduction to solid state physics. **1976**.

[4]　　White, R.M.; Geballe, T.H. *Long Range Order in Solids: Solid State Physics*; Elsevier, 2013.

[5]　　Ashcroft, N.W.; Mermin, N.D. *Solid State Phys.,* **1976**.

[6]　　Mizutani, U. *Introduction to the electron theory of metals*; Cambridge university press, 2001.
　　　http://dx.doi.org/10.1017/CBO9780511612626

[7]　　Merzbacher, E. Matrix methods in quantum mechanics. *Am. J. Phys.,* **1968**, *36*(9), 814-821.
　　　http://dx.doi.org/10.1119/1.1975154

[8]　　Dirac, P.A.M. *Lectures on quantum mechanics*; Courier Corporation, 2001, Vol. 2.

[9]　　Bloch, F. Heisenberg and the early days of quantum mechanics. *Phys. Today,* **1976**, *29*(12), 23-27.
　　　http://dx.doi.org/10.1063/1.3024633

[10]　Robinson, J.W. *Atomic spectroscopy*; CRC Press, 1996.

[11]　Sakurai, J.J.; Napolitano, J. *Modern quantum mechanics*; Cambridge University Press, 2017.
　　　http://dx.doi.org/10.1017/9781108499996

[12]　http://www2.ece.rochester.edu/projects/bdt/files/Landauer_buttiker_formalism.pdf, **2021**. Accession date: 15/17/2021

[13]　https://www1.itp.tu-berlin.de/skripte/cemary/nanoskript.pdf, **2021**. Accession date: 15/17/2021.

CHAPTER 3

Basic Microscopic Techniques to Characterize Nano Materials

Abstract: It is not surprising that materials with nanoscale dimensions have existed since the creation of the universe. The reason is simple since all the materials are composed of different atoms or molecules, so an assembly of a few molecules can give rise to what one calls "nanomaterials". The concept of nano, or more specifically, nanoscience and technology, is relatively new, as it requires a long journey of technological advancement to develop distinct optical devices that can see materials with dimensions of 10-9 metres and thus manipulate them for greater purposes. These special devices are commonly known as microscopes; however, they are not the same as traditional microscopes, which have a maximum resolution of 10-6 (micro) meters. In this chapter, the basic constructions and working principles of the more commonly used microscopes, rather than nanoscopes, will be discussed. Discussions on field emission scanning electron microscope (FESEM), high-resolution transmission electron microscopes (HRTEM), and scanning tunnelling microscope (STM) will also be done. Besides, the name and the main working principle of some other microscopic techniques will be mentioned. Apart from imaging, some other uses (if any) of these devices would also be mentioned.

Keywords: AFM, FESEM, Microscopy, Resolving power, Resolution, SPM, STM, TEM.

WAVE NATURE OF LIGHT

According to the ancient corpuscular theory of light, a source of light emits light in all directions in terms of a stream of small particles. These particles are assumed to be so small that when two such streams of particles overlap or are superposed, any collisions between these particles are really hard to occur. These assumptions give rise to a separate and rather an earlier branch of optics called geometrical optics. This theory can satisfactorily explain the basic optical phenomenon like reflection, refraction, *etc.*, but fail to explain some other practical phenomenon like the light entering into the geometrical shadow region of a sharp object (diffraction) or production of dark fringe by superposition of two light sources (interference). All these phenomena can be well understood only if one assumes that light has some wave character. The corresponding branch of optics is called physical optics. Based on this, Huygen was first able to give a satisfactory explanation for the propagation

Diptonil Banerjee, Amit Kumar Sharma and Nirmalya Sankar Das

of light. According to him, every point on a wavefront may be considered a source of secondary spherical wavelets which spread out in the forward direction at the speed of light. The new wavefront is the tangential surface to all of these secondary wavelets. According to Huygens' principle, a plane light wave propagates through free space at the speed of light, c. The light rays associated with this wavefront propagate in straight lines, as shown in Fig. **(3.1)**. It is also fairly straightforward to account for the laws of reflection and refraction using Huygens' Principle. Though according to the principle, there would be a backward propagation of light, which has to be neglected.

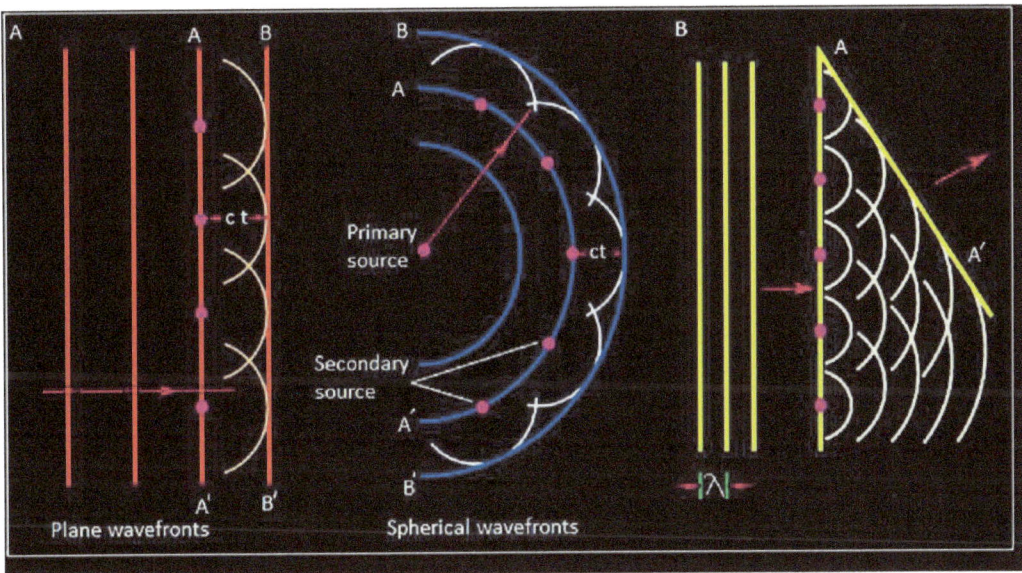

Fig. (3.1). Propagation of light according to Huygens' Principle

This theory can successfully explain the previously mentioned phenomena like interference diffraction, *etc.*, all of which originate from the basic principle of superposition of light waves.

Superposition Principle

The principle of superposition states that when two waves interact, the resulting wave function is the sum of the two individual wave functions. The superposition may be constructive or destructive, as shown in the side by Fig. **(3.2)**. When the crests overlap, the superposition wave reaches a maximum height. This height is the sum of their amplitudes (or twice their amplitude, in the case where the initial waves have equal amplitude). The same happens when the troughs overlap, creating a resultant trough that is the sum of the negative amplitudes. This sort of interference is called constructive interference because it increases the overall amplitude.

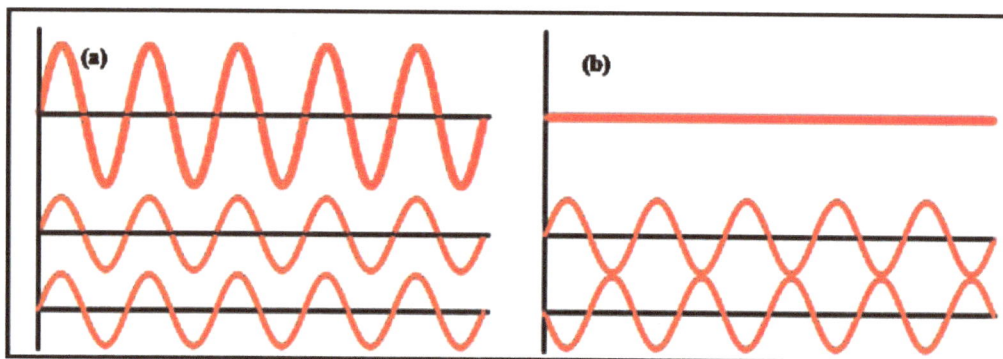

Fig. (3.2). Superposition of light (**a**) constructive (**b**) destructive interference.

Alternately, when the crest of a wave overlaps with the trough of another wave, the waves cancel each other out to some degree. If the waves are symmetrical (*i.e.*, The same wave function, but shifted by a phase or half-wavelength), they will cancel each other completely. This sort of interference is called destructive interference. Thus, interference can be considered broadly to be the modification of resultant lights due to the superposition of primary light waves coming from two different coherent sources (*i.e.*, the waves emitted from them have the same frequency and are 'phase-linked'; that is, they have a zero or constant phase difference).

The phenomenon of diffraction (which can be considered a special kind of interference), on the contrary, is nothing but the interaction of light coming from the different points, *i.e.*, secondary wavelets of the exposed part of the same wave fronts. Due to this kind of interaction, light gets bent around the sharp corners of an obstacle or slit (having a size comparable to the wavelength of light) and spreads into the regions of the geometrical shadow. The whole state of affairs is shown in Fig. (**3.3**).

It is to be noted that the phenomenon of diffraction is more closely related to the basic working principle of an optical microscope since the image of a point object is basically always a circular patch of light with centre illuminated as bright surrounded by alternatives dark and bright fringes. So when the pattern of two very close objects comes together and superposed it becomes difficult for an observer to distinguish between these two objects. So an optical instrument should have a good resolution to distinguish between two nearby objects. This minimum distance is called the optical resolution and its reciprocal is called the resolving power of an optical instrument.

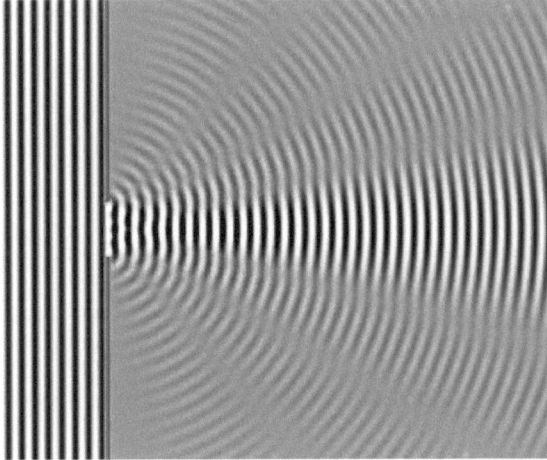

Fig. (3.3). Diffraction of light.

Resolving Power

The ability of an optical instrument, expressed in numerical measure, to resolve the image of two nearby points is termed its resolving power. In the case of a prism or a grating, the term resolving power refers to the ability of the prism or grating to resolve two nearby spectral lines so that the two lines can be viewed or photographed as separate lines.

To express the resolving power of an optical instrument as a numerical value, Lord Rayleigh proposed an arbitrary criterion. According to him, two nearby images are said to be resolved if the position of the central maximum of one coincides with the first secondary minimum of the other and *vice-versa*. The same criterion can be conveniently applied to calculate the resolving power of a telescope, microscope, grating, prism, *etc.* Also, the Rayleigh criterion can alternatively be stated as the two nearby objects can be said to be just resolved if the intensity of the resultant pattern is just 0.81 times of the intensity of the central maxima of the individual pattern, as shown in Fig. (**3.4**).

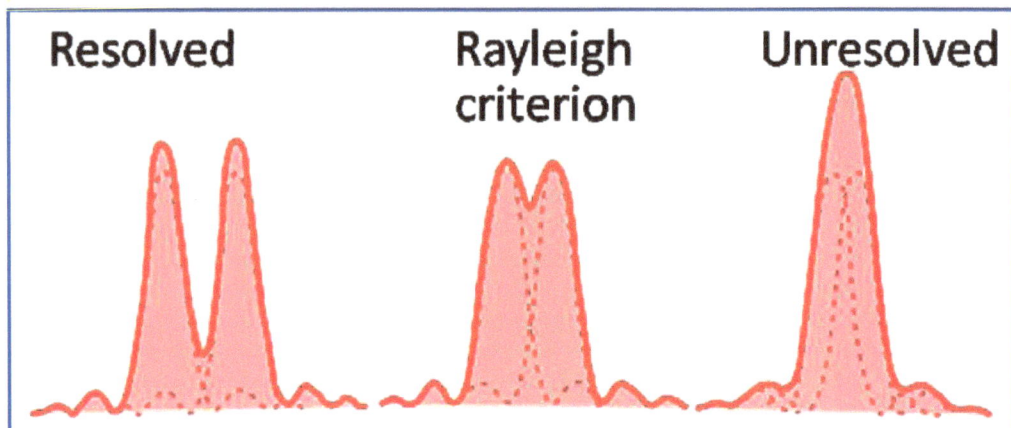

Fig. (3.4). Rayleigh criterion of optical resolution.

BASICS OF OPTICAL MICROSCOPES

A Brief History

The history of the microscope started long back in the early first century when with the invention of glass, the Roman's used to investigate the appearance of different objects and how they appeared larger. Then, in the 13th Salvino D'Armate from Italy, made the first eye glass, providing the wearer with an element of magnification to one eye. The earliest simple forms of magnification were magnifying glasses, usually about 6x - 10x and were used for inspecting tiny insects such as fleas, hence these early magnifiers called "flea glasses".

The first concept of the actual microscope came in the year 1950 when two Dutch spectacle makers, Zacharias Jansen and his father Hans, did experiments with combinations of lenses. The important discovery made by them was to see the object near the end of the tube much magnified compared to that seen by any conventional simple magnifying glass.

The importance of the first microscope rests mainly on its novelty rather than its scientific and technological aspects since its maximum magnification was only around 9x, and the images were somewhat blurry.

These Jansen microscopes survived only for Dutch royalty and were described as being composed of "3 sliding tubes, measuring 18 inches long when fully extended, and two inches in diameter". The microscope was able to magnify objects three times when fully closed, and 9 times when fully extended.

Although ordinary magnifying glasses are basically a simple microscope, when we speak of the invention of the microscope, we really mean the "compound microscope". Compound microscopes feature two or more lenses, connected by a hollow cylinder (tube). The top lens, the one people look through, is called the eyepiece. The bottom lens is known as the objective lens. So today, when we say "microscope," we really mean "compound microscope".

There is a lens called "the objective, "which produces a primary magnified image. Then there is another lens called "the eyepiece" or "ocular," which magnifies that first image. In actual practice, there are several lenses used for both the objective and ocular, but the principle is that of two-stage magnification.

It is believed that Zacharias Jansen's father, Hans, helped him build the first microscope in 1595. Zacharias wrote to a Dutch diplomat, William Boreel, about the invention. When the physician of the French king inquired about the invention in the 1650's, Boreel recounted the design of the microscope. The image of Jansen's microscope is shown in Fig. (**3.5**).

Dutch draper and scientist Anton van Leeuwenhoek (1632-1723) can be considered to be the pioneer of real microscopy techniques as in the late 17th century; he became the first man to make and use a real microscope. The microscope made by him could be able to magnify things 270 times and could view objects one-millionth of a meter (other microscopes of the time were lucky to achieve 50x magnification).

With this discovery, Van Leeuwenhoek became the first man to see and describe bacteria, yeast plants, the teeming life in a drop of water, and the circulation of blood corpuscles in capillaries. He devoted a prolonged period of his life to study many fascinating phenomena applicable both in the living as well as the non-living world with his newly discovered microscope and reported his findings in over a hundred letters to the Royal Society of England and the French Academy. His work was verified and further carried forward by English scientist Robert Hooke, who published the first work of microscopic studies, Micrographia, in 1665. Robert

Hooke's detailed studies, mainly in the field of microbiology, advanced the field of biological science as a whole to a great extent.

Fig. (3.5). Jansen's microscope.

Robert Hooke though, extensively worked on the microscopic studies of the various living and non-living objects like snow, a needle, a razor, *etc.* but would be most remembered for his most significant observations were done on fleas and cork. He observed the fleas under the microscope and was able to observe the tiny hairs on the fleas' bodies (Fig. **3.6**). On the cork, he saw pores. Upon examination of the pores, he decided to call them "cells"; but was unaware of the fact that had just discovered plant cells.

Surprisingly, in spite of all these famous discoveries and observation, the propagation of technological development regarding microscopy techniques almost remain static for next 200 years, even though there were imperfections when viewing an object due to the different refraction of light.

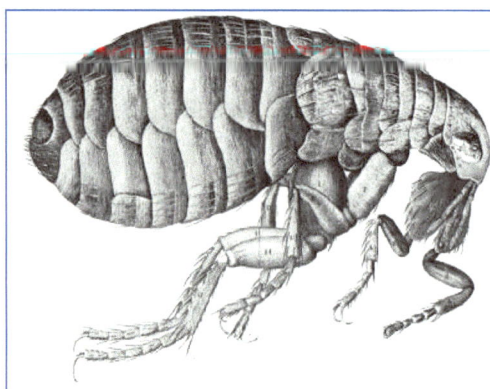

Fig. (3.6). The microscopic image of fleas as observed by Robert Hooke.

In the year of 1850s, German engineer Carl Zeiss started extensive research on the refinement of the different lenses used in the microscope and in 1880s, a glass specialist Otto Schott, was hired to conduct research on optical glass. His tireless effort contributed greatly to the improvement of the optical quality of the microscope.

In this regard, one should not forget the name of Ernst Abbe, who was hired by Zeiss to improve the manufacturing process of optical instruments. In a long and fruitful collaboration, Abbe carried out theoretical studies of optical principles, improving the understanding of the optical quality of a microscope.

Now a day, the theoretical minimum size able to be viewed by a modern optical microscope is 200 nm (as defined by Abbe), since optical microscopes are only able to focus on objects that are at least the size of a wavelength of light (usually, a wavelength of around 550 nm is assumed). An electron microscope, in contrast, can magnify images thousands of times smaller than a wavelength of light.

Basic Construction

Nowadays, the term optical microscope basically means an integrated microscope, which uses lenses and light that magnify the image and is alternatively called an optical microscope. The optical microscope is the simplest microscope capable of magnifying an object around 10 times. The integrated microscope has two magnifying lens systems:

1) an eyepiece lens that looks inwards and 2) an objective lens.

It is important to know the functions of each part.

Fig. (**3.7**) shows the image of an optical microscope, each part of which will be discussed one by one in this section.

Eyepiece Lens

This is the lens at the top through which the observers look. The lens has power ~ 10-15 X.

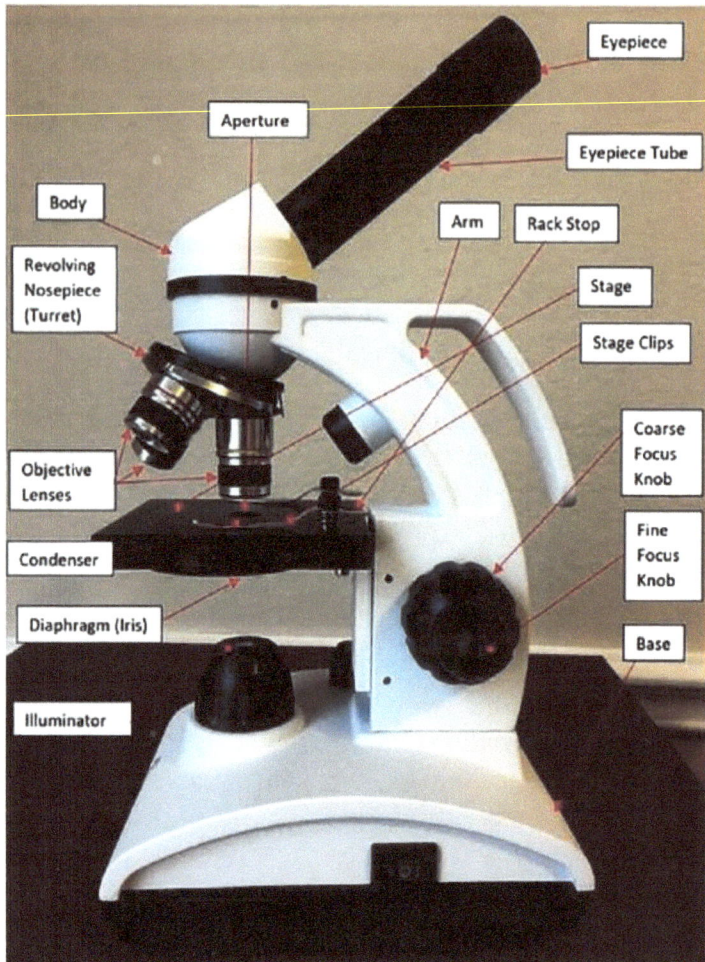

Fig. (3.7). Basic parts of an optical compound microscope.

Tube

This is basically a connecting element connecting the eyepiece to the objective lenses.

Arm

This component basically supports the tube and connects it to the base.

Base

As the name suggests, this part at the bottom of the microscope; basically bare the total weight of the instrument.

Illuminator

All the microscopes require intense illumination thus, a low voltage tungsten filament lamp with the filament wrapped in the form of a tightly-wound flattened grid is used here as an illuminator. It should be noted that the light emitted per unit area rather than the absolute values of power is important here. The combination of a suitable quartz-halogen bulb with a concave spherical reflector also serves as a good illuminator and is used commonly.

Stage

This basically holds the specimen under observation. There are arrangements of different sets of pins and clips which hold the specimen rigidly in the proper position. Sometimes the additional arrangement of mechanical stages is there, where with the help of two knobs, one becomes able to move the object according to the convenience. One moves it left and right, and the other moves it up and down.

Revolving Nose-piece or Turret

This is the part that holds two or more objective lenses and can be rotated to easily change the power.

Objective Lenses

Usually, there are 3 or 4 objective lenses associated with a specific microscope. Usually, they have the power of 4X, 10X, 40X and 100X thus, when coupled with another eyepiece lens mostly having power 10X, one gets total magnifications of 40X (4X times 10X), 100X , 400X and 1000X. It should be noted, however, that to have a good resolution of around 1000X, one would require a relatively sophisticated microscope having an Abbe condenser. As is obvious, the shortest lens has the lowest power; on the contrary longest one is the lens with the greatest power. The lenses are colour coded and if built to DIN standards, are interchangeable between microscopes. The high-power objective lenses are retractable (*i.e.*, 40XR). This means that if they hit a slide, the end of the lens will push in (spring-loaded), thereby protecting the lens and the slide. All quality microscopes have achromatic, parcentered (*i.e.* object in the centre of view will always remain in the centre when the objective is rotated), parfocal (refers to objectives that can be changed with minimal or no refocusing) lenses.

Rack Stop

This adjustment basically determines how close the objective lens can go to the specimen and this is basically set by the manufacturer. There is generally no need to readjust this unless one is working with a very thin slide and is not being able to focus on the specimen at high power.

Condenser Lens

The main utility of this lens is to focus the light onto the specimen. These kinds of lenses work more efficiently at the highest powers (400X and above). It has been seen that microscopes with condenser lenses perform much better than those with no lens. (at 400X). The convention is that a microscope with a maximum power of 400X would be most efficiently worked with a condenser lens rated at 0.65 NA or greater. 0.65 NA condenser lenses may be mounted on the stage and work quite well. Most 1000X microscopes use 1.25 Abbe condenser lens systems which can be moved up and down. It is set very close to the slide at 1000X and moved further away at the lower powers.

Diaphragm or Iris

In maximum Microscopes, there is a rotating disk under the stage having holes of different sizes. The function of these holes is to allow the light cones of different sizes and thus to vary the intensity of light projected upward into the slides. There is no set rule regarding which setting to use for a particular power. Rather, the setting is a function of the transparency of the specimen, the degree of contrast you desire and the particular objective lens in use.

How to Focus your Microscope

While focussing, some conventions are as follows: It is best to start with the objective lenses of lowest power, which should be made to move as close as possible to the specimen without touching it. Then, seeing through the objective lens, one should make it move upward only until the image is sharp. This process can be repeated several times if the focus is not achieved on the first attempt. Once the image is sharp with the low-power lens, one should be able to use the next power lens only after minor adjustments with the focus knob. If the microscope has a fine focus adjustment, turning it a bit should be all that's necessary. Continue with subsequent objective lenses and fine focus each time.

Different Kinds of Errors and Aberrations in Imaging through a Microscope

Optical aberration can generally be considered as the deviation or departure of an optical system from the basic principle of paraxial optics. The aberration generally occurs when light rays, after transmitting through the system, do not converge into (or do not diverge from) a single point. Aberrations occur because the simple paraxial theory is not a completely accurate model of the effect of an optical system on light, rather than due to flaws in the optical elements. These errors belong to both chromatic as well as non-chromatic classes producing images that are hazy, blurred and distorted. The basic aberrations that are observed in all the optical microscopes are the following:

1. Chromatic aberration,

2. Monochromatic aberration.

Chromatic Aberration

It is a well-known fact that the focal length of the lens depends upon the colour of light, and due to the prismatic action of the lens, the focal length for violet ray is smaller compared to that for red ray. Therefore a single lens forms one image of an object point, but a series of colour images at different distances from the lens. This defect due to the phenomenon of optical dispersion is called Chromatic aberration.

The second type can further be classified into:

- Spherical aberration

- Coma

- Astigmatism

- Field curvature

- Image distortion

Spherical Aberration

It is seen that when a set of parallel rays is incident on the lenses or mirrors, then after reflections or refractions, they do not meet in the same place. This is because of the fact that peripheral rays and axial rays have different focal points. This causes the image to appear hazy or blurred and slightly out of focus. This is very important in terms of the resolution of the lens because it affects the coincident imaging of points along the optical axis and degrades the performance of the lens. Now due to this phenomenon, if a screen is held perpendicular to the axes, then a circular patch is obtained. This circular patch of minimum diameter is known as the circle of least kind, and the value of the corresponding diameter is the measure of the transverse spherical aberrations. On the contrary, the extension between the focal length of the peripheral rays and axial rays is called the longitudinal spherical aberration.

Coma

The aberration, known as coma, occurs for the rays which come from the object situating off the lens axis. It is because of the fact that for these non-axial objects, the lateral magnification produced by different zones of the lens is different, thus there is an overall change of lateral magnifications with the height of the narrow circle zone through which the refraction takes place.

Astigmatism

This defect is due to the aberration of an oblique pencil of rays coming from the object point situating off-axis and passing through the centre of the lens.

Here the refracted or reflected pencil of rays never passes through a single point, but meets in two-line perpendicular to each other, called focal line. Such pencils of rays are called astigmatic pencil and the phenomenon is called astigmatism.

Curvature

When a flat object is placed perpendicular to the axis of a lens, the image, thus formed, is curved (even when the optical system is corrected for all other kinds of defects). The central part of the image is, which is formed by direct centric pencil of rays, is flat, but the edges of the image (formed by an oblique centric pencil) are curved. This is known as curvature and is because of the fact that the focal length of the lens for oblique rays is smaller compared to that of direct rays.

Distortion

This defect arises due to the variation in lateral magnification of different parts due to the difference in the distances of the object point from the axis. This produces a lack of similarity between the object and the corresponding images. This is known as distortion.

There are two kinds of distortions named barrel distortion and pincushions, and as the name suggests, the first occurs when the magnification of the central part is much more compared to the edges, and in the second case, the magnifications in the edges are more compared to that of the central parts.

The first type occurs when a rectangular wire gauge is viewed through a convex lens, and the image is real. The second error occurs when the image is virtual. Fig. (**3.8**) schematically shows all these aberrations.

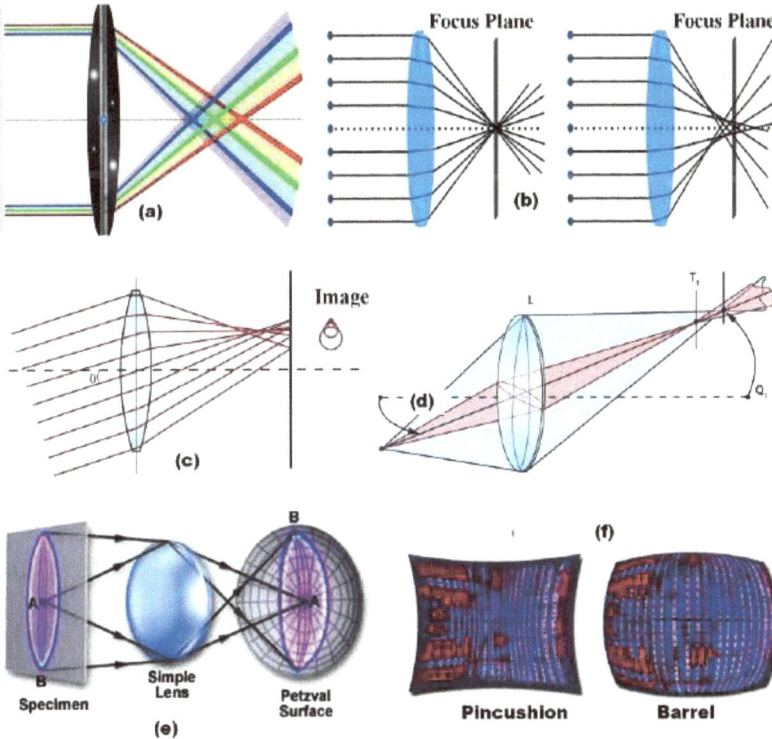

Fig. (3.8). Schematics of different aberrations (**a**) chromatic; (**b**) spherical; (**c**) coma; (**d**) astigmatism; (**e**) curvature and (**f**) distortion.

Resolving Power of Microscope

Broadly speaking, the resolving power of an optical microscope is basically defined as the reciprocal of the least angular separation between the two close objects, which makes them visible as a separate objects when seen through the microscopes.

Let us consider the following Figs. (**3.9a** and **b)**, where A and B are two self-luminous point objects separated by distance d and MN is the objective of the microscope. The point A_1 and B_1 are the central maxima of the Fraunhofer diffraction pattern corresponding to point A and B that are produced by the circular aperture of the periphery of the objectives. The Rayleigh criterion demands that the two points are said to be just resolved when central maximum of one point coincides with the first minima of the other. To obtain the required condition, let us calculate the path difference between two extreme rays, BNA_1 and BMA_1. *i.e.*

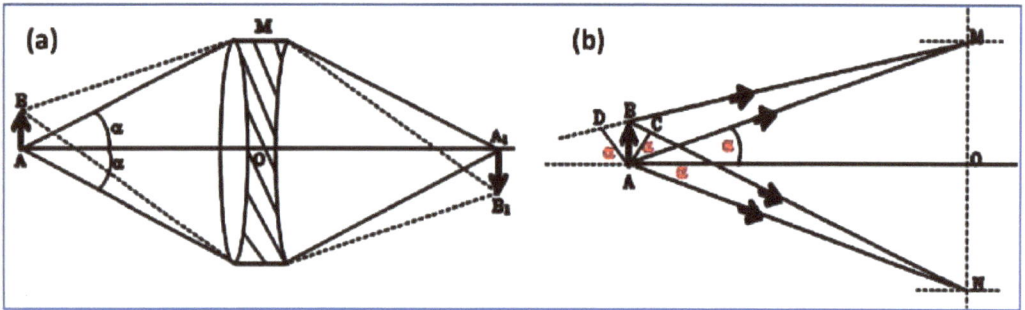

Fig. (3.9). (**a**) Resolving power of a microscope and (**b**) corresponding magnified view.

$$\delta = (BNA_1 - BMA_1)$$

$$= (BN + NA_1) - (BM + MA_1)$$

$$= BN - BM \text{ (as } NA_1 = MA_1)$$

Now, from Fig. (**3.9b**)

$$\delta = BN - BM = (BC + CN) - (DM - DB)$$

$$= BC + DB = d\sin\alpha + d\sin\alpha \text{ (As from Fig. (3.9b))}$$

$= 2d\sin\alpha$

If the path difference $2d\sin\alpha = 1.22\,\lambda$ then A_1 corresponds to the first minimum of the image B_1 and the two images are said to be just resolved.

thus

$2d\sin\alpha = 1.22\,\lambda$

or

$$d = \frac{\lambda}{2\sin\alpha} \tag{3.1}$$

Equation 4.1 gives the limit of resolution of the microscope when the objects are self-luminous emitting light of wavelength λ. It should be noted that generally, the objects that are seen through the microscope are not self-luminous. Generally, the high resolving powered microscope is an oil immersion type microscope where the space between the object and objective of the microscope is filled with suitable liquid of refractive index μ and thus, equation 3.1 would be changed to

$$d = \frac{\lambda}{2\mu\sin\alpha} \tag{3.2}$$

$\dfrac{1}{d} = \dfrac{2\mu\sin\alpha}{\lambda}$ Thus resolving power of the microscope can be formulated as

R.P. $\tag{3.3}$

The quantity $\mu\sin\alpha$ is called the numerical aperture (N.A.) of the objective.

$\dfrac{1}{d} = \dfrac{2(\text{N.A.})}{\lambda}$ Thus R.P. $\tag{3.4}$

It is seen from the above equation that resolving power of the optical instrument depends directly on the refractive index of the liquid used and inversely upon the wavelength. The last fact makes researchers interested in developing microscopes where different energy sources like ultra-violet ray or beams of electrons have wavelengths much shorter than that of ordinary light. Depending upon the nature of the wave used, these are said to respectively ultra-microscope and electron microscope.

Electron Microscope

Electron microscope (EM) basically uses an electron beam for illumination of the specimen in order to produce a magnified image of the same. It has a resolving power much higher than an ordinary light microscope and can reveal the structure of smaller objects even at the atomic level. This is because of the fact, as stated before, that electrons have wavelengths about 100,000 times shorter than visible light photons. It should be noted that an advanced electron microscope is able to magnify the image of an object by about 10000000 and possesses 50 Pico meters.

The history of the electron microscope started over 90 years back when in the year of 1926, Hans Busch first developed the electromagnetic lens then, in the year of 1928, he was convinced by the certain physicist named Leó Szilárd to develop an electron microscope and filed a patent. However, the development of the first prototype electron microscope is assigned to the name of German physicist Ernst Ruska and the electrical engineer Max Knoll in the year of 1931. Again two years later, the same group led by Ruska developed a much more advanced microscope having a resolution comparable to or even more than an optical microscope. Next year, Ernst Lubcke of Siemens & Halske were able to obtain images from their own developmental prototype microscope. The next breakthrough came almost five years later when Ernst Ruska and Bodo von Borries, and newly employed Helmut Ruska (Ernst's brother) were financed to develop microscopes mainly to study biological samples.

Though Manfred von Ardenne was first to develop a real scanning electron microscope in the year of 1937, it was Eli Franklin Burton at the University of Toronto and his students Cecil Hall, James Hillier, and Albert Prebus; who first constructed the practical electron microscope in the year of 1938, and Siemens produced the first commercial transmission electron microscope (TEM) in 1939. It should be noted that though now a day contemporary electron microscopes can magnify a suitable sample up to two million-power magnification, as scientific instruments, they remain based upon Ruska's prototype.

There are mainly two kinds of electron microscopes are used to characterize the nanomaterial microscopically depending upon the nature of the electron beam with the sample under investigation; scanning electron microscope (SEM) (field emission scanning electron microscope in a sophisticated version (FESEM)) and transmission electron microscope (TEM) (high-resolution transmission electron microscope (HRTEM)).

High-Resolution Transmission Electron Microscope

The basic parts of a transmission electron microscope consist of the following:

Electron gun that produces a beam of electrons under the application of suitable forms of energy; these emitted electrons are subsequently converged to a highly collimated beam by a set of several magnetic lenses, stacked vertically to form a lens column. All these are basically the parts of the illumination system. The diameter of the electron beam, often called as "illumination" can be modified by suitable adjustment of these basic parts. Also, it has a profound effect on the intensity level in the final TEM image.

The next part is known as the specimen stage, and the main function of this stage is to hold the specimen either stationary or to make it move according to convenience. It also helps the system to make the sample inserted into or withdrawn from the instrument. It should be noted here that the mechanical stability of the specimen stage is an important factor that determines the spatial resolution of the TEM image and a number of precautions are taken to make the instrument free of vibration.

The next and last part is the imaging system containing a lens system of at least three lenses. The objective of this part is to produce a magnified image (or diffraction pattern) of the sample under consideration, on a fluorescent screen, on photographic film, or on the monitor screen of an electronic camera system. The magnification of the TEM images is determined by the working of this imaging system, while the spatial resolution of the system can be characterized by the design of the imaging lenses.

The Electron Gun

As stated before, the electron gun, under the application of suitable forms of energy, produces an electron which is made to accelerate in order to impart sufficient energy so that the electron can transmit the specimen (hence the name transmission electron microscope). The electron gun is nothing but a suitable electron source named cathode as it is at a high negative potential, and an electron-accelerating chamber. There are several types of electron source, operating on different physical principles. They are thermionic emissions where the electrons are emitted by the application of sufficient heat so that the temperature of the filament remains around 2700 K.

$J = AT^2 \exp\left(\frac{-\Phi}{K_B T}\right)$ The rate of electron emission can be formulated with the help of the famous Richardson equation discussed in detail with mathematical treatment as well as in the energy diagram (Here Richardson constant A though slightly dependent upon the cathode material but is independent upon the temperature).

Here kT is approximately the mean thermal energy of an atom (or of a conduction electron, if measured relative to the Fermi level). The work function Φ is generally expressed in eV and should be converted to joule by multiplying it 1.6×10^{-19} for use in Richardson equation. Now, as can be seen from the Richardson equation that despite the T^2 factor, the presence of temperature term in the exponent factor mainly dominate magnitude of the thermal current. So, though temperature increases J does not show considerable value unless the product kT approaches a few percent of the work function.

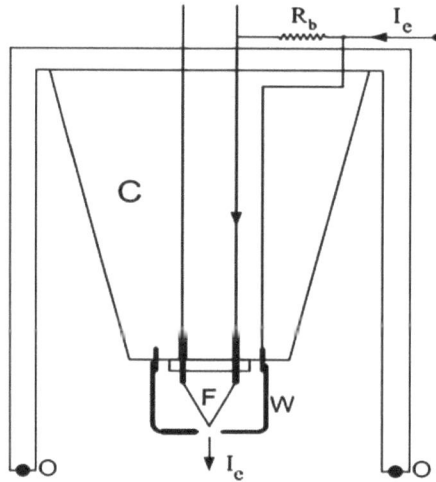

Fig. (3.10). F is tungsten filament attached with the thermionic electron gun, W is Wehnelt electrode, C, ceramic high-voltage insulator, and O is o-ring seal connected to the lower part of the TEM column. A potential difference between W and F is created by auto bias resistor R_b, which in turn controls the electron-emission current I_e.

Also the temperature is highest at the tip of the V-shaped filament as shown in Fig. (**3.10**), making it the most promising site for electron emission and most of the emission occurs in the immediate vicinity of the tip.

Tungsten is generally used as filament material as it has a very high melting point (~ 3650 K) and a low vapor pressure. The combined effect of these two lets it be maintained at a temperature of 2500-3000 K in vacuums. Though the work function of this material is high, but the high working temperature makes the value Φ/it lower enough to allow sufficient electron emission making it conducting. The additional advantages of using this material are resting in its chemically inertness and the possibility of giving it wire-like shapes.

Alternatively, one can use a material like lanthanum hexaboride (LaB_6; $\Phi= 2.7$ eV), having a lower work function and thus need not to be heated much. The LaB_6 crystal in wire shape is generally heated (1400 - 2000 K) by mounting it between the wires or onto a carbon strip through which a current is passed. The disadvantage of using this material is that it is highly sensitive to the trace amount of oxygen present in the vacuum chamber and thus gets poisoned. So the use of this material implies that one should maintain a very high vacuum level for better performance of the electron gun in TEM. This material is also expensive compared to tungsten but has the advantage of its long durability. When used, it is heated slowly in order to avoid resulting thermal shock and the resulting mechanical fracture. The magnitude of the emission current obtained is comparable to that obtained from tungsten filament but from a comparably smaller cathode area. This in turn, enables the electron beam to be focused onto a smaller area of the specimen. The resulting higher current density helps the instrument to provide a brighter image at the viewing screen (or camera) which is indeed desirable.

There is another useful component (Fig. **3.10**) associated with electron gun and this is Wehnelt cylinder. This is nothing but a metal electrode capable of being removed easily, but normally surrounds the filament completely except for a small (< 1 mm diameter) hole through which the electron beam emerges. Wehnelt electrode basically controls the emission current of the electron gun and because of the very purpose, its potential is made more negative than that of the cathode. The negative potential does not allow the emission of excess electrons unless they are emitted from a region near its tip. Increasing the magnitude of the negative bias reduces both the emitting area and the emission current.

Schottky emission, again discussed in the previous chapter, on the other hand, lets the emitter emits electron at much lower temperatures (~1800 K) by applying an additional electric field. In practice, V-shaped tungsten filament coated with zirconium oxide to obtain work function much lower (2.8 eV) compared to the pure material. This results increase in emission current 10 times compared to the thermally emitted current.

When the electric field is sufficiently increased, the electron emission can occur in the absence of any additional heat applied by the phenomena of quantum mechanical tunneling and the process is known as **cold field emission** or simply **field emission**. Here the Wehnelt cylinder is replaced by an extractor electrode maintained at a positive potential relative to the tip to obtain the high electric field. How the high electric field value is achieved can be understood if we take the tip to be a sphere having radius r, which is very lesser than the distance to the extractor electrode. Now we are in a position to apply the electrostatic formula for an isolated sphere: $E = KQ/r^2$ to relate the surface electric field E to the charge Q on the tip, $K = 1/(4\pi\varepsilon_0)$ being the Coulomb constant. The electric field E just outside the tip can be measured by taking the gradient of the potential, thus E = -(dV/dr) and integrating, we obtain V = KQ/r = E.r. Thus against a normal applied potential considerable high electric field can be obtained if the tip radius is made sufficiently smaller.

As here, the tip is not heated, the durability of the tip is much longer, and the instrument can be operated at room temperature. It is though, needed to be heated after a certain specific interval in order to remove adsorbed gases and the process is known as "**flashing**". The disadvantage this process is that it requires ultra-high vacuum (UHV: pressure ~ 10^{-8} Pa) to achieve stable operation, requiring an elaborate vacuum system and resulting in substantially greater cost of the instrument.

The operating conditions and the main performance criteria for the four types of electron source are compared in Table **3.1**.

It is to be noted that the emission current decreases from thermionic to field emission, but the radius of the tip also decreases and with much faster rate and thus the overall current density resulting in an increase in the current density J. β (not to be confused with the enhancement factor defined the in previous chapter) is another important parameter here and is known as electron-optical brightness of the source.

This is defined as the current density divided by the *solid* angle Ω over which electrons are emitted: $\beta = Ie /(As\ \Omega) = Je /\Omega$. Table **3.1** shows that this is also increasing as one goes from thermionic to field-induced emission. Note that in three-dimensional geometry, solid angle replaces the concept of angle in two-dimensional (Euclidean) geometry.

Table 3.1. Operating parameters of four types of electron source (Φ is the work function, T the temperature, E the electric field, Je the current density, and β the electron-optical brightness at the cathode; ds is the effective (or virtual) source diameter, and ΔE is the energy spread of the emitted electrons).

Type of Source	Tungsten Thermionic	LaB6 Thermionic	Schottky Emission	Cold field Emission
Material	W	LaB6	ZrO/W	W
Φ (eV)	4.5	2.7	2.8	4.5
T (K)	2700	1800	1800	300
E (V/m)	low	low	$\sim 10^8$	$> 10^9$
J (A/m2)	10^4	10^6	10^7	10^9
β (Am^{-2}/sr^{-1})	10^9	10^{10}	10^{11}	10^{12}
d_s (μm)	40	10	0.02	0.01
Vacuum (Pa)	$< 10^{-2}$	$< 10^{-4}$	$< 10^{-7}$	$< 10^{-8}$
Lifetime (hours)	100	1000	10000	10000
ΔE (eV)	1.5	1.0	0.5	0.3

Another important parameter ΔE has been tabulated in the last row and is called the energy spread, which is basically the variation in kinetic energy of the emitted electrons. For thermionic or field-induced thermionic (Schottky) emission, the energy spread can be taken to be a reflection of the statistical variations in thermal energy of electrons within the cathode and depends upon the temperature of this. In case of field emission, some electrons are always emitted below the Fermi energy level. It is to be noted that in both cases, ΔE increases with increasing emitting current due to the electrostatic interaction between electrons at "crossovers" where the beam has a small diameter and the electron separation is relatively small. The effect is known as The Boersch effect. Larger ΔE leads to increased chromatic aberration and a loss of image resolution in both the TEM and SEM.

Electron Acceleration

After emission from the cathode, electrons are accelerated to their final kinetic energy E0 by means of an electric field parallel to the optic axis. This field is generated by applying a potential difference V0 between the cathode and an **anode**, a round metal plate containing a central hole (vertically below the cathode) through which the beam of accelerated electrons emerges. Many of the accelerated electrons are absorbed in the anode plate and only around 1% pass through the hole, so the beam current in a TEM is typically 1% of the emission current from the cathode. To produce electron acceleration, it is only necessary that the anode be positive relative to the cathode. This situation is most conveniently arranged by having the anode (and the rest of the microscope column) at ground potential and the electron source at a high negative potential (- V0). Therefore the cathode and its control electrode are mounted below a high-voltage insulator made of a ceramic (metal-oxide) material, with a smooth surface and long enough to withstand the applied high voltage, which is usually at least 100 kV.

Because the thermal energy kT is small (<< 1 eV), we can take the kinetic energy (KE) of an electron to be zero before the acceleration and its potential energy (PE) to be the product of its electrostatic charge (- e) and the local potential (- V0). After acceleration, the KE of the electron is E0 and its PE is zero. Applying the principle of conservation of total energy (KE + PE):

$$0 + (-e)(-V_0) = E_0 + (-e)(0) \tag{3.5}$$

The final kinetic energy of the electron (in J) is, therefore, $E_0 = (e)V_0$. Expressed in electron volts, it is $(e)V_0/e = V_0$. In other words, the electron energy, in eV units, is equal to the magnitude of the accelerating voltage. Note that a similar statement would hold for protons (charge = +e) but not for alpha particles (charge = +2e) or any ion whose charge differs from -e. According to classical physics (Newtonian mechanics), we could deduce the speed v of an accelerated electron by equating E_0 to $mv^2/2$. However, this simple expression becomes inaccurate for an object whose speed is a significant fraction of the speed of light in a vacuum (c). In this case, we must make use of Einstein's Special Theory of Relativity, according to which the energy of a material object can be redefined so that it includes a rest-energy component m_0c^2, where m_0 is the familiar rest mass used in classical physics. If defined in this way, the total energy E is set equal to mc^2, where $m = \gamma m_0$ is called the relativistic mass and $\gamma = 1/(1-v^2/c^2)^{1/2}$ is a relativistic factor that represents the apparent increase in mass with speed. In other words,

$$E = mc^2 = (\gamma m_0) c^2 = E_0 + m_0 c^2 \qquad\qquad (3.6)$$

and the general formula for the kinetic energy E_0 is therefore:

$$E_0 = (\gamma - 1) m_0 c^2 \qquad\qquad (3.7)$$

Applying Eq. **(3.7)** to an accelerated electron, we find that its speed v reaches a significant fraction of c for the acceleration potentials V_0 used in a TEM, as shown in the third column of Table **3.2**. Comparison of the first and last columns of this table indicates that, especially for the higher accelerating voltages, use of the classical expression $m_0 v^2/2$ would result in a substantial underestimate of the kinetic energy of the electrons. The accelerating voltage ($-V_0$) is supplied by an electronic high-voltage (HV) generator, which is connected to the electron gun by a thick, well insulated cable. The high potential is derived from the secondary winding of a step-up transformer, whose primary is connected to an electronic oscillator circuit whose (ac) output is proportional to an applied (dc) input voltage. Because the oscillator operates at low voltage, the large potential difference V_0 appears between the primary and secondary windings of the transformer. Consequently, the secondary winding must be separated from the transformer core by an insulating material of sufficient thickness; it cannot be tightly wrapped around an electrically conducting soft-iron core, as in many low-voltage transformers. Because of this less-efficient magnetic coupling, the transformer operates at a frequency well above the mains frequency (60 Hz or 50 Hz). Accordingly, the oscillator output is a high frequency (~10 kHz) sine wave, whose amplitude is proportional to the input voltage V_i.

In order to deliver direct high voltage, the current from the secondary of the transformer is rectified with the help of a series of solid-state diodes allowing electrical current flow along single direction *i.e.* alternating component is removed (ripple). Smoothing is achieved largely by the HV cable, which has enough capacitance between its inner conductor and the grounded outer sheath to short-circuit the ac-ripple component to ground at the operating frequency of the transformer output. Because the output of the oscillator circuit is linearly related to its input and rectification is also a linear process, the magnitude of the resulting accelerating voltage is given by:

Table 3.2. Speed v and fractional increase in mass (γ) of an electron (rest mass m₀) for four values of the accelerating potential V₀. The final column illustrates that the classical expression for kinetic energy no longer applies at these high particle energies.

V_0 (kV) or E_0 (keV)	γ	v/c	$m_0 v^2/2$ (keV)
100	1.20	0.55	77
200	1.39	0.70	124
300	1.59	0.78	154
1000	2.96	0.94	226

$$V_0 = G \, V_i \qquad\qquad (3.8)$$

where G is a large amplification factor (or gain), dependent on the design of the oscillator and of the step-up transformer. Changing V_i (by altering the reference voltage $V+$ in Figs. **3-6**) allows $V0$ to be intentionally changed (for example, from 100 kV to 200 kV). However, V_0 could drift from its original value as a result of a slow change in G caused by drift in the oscillator or diode circuitry, or a change in the emission current I_e for example. Such HV instability would lead to chromatic changes in focusing and is generally unwanted. To stabilize the high voltage, a **feedback resistor** R f is connected between the HV output and the oscillator input. If G were to increase slightly, V_0 would change in proportion, but the increase in feedback current I_f would drive the input of the oscillator more negative, opposing the change in G. In this way, the high voltage is stabilized by negative feedback, in a similar way to stabilize of the emission current by the bias resistor. To provide adequate insulation of the high-voltage components in the HV generator, they are immersed in transformer oil (used in HV power transformers) or in a gas such as sulfur hexafluoride (SF₆) at a few atmospheres pressure. The high-voltage "tank" also contains the bias resistor Rb and a transformer that supplies the heating current for a thermionic or Schottky source. Because the source is operating at high voltage, this second transformer must also have good insulation between its primary and secondary windings. Its primary is driven by a second voltage-controlled oscillator, whose input is controlled by a potentiometer that the TEM operator turns to adjust the filament temperature. It is to be noted that the direction of the electron accelerating electric field is along the optic axis; it gets curved in the nearby region of the Wehnelt control electrode and the reason of this curvature is the lesser negative polarity of the electron source than the electrode. Curvature gives rise to an electrostatic lens action equivalent to a convex lens favoring the convergence of electrons. On the contrary, the curves above the hole near to the anode offer a diverging effect. This makes the electrons exist in the lens column look like they are coming from a virtual source.

The Specimen Stage

Sample preparation and specimen stage have always been very crucial parameters in TEM operation. The sample should be thin enough in order to pass the electrons through the sample or, in other words, transmit through this. The specimen stage is made in such a way that it should be unmovable and vibration-free as much as possible. The reason for this is that the slightest movement of the specimen stage may get significantly enlarged at the final stage, destroying its spatial resolution significantly. However, it should also be noted that to have a view of all the parts of the sample, it sometimes becomes necessary to move the specimen stage at most around 3 mm. Care has also been taken that during insertion of the sample through the specimen stage, the vacuum condition of the system should remain undisturbed. The latter is done by inserting the sample through an airlock which is nothing but a chamber through which the sample is placed into the TEM and before lacing the sample may be evacuated. The mechanical part of the airlock, as well as the specimen stage, is a bit complex, and there are two mechanisms or designs for the latter, *i.e.*, side-entry and top-entry.

In the first configuration, *i.e.*, the side-entry stage, the sample is inserted through the specimen holder horizontally through an airlock, as has been shown in Fig. (**3.11a**). The airlock-evacuation valve, as well as the high-vacuum valve (entry point of TEM column), is made activated consecutively by the act of rotation. Here the advantage stays in the fact that here the precision motion is easy to control.

Sometimes the movement along x, y or z direction is achieved by applying proper movement to an end stop that is kept in touch with the pointed end of the specimen holder. Sometimes in a microscopic study, it gets necessary to tilt the sample may be to have a clear idea about the shape of the sample or to get an insight into the microscopic defects present in the sample. Another significant advantage associated with the side entry stage is the possibility to set a heating arrangement that can help us to study the *in situ* structural change.

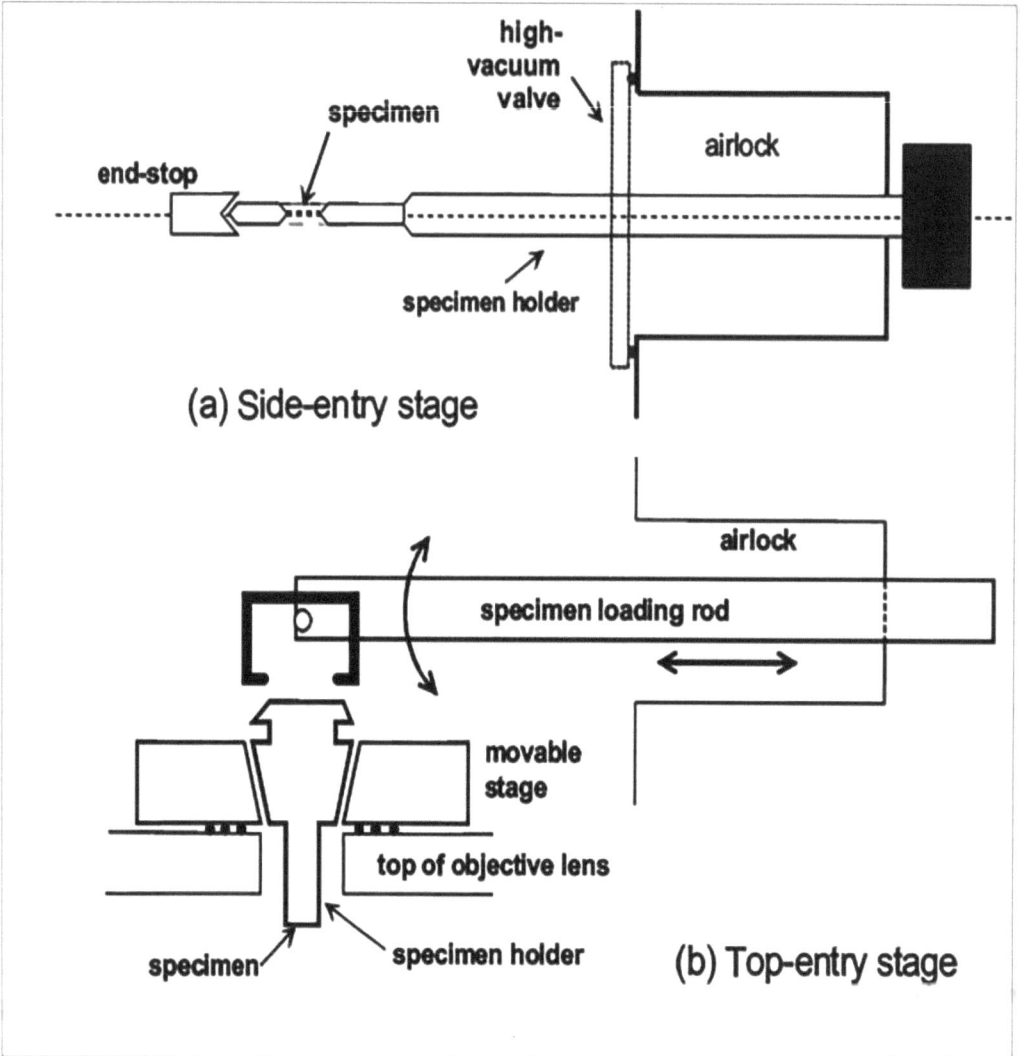

Fig. (3.11). Schematic diagrams of (a) a side-entry and (b) a top-entry specimen holder.

Specimen cooling can also be achieved, by incorporating (inside the side-entry holder) a heat-conducting metal rod whose outside end is immersed in liquid nitrogen (at 77 K). If the temperature of a biological-tissue specimen is lowered sufficiently below room temperature, the vapor pressure of ice becomes low enough that the specimen can be maintained in a hydrated state during its examination in the TEM.

One disadvantage of the side-entry design is that mechanical vibration picked up from the TEM column or acoustical vibrations in the external air, are transmitted directly to the specimen. In addition, any thermal expansion of the specimen holder can cause drift of the specimen and the TEM image. These problems have been largely overcome by careful design, including the choice of materials used to construct the specimen holder. As a result, side-entry holders are widely used, even for high-resolution imaging. In a top-entry stage, the specimen is clamped to the bottom end of a cylindrical holder that is equipped with a conical collar (shown in Fig. **3.11b**).The holder is loaded into position through an airlock by means of a sliding and tilting arm, which is then detached and retracted. Inside the TEM, the cone of the specimen holder fits snugly into a conical well of the specimen stage, which can be translated in the (x and y) horizontal directions by a precision gear mechanism. The major advantage of a top-entry design is that the loading arm is disengaged after the specimen is loaded, so the specimen holder is less liable to pick up vibrations from the TEM environment. In addition, its axially symmetric design tends to ensure that any thermal expansion occurs radially about the optic axis and therefore becomes small close to the axis. However, it is more difficult to provide tilting, heating, or cooling of the specimen. Although such facilities have all been implemented in top-entry stages, they require elaborate precision engineering, making the holder fragile and expensive. Because the specimen is held at the bottom of its holder, it is difficult to collect more than a small fraction of the x-rays that are generated by the transmitted beam and emitted in the upward direction, making this design less attractive for high-sensitivity elemental analysis.

The Imaging System

Sample in the TEM is placed in front of the objective lens in such a way (preferably a laminar way) that an electron beam from the source gets able to transmit through it (Fig. **3.12**). The objective lens, in turn, makes an image of the sample based on the electron distribution at the existing surface. On the back focal plane of the objective lens, a diffraction pattern gets produced, resulting in an image formation at the objective lens on its image plane. There are several steps and factors that convert the electron output into the signal in the visible range. The diffraction pattern produces, projection and set of intermediate lenses below the objective lens all collectively help the system to enlarge the diffraction pattern or the image. TEM is capable of producing both the diffraction pattern and as well as the microstructure that capable of conveying crystallographic as well as structural information.

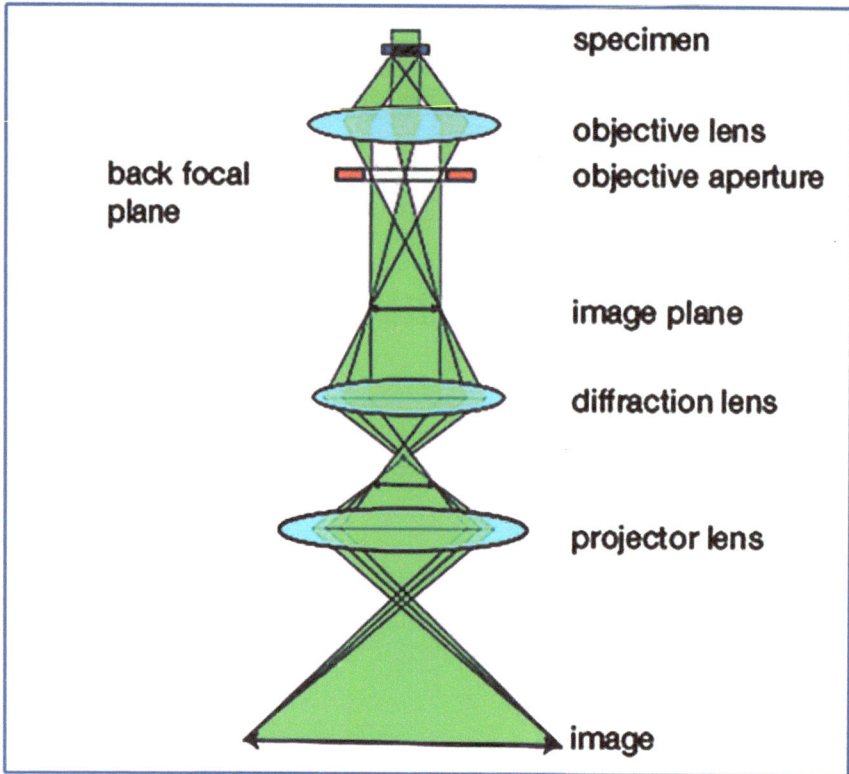

Fig. (3.12). Schematic of lens system used in imaging part of a transmission electron microscope.

Fig. (**3.12**) may explain the features of both the diffraction and image modes using the basic geometrical optics connected to the objective lens. One can see in Fig. (**3.12**) how the electron beams parallel to the optic axis come and falls upon the sample. When both the transmitted and the diffracted beam gets concentrated on the back focal plane of the objective lens, one can see the picture of the specimen under consideration. Thus, it is not surprising that as both diffraction patterns and the image of the specimen are produced by the objective lens, the further magnification is obtained on the screen under view by focusing the next lens in the magnification system. The scattering and absorption of the electrons by the specimen basically results in an intensity variation in the sample exit surface that gets magnified and in turn, produces images on a fluorescent screen.

It is to be noted that apart from diffraction, there are three other factors that play a major role in producing image contrast in TEM. The parameters are as follows:

Mass-Thickness Contrast

This phenomenon is rather the consequence of the Rutherford scattering that the electrons in the beam suffer while moving through the specimen. They are incoherent, off-axis and clastic scattering. The extent of scattering is proportional to the atomic number of the target and the higher the atomic number more electrons will scatter producing higher differential intensity in the image formed from the thicker region of the specimen. This mass thickness contrast plays a major role in image formation, especially in biological samples and is highly dependent upon the objective aperture size and accelerating voltage.

Diffraction Contrast

Here, the main mechanism is based on the fact that only one diffracted beam leaving the lower surface of the sample is allowed to form the image, and the rest are chopped out. This finds its wide use in describing the detail of a crystalline material having a diameter over 15 Å especially studying their defects even without reaching the maximum resolution.

Bright field (BF) and centred dark field (CDF) are the two most common diffraction contrast modes where diffracted and transmitted beams are blocked by changing the objective aperture position. However, due to the use of optic axis path, CDF suffers errors like spherical aberration or astigmatism, and the image qualities are comparatively poor. However, one may use a tilting of the electron gun (older microscope) or electromagnetic tilting device (modern microscope) in order to keep the diffracted beam lying along the optic axis and thus to retain the resolution of BF mode intact. This method is called as centred dark field (CDF) configuration and is used when detailing in DF image is needed. The whole state of affairs may be understood from Fig. (**3.13**). In Fig. (**3.13A**) transmitted and incident beam rests on the optic axis, but the diffracted ray does not and thus gets eliminated (BF mode). In the second case, where the diffracted ray is allowed, but away from optic axis, one gets displaced aperture DF (Fig. **3.13B**). The last is the modified version of the second, where using a suitable tilting device can bring the displaced diffracted beam onto the centre (Fig. **3.13C**).

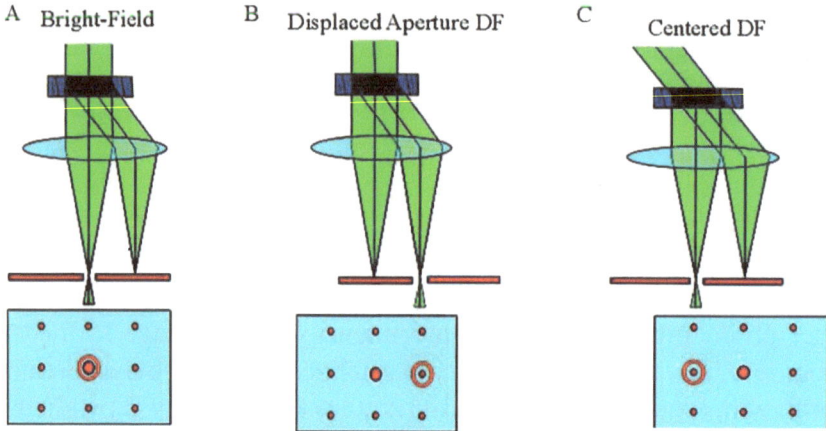

Fig. (3.13). Different imaging modes used in transmission electron microscopes.

Here the diffraction is nothing but a reflection from the set of crystal planes oriented at a particular angle w. r. t. incident ray. This angle is well known Bragg's angle.

Phase Contrast

As the name suggests, this method utilizes the differences in phases of the electron beams scattered by the thin specimen. Unlike the convention bright or dark field TEM, where only a central or single diffracted electron beam is used, the phase-contrast mechanism uses at least two and usually many diffracted beams, which form the image of interference and are usually capable of giving sub-unit cell details. Though the main principle of this method is the differences in phase of the electron beams scattered from the specimen, to convert this signal into the image, one should additionally supply external phase shifts in suitable ways and means. As this method can supply the sub-unit cell detail, it is the basis of the working of HRTEM. Phase contrast carries very important information regarding the periodicity *i.e.* the crystal structure as well as microstructures as it depends on both the phase information as well as information regarding the Bragg's criterion. For a periodic structure, it is capable of giving complete lattice imaging.

Sample Preparation

Since the TEM relies on the information that comes from the transmitted electron, the sample preparation here is very crucial. Firstly, to retain the high vacuum in the TEM chamber, the volatile sample should be avoided for the possible de-gassing.

The sample should be small enough, not more than 3 mm in diameter, which makes them enable to be inserted into the vacuum chamber. Also, in order to transmit sufficient numbers of the electron, the sample should be thin enough, not more than 20 nm, for ideal high-resolution operation and the biological sample may be around 300 nm thick. However, biological samples should go under a special preparation technique that keeps the cell structure intact while removing the entire water content. It is customary to use either a carbon-coated (Preferably) or non-coated copper mesh onto which a few drops of alcohol suspension of the sample are placed upon evaporation of the alcohol the sample is ready to be studied under an electron microscope. However, one should remember that electron microscope study is destructive in nature, especially for thin sheets or biological samples where hitting the sample with a very high energy electron beam may destroy the sample property.

A digital image of a high-resolution transmission electron microscope (JEOL-JEM-2100) is shown in Fig. (**3.14**).

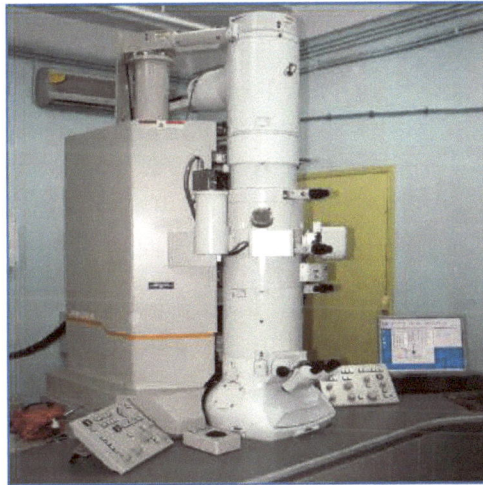

Fig. (3.14). The photographs of HRTEM.

Scanning Electron Microscope (SEM) and Field Emission Scanning Electron Microscope (FESEM)

The main drawback of the transmission electron has been made the main working tool of another electron microscope called scanning electron microscope (SEM), capable of the relatively thick sample, thus working equivalent to metallurgical light microscopes with further advantages of much higher spatial resolution.

When a high energetic incoming electron beam gets obstructed by a heavy target (sample), it may either get reflected (called back scattered) or it can impart its energy to the atomic electron of the sample causing secondary emission, which may be used to form an image of the sample. However, a special technique has to be taken as the secondary electrons are emitted with a range of energies and thus, it gets more difficult to focus them into an image by simply the means of electron lens. However to overcome the issue one has to make use of the fact that the direction of electron beam travelling may be altered by application of electrostatic or magnetic fields at perpendicular. Here the primary electrons are concentrated into an electron probe that remains scanning on the sample. A raster *i.e.* a square or rectangular area of the sample can be covered by simultaneous scanning in two perpendicular directions. The image now gets formed after collection of electron from each point of the specimen.

In the raster scan unlike the TEM or optical microscope the image is generated point wise not simultaneously.

Initially after the development of secondary electron-based scanning electron microscope at RCA laboratory, New Jersey, older version used cold emission source (which was later replaced by a thermal gun), electrostatic lens for focusing and early version of the FAX machine as image formation means. In this way, the resolution had reached around 50 nm, almost 10 times than the ordinary light microscope used.

Fig. (**3.15**) shows one example of a modern instrument where the image is stored in a computer and images appear in the associated display monitor. The best quality SEM (FESEM) may provide images having a resolution as high as 10 nm, but this is still inferior compared to the TEM here, the field of view (*i.e.*, the scan area) is much higher than the TEM, and thus it is more suitable for large area information. The basic components of SEM are almost the same as that of TEM, as both are electron microscopes. They both have electron gun, accelerator, lens system, vacuum system and others, as can be seen from the block diagram shown in Fig. (**3.15**).

In SEM system, the electron columns consist of the electron gun and focusing electromagnetic lenses, both kept at a high vacuum. Electrons that get generated from a source, either thermally or by cold emission, get accelerated under a potential difference of 1-40 KeV and simultaneously get converged to a narrow electron beam (electron probe) with a spot size less than 10 nm in diameter. However, it still carries current sufficient to form the image. The electron probe is mainly determined by three parameters; first is the probe diameter d (1 nm – 1μm), probe current (pA-μA) and lastly, probe convergence (10^{-4} to 10^{-2} radians).

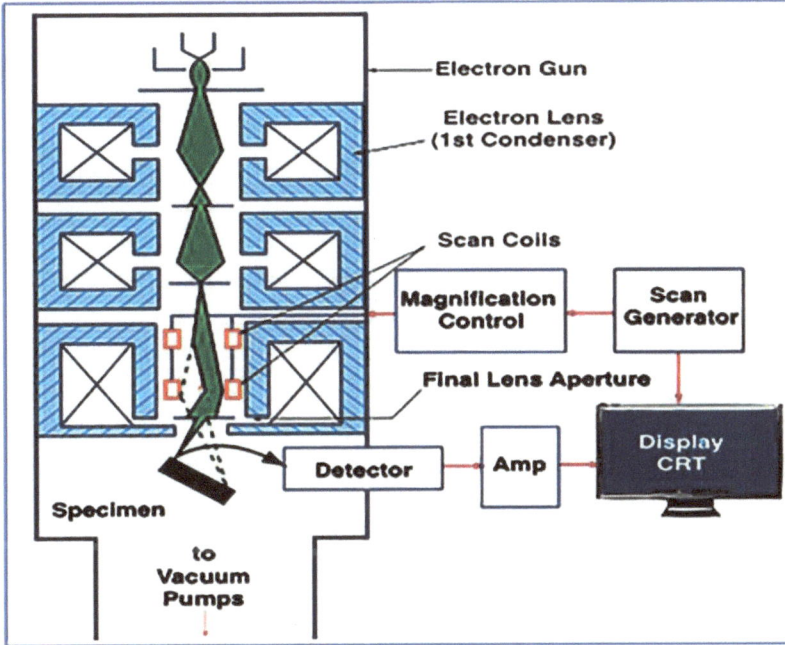

Fig. (3.15). Block diagram of Scanning Electron Microscope.

When the electron beam is focused on a final probe, which continuously scans the surface by means of a scanning coil, each point of the sample emits electromagnetic signals due to either secondary or backscattered electrons. The signal is then collected by the detector and converted to the image that, in turn, gets displayed on the monitor. The SEM image is rather straightforward to interpret so far as the topographical information is concerned. However, one may further concentrate on complex interactions between the beam and the sample in the entire penetration depth of around 1 μm to interpret further complex information out of the scope of this book.

Lenses in the SEM

The combinations of lenses are for producing converging electron beams having a particular crossover diameter. The lenses are made up of a metal cylinder with a cylindrical aperture inside into which magnetic fields are generated to control the focusing (or defocussing) of the electron beam. In general, there are total three condenser lenses that are used to lower the crossover diameter to the desired value. The first two monitor the amount of demagnification and the third one focuses the probe onto the sample. It should be remembered that as heavy current flows through the lenses thus, there remains an acute heating effect which may only be minimized by a suitable chiller arrangement.

In the final lens, it may be designed in three different ways, all having their own advantages and drawbacks. They are pinhole or conical lens, immersion lens and snorkel lens. Of which the pinhole configuration has advantages in two ways, *i.e.*, Firstly, here, the sample size is determined by the sample chamber only, not by the lens and secondly, as here the working distance is variable user has control over the scan area. Whereas due to the very short focal distance immersion lens offers the least aberration, probe size and thus in turn, the highest resolution. The third one, *i.e.*, Snorkel lens, has the best features of the previous two. It has low aberration as well as no restriction in sample size. However, the magnetic sample is difficult to study with this configuration since here, the magnetic fields may come out of the lens and reach the sample and thus may bring negative consequences both for the sample as well as for the instrument.

The general approach for efficient working of an SEM is minimizing the probe diameter and maximizing the probe current. The first parameters are functions of variables like accelerating voltage, electron gun size, spherical aberration and optical brightness. The brightness depends on the design and performance of the electron gun and it increases linearly with increasing accelerating voltage. If we increase the accelerating voltage the energy of the accelerated electron beam also gets results in less wavelength of the electron beam. This lesser wavelength has a regular consequence to decrease the probe size.

The easiest way of increasing beam current is to increase the accelerating voltage. However, increasing voltage beyond a certain limit (typically 30 kV) results in lesser resolution and thus should be avoided. Decreasing the spherical aberration is a rather more wise approach to achieving high current, which is done more efficiently in immersion-type lenses.

Interaction Volume

Interaction volume is associated with the interaction of accelerating electrons with the sample and is of extreme importance in interpreting the SEM data/image. This is basically the combined volume of the primary incident beam and the several secondary radiations that include back scattered electron, secondary electron, characteristic as well as bremsstrahlung x-rays, and sometimes cathode-luminescence. This parameter is one of the main determining factors of the image quality as well as resolution in SEM. Different elastic and inelastic scattering dominates the penetration depth of the incoming electron beam into the solids and it is that region over which incoming electrons interact with a sample which is called as interaction volume. For lighter targets, the volume has a pear shape, whereas for intermediate and heavy targets, the shape becomes hemispherical. The interaction volume varies directly with incident beam energy and, inversely average atomic number of the target or sample.

Image Formation

The SEM produced basically no true images, unlike TEM or optical microscope. It is an intensity mapping of the signal that gets generated due to sample-beam interaction in analog or digital domain. Here every pixel on display somehow is mapped to a particular point in the sample and is associated with the signal intensity detected by the detector from that particular point (Fig. **3.16**). As mentioned before, placing film anywhere and recording the photograph in SEM is impossible. Here images are both generated and displayed electronically.

For an analog scanning system, the incoming electron beam is allowed to move rapidly and continuously along the X-axis (line scan) combined with a relatively slow scan along the Y axis at predefined line numbers. The scan time of a single line multiplied by line numbers gives the parameter called frame time. When the question comes to a digital image where only discrete beam locations are allowed, the beam is allowed to stay at a particular point for a fixed dwell time and then goes to the next point. Here the steps are as follows:

When the beam comes and interacts at a particular point in the sample, the detector detects the analog signal produced as a result of the interaction. The signal detected by the detector is then amplified and gets converted to digital form, and stored at a specific location in the computer. The intensity most commonly is digitized into 8,

12 or 16 bits. One can see the image when the values stored in the computer memory get converted into an analog version for the purpose of display.

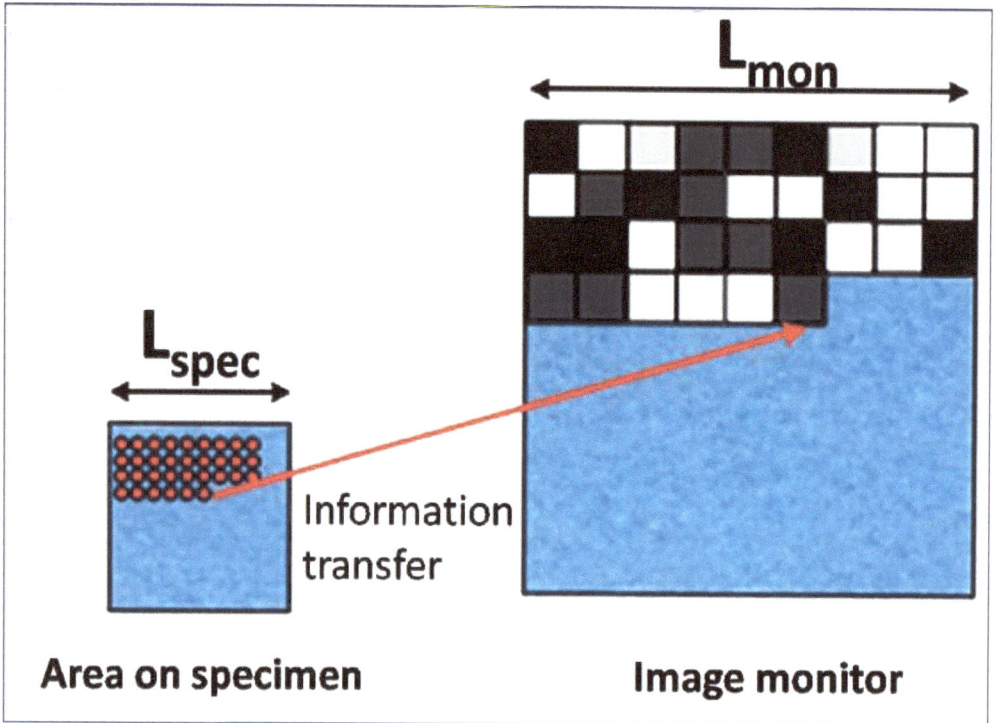

Fig. (3.16). Schematic of mapping in image formation process occurs in SEM.

Magnification in the SEM

Here the magnification is achieved by changing the scan area (lesser than the area of the display). As the display area is fixed, magnification is achieved by simply by varying the length of the scan. Magnification is done by the excitation of the scan coils and not by excitation of the objective lens. Quantitatively the magnification (M) is determined by the ratio of length of the monitor ((L_{mon}) over that of the scan ((L_{specs}) *i.e.* $M = L_{mon} / L_{spec}$. *i.e.*

Image Quality

The image quality is the manifestation of the signal-to-noise ratio, where the signal is nothing but the number of electrons reaching detectors from each point of the sample due to interaction with an incoming beam. When the number of counts from each point increases the signal-to-noise ratio automatically increases as there would be a lesser chance for the random fluctuation to get more prominent. Contrast has a value that lies between 0 to 1, and thus it is a positive definite quantity. A threshold of signal may be defined in terms of beam current, frame time, and efficiency of signal collection. The utility of the parameter is that if an object produces a signal lesser than the threshold, it cannot be separated from the signal. Each beam current is associated with a particular contrast value below which nothing can be observed, and once one knows the required beam current for imaging a particular contrast level, one can calculate the associated probe size considering all the aberrations. The desired contrast quality may also be achieved by optimizing the frame time, *i.e.*, primarily optimizing the scan speed. The image quality is also a function of the mechanical stability of the instrument as well as the external electromagnetic interference.

Signals and Imaging Modes in the SEM

The main two probes that are used in SEM imaging are SE and BSE of which the first particle has energy around 5 eV; the BSE has energy around a couple of thousand eV and thus, the difference is huge.

Compositional Contrast: This exists due to the differences in signal comes from the region of the specimen having different atomic numbers. This may be efficiently imaged by using BSE as the BSE coefficient η increases almost linearly with atomic number. Thus, simply the region with high Z will look brighter than that with low Z giving possibilities of the predicting compositional contrast. If one uses a detector which is sensitive to BSE numbers, then the signal detected is proportional to η. Thus the contrast (C) between two regions at 1 and 2 having back scattered coefficients η_1 and η_2 may be quantified as:

$$C = (\eta_1 - \eta_2)/ \eta_2$$

Topographic Contrast: Topographic contrast is associated with predicting the size, shape and morphology of the specimen, which is actually the major utility of the SEM. This kind of contrast arises due to the fact that the path of the BSE and SE depends on the angle at which the electron beam falls on the local surface of the specimen. The angle varies due to the variation of the local structures of the sample.

The contrast is a strong function of the type of detector and its relative position w.r.t. the sample and also its inclination.

Defects in SEM Imaging

Contamination is one of the major problems in SEM imaging. Contamination is basically the presence of unwanted artifacts (generally carbonaceous materials that get formed due to the breakdown of hydrocarbon) on the sample surface. The manifestation of the presence of impurities is associated with the reduction of the magnifications. This effect may be reduced by a gradual increase of magnification while starting from a low value. The contamination may become a serious issue, particularly for a high-resolution image. Using an ultra-high vacuum and use of the anti-contamination device is another way to get rid of this problem where the last one cools the area in the nearby region of the specimen and freezes the hydrocarbon migration. Before SEM study, the sample may adequately be exposed under UV light that can fix the contamination at the initial position.

Charging

Sometimes it is seen, especially when one deals with the insulating sample, that some parts of the field view appear extremely bright, and the rest is extremely dark. This happens when there is obstruction of the flow of electrons and charges get start to accumulate on the sample. This phenomenon is technically known as the charging effect. During the sample electron interaction, most of the electrons get captured by the sample and later, they are grounded through a proper path. However, sometimes this passage may get hindered, say for an insulator sample, and charges start to get accumulated and the surface potential rises. If there is significant local charging, it may affect the field lines associated with the detector potential, which in turn greatly affects the collection of secondary electrons. In this way, the voltage contrast that gets developed make some part of the sample look brighter and some part are black. This drawback may be eliminated by coating the sample with conducting particles (metals). The coating is so thin that it only makes the sample conduct without compromising its microstructure. Rapid scanning, using BSE or applying a less accelerating voltage are another few remedies.

Depth

The very small angular aperture of the electron probe forming system permits a large depth; the optical microscope will have a depth of field all in focus at once.

At a resolution of 1 µm, the optical microscope will have a depth of field of the same order, about 1 µm. The scanning microscope will have a depth of field of up to 7000 pm at this resolution. As a result, very rough surfaces such as pollen grains, microfossils, bone and tooth surfaces may be seen in focus across the whole specimen. At higher magnification, where a resolution of 10 nm or better from solid surfaces has been reached, the depth of field is still some hundred times greater than the value of the point-to-point resolution.

Contrast

The third feature of the scanning electron microscope, in addition to high resolution and great depth of field with respect to the optical microscope for solid specimens, is contrast. Scanning micrographs of solids inclined at an angle to the electron beam appear as if the eye were placed along the electron beam axis and illumination fell on the specimen from the electron collector placed to one side. This enhances the three-dimensional effect seen even when looking at one micrograph alone. The specimen may be tilted with respect to the beam, or *vice versa*, and two micrographs recorded will combine to give a stereographic pair. The third dimension can be measured with accuracy by using photogrammetric techniques from aerial photography (Boyde 1970). A typical image of SEM (JEOL-JSM-6360) and FESEM (Hitachi S-4800) are shown in Figs. (**3.17a** and **b**), respectively.

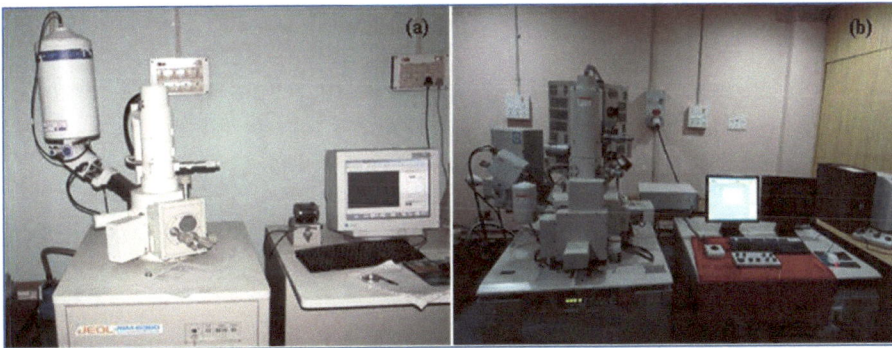

Fig. (3.17). The photographs of (**a**) SEM and (**b**) FESEM with EDX attachment.

Scanning Probe Microscopy

Scanning probe microscopy is a faster method to get the surface topology of a relatively smooth surface where a sharp tip, which we call the probe, scans the surface in a plane (say X-Y plane) in closed proximity to the surface and stores the z information (3-D) by means of particular patterns of interaction (may be mechanical or tunnelling). Fig. (**3.18**). There are mainly two kinds of SPM techniques that are:

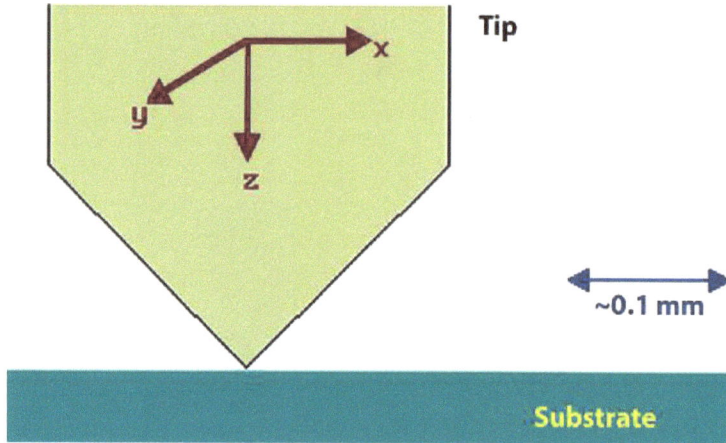

Fig. (3.18). Schematic of interaction between tip and surface in SPM techniques.

Scanning Tunnelling Microscopy (STM) - is based on the tunnelling current between a metallic tip and a conducting substrate which are in very close proximity but **not** actually in physical contact.

Atomic Force Microscopy (AFM) - is based on the van der Waals force between the tip and the surface; this may be either the short-range repulsive force (in contact-mode) or the longer-range attractive force (in non-contact mode).

Scanning tunnelling microscope is the first way and means in the world of microscopy that was able to give a surface topology with really atomic-scale resolution. Here a conducting tip is brought very close to the sample (conducting surface with relatively less roughness) of the order of 1 nm. A biasing voltage of say 1 V is applied. As a result of this depending upon the Fermi level of the material of tip and sample a very small tunnelling current flows between tip to sample or sample to tip. Here the sample and the tip comes very close to each other, but it never touches each other, and thus, the flow of electron occurs solely due to the quantum mechanical phenomenon called "tunnelling". To have an idea about the sharpness of the tip, one may study the following schematics (Figs. **3.19a-c**).

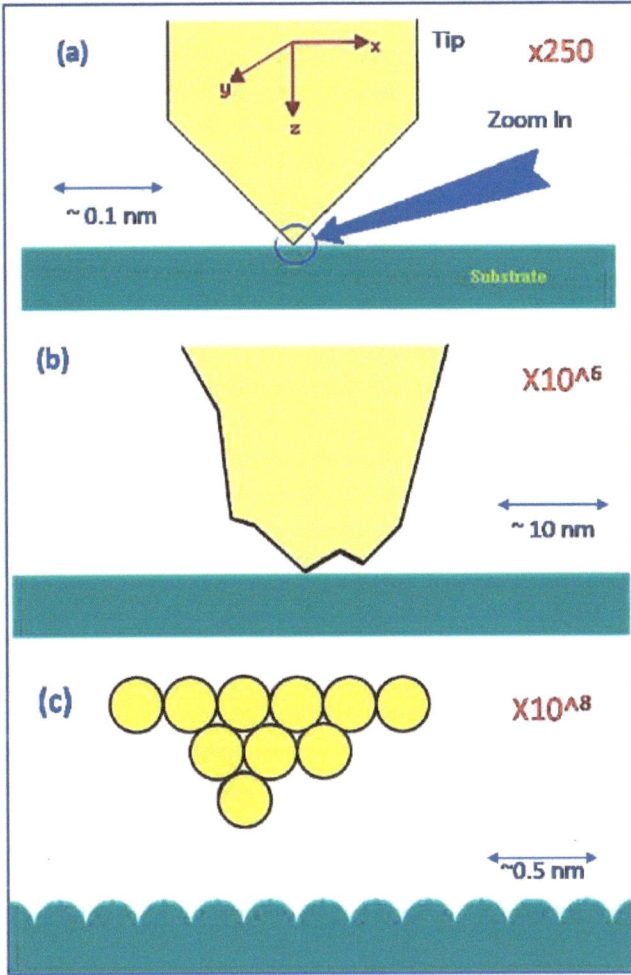

Fig. (3.19). Evolution of a sharp tip structure with increasing magnification.

Let us look at the region where the tip approaches the surface in greater detail. If the tip is biased with respect to the surface by the application of a voltage between them, then electrons can tunnel between the two, provided the separation of the tip and surface is sufficiently small - this gives rise to a tunnelling current. The direction of current flow is determined by the polarity of the bias. If the sample is biased -view with respect to the tip, then the electrons will flow from the surface to the tip as shown above, whilst if the sample is biased +ve with respect to the tip, then the electrons will flow from the tip to the surface as shown below as shown in Fig. (**3.20**).

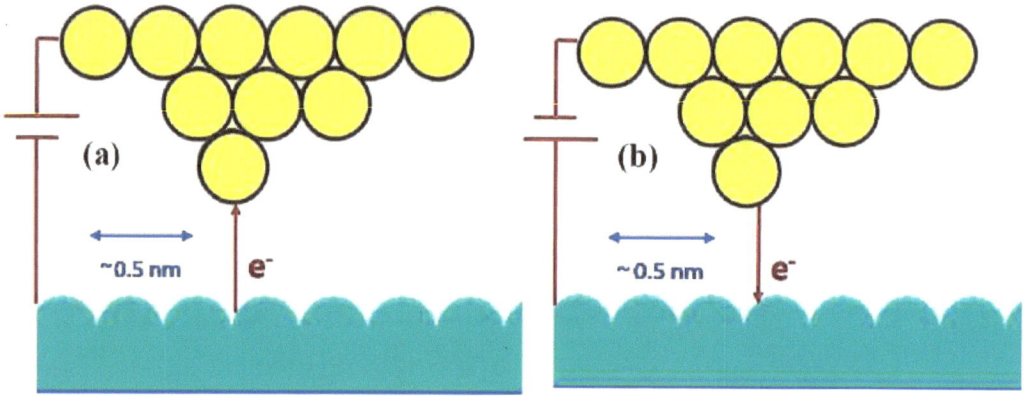

Fig. (3.20). Direction of electron flow between tip-surface when sample is biased (**a**) – ve and (**b**) + ve with respect to tip.

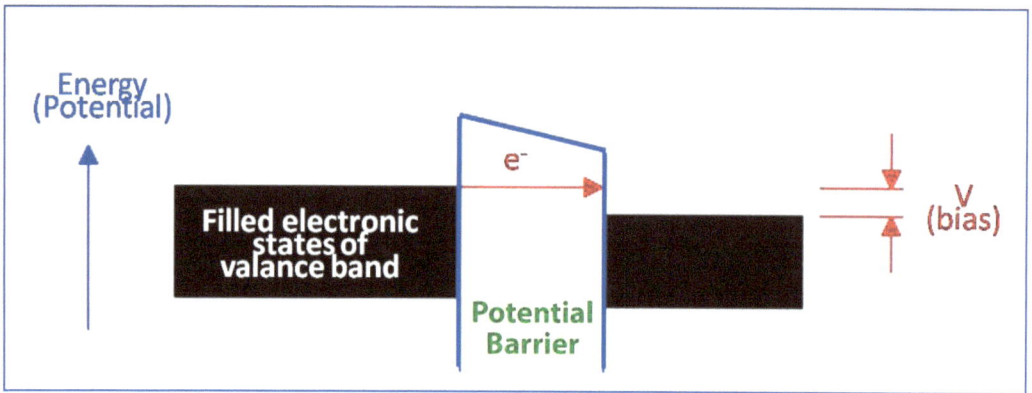

Fig. (3.21). Schematic of tunnelling phenomena.

The name of the technique arises from the quantum mechanical *tunnelling*-type mechanism by which the electrons can move between the tip and substrate. Quantum mechanical tunnelling permits particles to tunnel through a potential barrier which they could not surmount according to the classical laws of physics - in this case, electrons are able to traverse the classically-forbidden region between the two solids as illustrated schematically on the energy diagram below in a simplified way (Fig. **3.21**).

It is to be noted that imaging of the surface topology may then be carried out in one of two ways:

1. in constant height mode (in which the tunnelling current is monitored as the tip is scanned parallel to the surface).

2. in constant current mode (in which the tunnelling current is maintained constant as the tip is scanned across the surface).

If the tip is scanned at what is nominally a constant height above the surface, then there is actually a periodic variation in the separation distance between the tip and surface atoms. At one point, the tip will be directly above a surface atom, and the tunnelling current will be large whilst at other points, the tip will be above hollow sites on the surface, and the tunnelling current will be much smaller.

In practice, however, the normal way of imaging the surface is to maintain the tunnelling current constant whilst the tip is scanned across the surface. This is achieved by adjusting the tip's height above the surface so that the tunnelling current does not vary with the lateral tip position. In this mode, the tip will move slightly upwards as it passes over a surface atom, and conversely, slightly towards the surface as it passes over a hollow.

A plot of the tunnelling current v's tip position, therefore, shows a periodic variation that matches that of the surface structure - hence it provides a direct "image" of the surface (and by the time the data has been processed, it may even look like a real picture of the surface!). The image is then formed by plotting the tip height (strictly, the voltage applied to the z-piezo) v's the lateral tip position. Both the state of affairs have been shown Fig. (**3.21**).

Maintaining a tip within1 nm of a surface (without touching) requires great mechanical precision, an absence of vibration, and the presence of a feedback mechanism. As the tunnelling current increases dramatically with decreasing tip-sample separation, a motorized gear system can be set up to advance the tip towards the sample (in the z-direction) until a pre-set tunnelling current (*e.g.*, 1 nA) is achieved; see Fig. (**3.22**) a. The tip-sample gap is then about 1 nm in length, and fine z-adjustments can be made with a piezoelectric drive (a ceramic crystal that changes its length when voltage is applied). If this gap were to decrease due to thermal expansion or contraction, for example, the tunneling current would start to increase, raising the voltage across a series resistance (see Fig. **3.22a**).

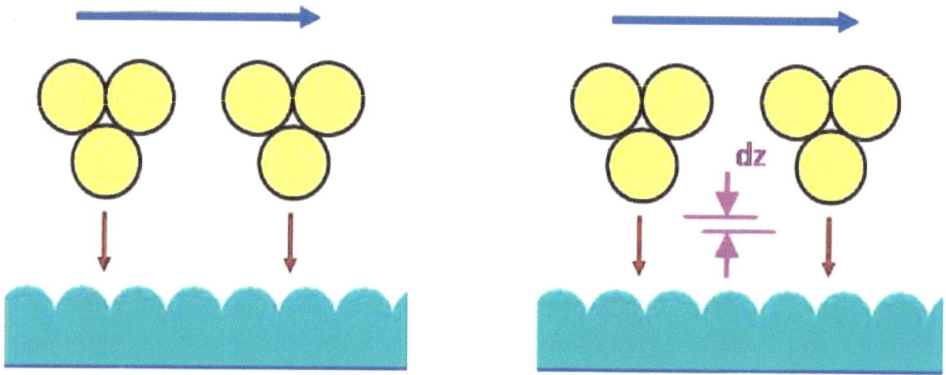

Fig. (3.22). Different modes of STM (**a**) constant height and (**b**) constant current mode.

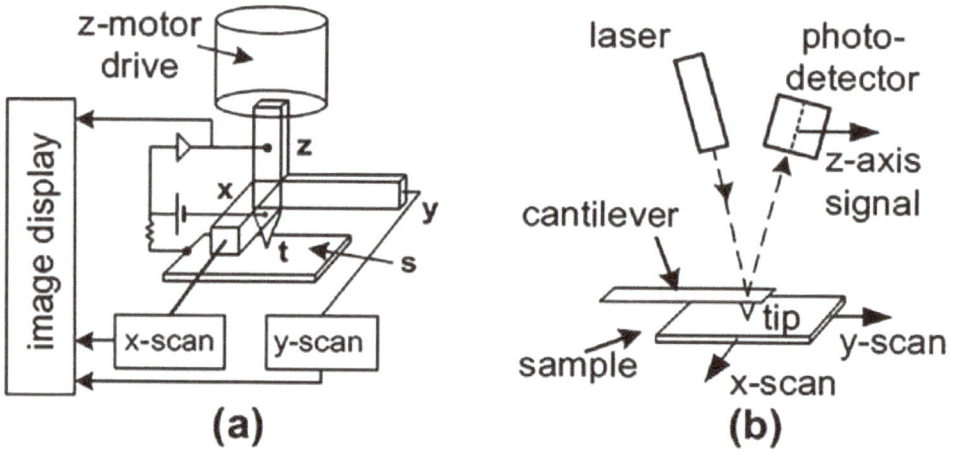

Fig. (3.23) (**a**) Principle of the scanning tunnelling microscope (STM); x , y, and z represent piezoelectric drives, t is the tunnelling tip, and s is the specimen. (**b**) Principle of the scanning force (or atomic force) microscope. In the x- and y-directions, the tip is stationary and the sample is raster-scanned by piezoelectric drives.

The change of voltage, as mentioned, then gets amplified and applied back to the piezo, thus decreasing the gap, which in turn maintains the constancy in the current. This is called the negative feedback mechanism, as here, the tip to sample distance is fed back to the electrochemical system to keep this gap constant.

The principle of the image formation technique is rather simple. Here the probe goes on scanning the surface along the x and y direction again and again. During scanning, the z directional movement of the probe follows exactly the same undulation as that on the surface, thus giving the surface topography (Fig. **3.23a**).

As the tunnelling current is extremely sensitive to the gap between tip and sample and varies exponentially at the same, the resolution that can be achieved with STM is extremely high and may go as high as 0.01 nm that is up-to atomic resolution. However, it suffers different inherent drawbacks, also. Firstly, it is applicable for a very smooth surface as it can very easily be assumed that if there is a huge difference in the heights of adjacent sites in the sample, the movement cannot be made flexible beyond a certain limit, and thus probe gets broken or damaged by collision. Also here, the scan time is very long and thus, it requires a considerable duration to scan an area as less as a micro-meter range, and at the same time, one has to keep the entire system absolutely vibration-free (mere talking near the system may create vibration which is sufficient to introduce error in the measurement). Unlike the previous case where the current remained unchanged, the constant height mode may also employ in the STM operation where the modulation of tunnelling current with local roughness would produce the topology of the sample. However, in this mode, there would be a higher chance that during movement, the tip crashes to the surface and gets damaged permanently.

The huge success of this technique for the conducting surface inspired other workers to develop different kinds of scanning probe microscopes most prominently atomic force microscope (AFM), which works on the interatomic forces between a tip fixed at the endpoint of a piezoelectric flexible cantilever and the surface atoms/molecules of our interest. The working of the AFM may also be classified into two types that include contact mode and semi-contact mode, where the nature of interaction respectively becomes repulsive and attractive. The associated interaction energy has best been given by well-known Lennard Jones potential. Here a LASER light is focussed on the tip of the cantilever, and the reflected LASER beam is caught on the detector. As the cantilever scans the surface, the tip of the cantilever moves ups and down due to its flexibility and piezoelectric property. This would produce a shifting of reflected LASER light in the detector. The deflection is the measure of the surface irregularity and thus produces the topography of the surface (Fig. **3.23b**). It should be noted that by changing the nature of the tip, several parameters like a magnetic field, entropy, roughness, histogram, *etc.*, may also be noted obtained.

AFM has inherent drawbacks compared to STM with reference to the atomic resolution, but that can be compromised with the advantage that it does not require any external vacuum system. Another very interesting fact about the AFM is that the corresponding measurement may be done in a liquid medium, thus making it suitable for being used in the case of biological samples.

For this, versatile application coupled with the relatively low cost has made SPM suitable for the application in a few cases where initially SEM and TEM were used.

There are still two other drawbacks related to SPM measurement one is obviously as mentioned before, related to the excess time consumption in mechanical scanning. It is also not suitable for varying the magnification dynamically.

The technique, unlike SEM or TEM, can give only a little or actually no idea regarding the chemical composition of the sample. STM, when used in spectroscopy mode, though it can give little idea about the valence electron and thus, it is not suitable for chemical composition analysis. SPM, above all, is an excellent measurement technique, but when the measurement is associated with surface properties only.

Fig. (**3.24**) shows the image of an atomic force microscope (AFM-NT-MDT, Solver Pro., Russia).

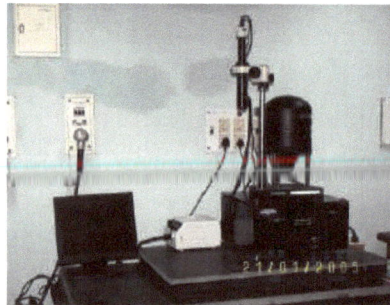

Fig. (3.24). Photograph of Atomic Force Microscopes.

CONCLUSION

This chapter has dealt with one of the most fundamental aspects of nanoscience and technology, *i.e.*, how to see nanomaterials. It has been concluded that to study and characterize nanomaterial, it is most obvious to develop a new kind of microscope that can magnify the objects several thousand times to see them clearly, and this was the main problem in the early days of nanotechnology. In this consequence, the basic properties of light have been discussed, and it is mentioned that light can have both rays as well as wave characters.

While discussing the resolution of the microscope, the chapter confirms that in an optical microscope, the diffraction pattern of the two nearby objects superpose into each other and hinders their visibility as separate objects. The ability to see two nearby objects separately is called the resolving power of the optical instrument. The resolving power of a microscope depends upon the wavelength of the incident radiation, and this is the basis of all the sophisticated electron microscopes developed so far.

The construction of the electron microscope was first initiated in 1926 by Hans Busch and Albert Prebus; the practical electron microscope was constructed in 1938, and Siemens produced the first commercial transmission electron microscope in 1939.

The very basic working principles of electron microscopes like TEM or SEM have been discussed, and it is concluded that the limitation of the TEM is, unless the specimen is made very thin, electrons are strongly scattered within the specimen, or even absorbed rather than transmitted. This constraint has provided the incentive to develop electron microscopes capable of examining relatively thick (so-called bulk) specimens.

The scanning electron microscope, though it provides much lower magnification than TEM, can be a good tool to see the structure of the nanomaterial. This uses secondary electrons instead of transmitting electrons, and when the electron gun emits electrons due to the applied electric field, it is called a field emission scanning electron microscope.

There are mainly two modes in scanning probe microscopy; in STM, it is the tunnelling current between a metallic tip and a conducting substrate which are in very close proximity but **not** actually in physical contact, that basically provides the information of the substrate, whereas in AFM, it is the van der Waals force between the tip and the surface; this may be either the short-range repulsive force (in contact-mode) or the longer-range attractive force (in non-contact mode) that gives the topological information.

BIBLIOGRAPHY

[1]   Lee, W.M.; Reece, P.J.; Marchington, R.F.; Metzger, N.K.; Dholakia, K. Construction and calibration of an optical trap on a fluorescence optical microscope. *Nat. Protoc.,* **2007**, *2*(12), 3226-3238.
        http://dx.doi.org/10.1038/nprot.2007.446 PMID: 18079723

[2]   Tarrach, G.; Bopp, M.A.; Zeisel, D.; Meixner, A.J. Design and construction of a versatile scanning near-field optical microscope for fluorescence imaging of single molecules. *Rev. Sci. Instrum.,* **1995**, *66*(6), 3569-3575.
        http://dx.doi.org/10.1063/1.1145471

[3]   Lin, H.N.; Chen, S.H.; Lee, L.J.; Tsai, D.P. Construction of a dual mode scanning near-field optical microscope based on a tapping mode atomic force microscope. *Rev. Sci. Instrum.,* **1998**, *69*(11), 3840-3842.
        http://dx.doi.org/10.1063/1.1149187

[4]   Carter, A.R.; King, G.M.; Ulrich, T.A.; Halsey, W.; Alchenberger, D.; Perkins, T.T. Stabilization of an optical microscope to 01 nm in three dimensions. *Appl. Opt.,* **2007**, *46*(3), 421-427.
        http://dx.doi.org/10.1364/AO.46.000421 PMID: 17228390

[5]   Sridhar, M.; Basu, S.; Scranton, V.L.; Campagnola, P.J. Construction of a laser scanning microscope for multiphoton excited optical fabrication. *Rev. Sci. Instrum.,* **2003**, *74*(7), 3474-3477.
        http://dx.doi.org/10.1063/1.1584079

[6]   Steidtner, J.; Pettinger, B. High-resolution microscope for tip-enhanced optical processes in ultrahigh vacuum. *Rev. Sci. Instrum.,* **2007**, *78*(10), 103104.
        http://dx.doi.org/10.1063/1.2794227 PMID: 17979403

[7]   Hogg, J. *The Microscope: Its History, Construction, and Applications*; Illustrated London Libr, 1854.

[8]   Matthews, H.J. **1987**.Applications of Image Processing in the Scanning Optical Microscope.

[9]   Lewis, A.; Isaacson, M.; Harootunian, A.; Muray, A. Development of a 500 Å spatial resolution light microscope. *Ultramicroscopy,* **1984**, *13*(3), 227-231.
        http://dx.doi.org/10.1016/0304-3991(84)90201-8

[10]   Kim, D.U.; Song, H.S.; Song, U.; Kim, D.Y. Construction of Confocal Optical Microscope based on Resonant Scanning Mirror. *Proceedings of the Optical Society of Korea Conference,* **2008**, pp. 275-276.

[11]   Martínez-Tejada, H.V. General considerations for design and construction of transmission electron microscopy laboratories. *Acta Microscopica,* **2014**, *23*(1), 56-69.

[12]   Hinks, J.A. A review of transmission electron microscopes with *in situ* ion irradiation. *Nucl. Instrum. Methods Phys. Res. B,* **2009**, *267*(23-24), 3652-3662.
        http://dx.doi.org/10.1016/j.nimb.2009.09.014

[13]   Kuwahara, M.; Takeda, Y.; Saitoh, K.; Ujihara, T.; Asano, H.; Nakanishi, T.; Tanaka, N. Development of spin-polarized transmission electron microscope. *J. Phys. Conf. Ser.,* **2011,** *298*(1), 012016. []. IOP Publishing.].
http://dx.doi.org/10.1088/1742-6596/298/1/012016

[14]   De Graef, M. *Introduction to conventional transmission electron microscopy*; Cambridge university press, 2003.
http://dx.doi.org/10.1017/CBO9780511615092

[15]   Kimura, Y.; Yasuno, M.; Shimizu, R. Development of Coincidence Transmission Electron Microscope (I) Coincidence Image Construction System. *Microscopy,* **1995,** *44*(5), 295-300.

[16]   Wang, C.; Huddle, T.; Lester, E.H.; Mathews, J.P. Quantifying curvature in high-resolution transmission electron microscopy lattice fringe micrographs of coals. *Energy Fuels,* **2016,** *30*(4), 2694-2704.
http://dx.doi.org/10.1021/acs.energyfuels.5b02907

[17]   Flegler, S.L.; Heckman, J.W., Jr; Klomparens, K.L. *Scanning and transmission electron microscopy: an introduction*; Oxford University Press: UK, 1993, p. 225.

[18]   Kliewer, C.E.; Kiss, G.; DeMartin, G.J. *Ex situ* transmission electron microscopy: a fixed-bed reactor approach. *Microsc. Microanal.,* **2006,** *12*(2), 135-144.
http://dx.doi.org/10.1017/S1431927606060077 PMID: 17481349

[19]   Slayter, E.M.; Slayter, H.S. *Light and electron microscopy*; Cambridge University Press, 1992.

[20]   Muller-Reichert, T.; Verkade, P. *Correlative light and electron microscopy II*; Academic Press, 2014.

[21]   Haugstad, G. *Atomic force microscopy: understanding basic modes and advanced applications*; John Wiley & Sons, 2012.
http://dx.doi.org/10.1002/9781118360668

[22]   Bowen, R.; Hilal, N. *Atomic force microscopy in process engineering: An introduction to AFM for improved processes and products*; Butterworth-Heinemann, 2009.

[23]   Takeyasu, K., Ed.; *Atomic force microscopy in nanobiology*; CRC Press, 2014.

[24]   Braga, P.C.; Ricci, D., Eds.; *Atomic force microscopy: biomedical methods and applications*; Springer Science & Business Media, 2004, Vol. 242.

[25]   Salapaka, S.M.; Salapaka, M.V. Scanning probe microscopy. *IEEE Control Syst.,* **2008,** *28*(2), 65-83.
http://dx.doi.org/10.1109/MCS.2007.914688

[26]   Meyer, E.; Hug, H.J.; Bennewitz, R. *Scanning probe microscopy*; Springer: New York, 2003, Vol. 4.

[27]   Voigtländer, B. *Scanning probe microscopy: Atomic force microscopy and scanning tunneling microscopy*; Springer: Berlin, 2015, pp. 324-327.
http://dx.doi.org/10.1007/978-3-662-45240-0

CHAPTER 4

Structures and Basic Properties of Textile Dyes and their Impact on the Environment

Abstract: In today's world, the textile industry is one of the most important sectors, both economically and in everyday life. The textile industry is one of the most important commercial sectors that require a significant amount of water and chemical ingredients for several types of processing needed during the conversion of fibres to final textile products ready for sale. Textile dyes are an important topic to discuss from both a positive and negative perspective, as dyes are an unavoidable part of colouring clothes or papers, but when they end up in the environment as a waste product from the industries mentioned above, they have a significant negative impact on the environment and the water ecosystem. Thus, it becomes necessary to handle the dyes properly and develop ways and means to remove/reuse them. However, in order to do so, one must have an in-depth understanding of the structures and properties of various dyes, as well as the treatment that should be applied. Keeping this in mind, we have discussed the basics of textile dyes in this chapter. The classifications of textile dyes, as well as their chemical structures and qualities, were also covered in this chapter. The in-depth discussion on the fundamental of textile dyes may help the workers handle dyes in a controlled way to protect the environment from the associated toxicity.

Keywords: Dyes, Toxicity, Water pollution, Ecosystem, Chromophores, Environment.

INTRODUCTION

The textile dye industry has a history of more than 4000 years, but it was only around 150 years ago that people got capable of synthesizing dyes artificially [1]. This boom was a result of the work of William Henry, who discovered Mauveine accidentally, while working on Quinine synthesis [2].

This accidental discovery led to a community of a new generation of dyes, and subsequent research into the synthesis and applications of dyes and pigments began. It is noteworthy that there exists firm demarcation between dyes and pigments, mainly in terms of stability as well as molecular size. Dye molecules are soluble either in water or oil, lesser in molecular size, and thus more vulnerable to external perturbation compared to pigments [3]. Dyes basically are unsaturated organic

Diptonil Banerjee, Amit Kumar Sharma and Nirmalya Sankar Das

molecules with complex structures capable of absorbing light and subsequently giving colors in the visible region [4]. There are a variety of dyes with varying formulas, structures, and properties. As mentioned before, dyes are capable of absorbing part of the visible spectrum (chromophore) and thus, the color is associated with the part that is not absorbed. Dyes may be distinguished from one-another, from the point of view of their ability to absorb a specific part of the entire visible spectrum, *i.e.*, wavelength between 380 to 750 nm. In the process, a selective portion of white light gets absorbed by a group of atoms during reflection, transmission, diffusion, or other processes. The very group of atoms responsible for the absorption of colour is called chromophoric groups. This clearly suggests it is the chemical composition as well as the molecular structures that are responsible for the intensity of coloration. It basically takes certain luminous radiation in turn, giving their complementary colours [5]. As mentioned before, the use of natural dyes was rather common in ancient days when different natural materials like beetroot, other plants, insects, and animal minerals were used as natural dyes. The concept was environmentally friendly as all these natural dyes are low or null toxic, and the corresponding wastewater can be degraded by different biological means [6].

Nowaday, natural dyes are almost totally replaced by synthetic dyes due to the constantly growing demand for color in textile, paper and associated other industries. A recent survey says that now a day, around 10 % of the stuff is coming out as waste material from almost 10^8 tons of synthetic dyes produced every year [7]. It is also an alarming fact that almost 50 % of the dyes actually have nothing to do with textile but exist in the liquid phase with all its toxicity.

In a dye, the most important part is the chromophore which is actually nothing but the spatial arrangements of atoms responsible for absorbing colors. The absorption takes place due to the photo-induced excitation of electrons that exists in these chromophores. It is very clear that depending on the nature of dye; chromophores are representatives of different functional groups that include azo, Nitro, nitroso, carbonyl, thiocarbonyl, and alkenes or others [8]. The molecules containing chromophores are called chromogenic. However, they are not effective unless the remaining part of the molecule called auxochrome comes and joins together to modify the color of the dye and thus to open the dye possibilities. The auxochromes may be of both the types, *i.e.* acidic and basic. The remaining part of the molecule is the third part only. It is the existence of a stable functional group only that makes it difficult to handle them by conventional treatment [9].

If we discuss more the basic structures of dyes and their connection in the entire procedure, we have to remember that there are three parts consisting of chromophoric, auxochromic group and the basic skeleton aromatic rings (like benzene or anthracene or other). Depending upon the relative abundance of all the groups mentioned above, the conjugation of the unsaturated bonds increases, which in turn affects the activity of the electrons giving different colors [10]. Presently, we can hear around 40000 trade names that have been given to 8000 different synthetic dyes, having specific color index (CI) specifying color, class and order numbers [11]. Just for a better understanding of the reader, we have taken Reactive Black 5 dye as an example, which happens to be a diazo dye. The structure is shown in Fig. (**4.1**).

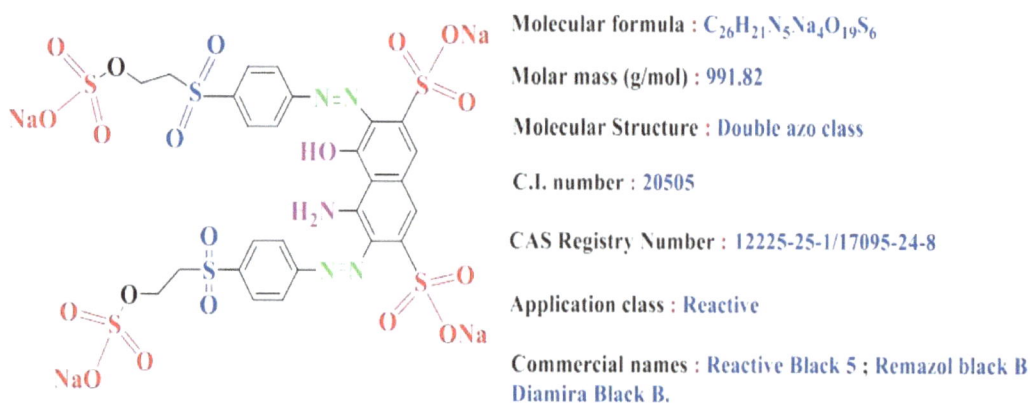

Molecular formula : $C_{26}H_{21}N_5Na_4O_{19}S_6$

Molar mass (g/mol) : 991.82

Molecular Structure : Double azo class

C.I. number : 20505

CAS Registry Number : 12225-25-1/17095-24-8

Application class : Reactive

Commercial names : Reactive Black 5 ; Remazol black B Diamira Black B.

Fig. (4.1). Reactive Black 5 as an example of how to classify and uniquely recognize a dye uniquely Representative example for the color index classification.

CLASSIFICATION OF DYES

As mentioned before, there are very few numbers of natural dyes that are available for use, thus, a simple nomenclature was enough to remember their detailed properties, structures and applications of them. With the advent of different industries like textile, papers and other demand for dyes found new exponential growth, which basically initiated the need for systematic classifications of dyes. The classifications have been done based on a number of parameters, mainly chromophore structures or color index and Industrial application.

Dyes Based on the Chromophores Present and CI

Triphenylmethane, azo, diphenylmethane, ox-azine, xanthene, *etc.*, are the different classes of dyes that possess different chromophores [12]. The chromphores are the main ones responsible for the dye colour shown by the chromogenic. It is the structure of chromophores and its interaction with auxochrome that gives rise to the color, and thus it is very natural that dyes will be classified in terms of the chromophore chemical structures, which, when combined with auxochrome and the base skeleton, give unique structure to a particular dye molecule. Table **4.1.** summarizes the classification of dyes as per their chromophore structure.

Table 4.1. Chromphore-induced classification of dyes [11].

Class	Chromophore	Dye Example
Azo		
Cyanine		
Xanthene		
Nitro		

(Table 4.1) cont.....

Quinione-imine		
Indigoid		
Acridine		
Oxazine		
Anthraquinone		
Triarylmethane		

(Table 4.1) cont.....

Phthalein		
Triphenylmethane dyes		
Nitroso		
Nitro		
Diarylmethane		

Not only chromophores but also the auxochrome, when combined with chromophores, take an important role to color the chromogenic. Few very common chromogenic are (–C = C-), (-C = N-), (-C = O-), (-N = N-), (–NO$_2$) and quinoid rings and on the other hand, few auxochromes are (–CO$_2$H), (–NH$_3$), (-SO$_3$H), and (–OH), *etc.* [13, 14].

Industrial Application Categories

Under this section, there would come a number of subgroups. We will see it one by one.

Disperse Dye

The first mention-worthy information about this class of dye is that it is only partially soluble in water and thus mostly applied to hydrophobic fibers. These classes of fibers have non-ionic characters. Generally, these dyes contain azo structures, but anthraquinone derivative-induced blue and violet colors are also not uncommon for them [15, 16]. These kinds of dyes are best suited to be used in polyester cloths. However, it is not uncommon to use in other fibers like acrylic, nylon, cellulose acetate and others, but maybe after compromising their wet-fastness properties. As mentioned before, though few anthraquinone-based dyes belong to the same disperse class of dyes, however more than 50 % disperse dyes alone contain azo group. Moreover, these dyes are stable dyes due to their recalcitrant attitude towards external perturbation and thus belong to the persistent class and are not biodegradable. As the dyes are mostly insoluble in water, the coloring process is mostly related to the dye bath of the fabric, where a carrier is also introduced in order to get better diffusion. And thus, sublimation temperature is one of the most important parameters for such dyes. Depending upon the molecular formula rather than molecular size, the sublimation temperature varies, and dyes are also classified as class A to class D dye, with sublimation temperature higher in class A dye and lowest for class D dye [16].

Just like the sublimation temperature, the stability of the dye also varies with system size, and the stability gets substantially better when the system size enters into nano-regime [17].

In this regard, Glover and coworkers reported important findings where it is reported that though dispersed dyes are absorbed into both polyamides based as well as polyester dyes, the fabric-dye bond strength is much stronger in the case of polyamide-based dyes because of the negligible or zero dissolution of dyes into such fabrics. However, the bonds may be easily broken under perturbation of a polar bond interrupter like water [18]. This explains why the fastness property in polyamide fabric and disperse dye system is not very appreciating. The structure of a few dispersed dyes along with corresponding CI is illustrated in Fig. (**4.2**).

Fig. (**4.3**) shows how different structures can be developed by changing functional groups and their position in the molecular skeleton structure.

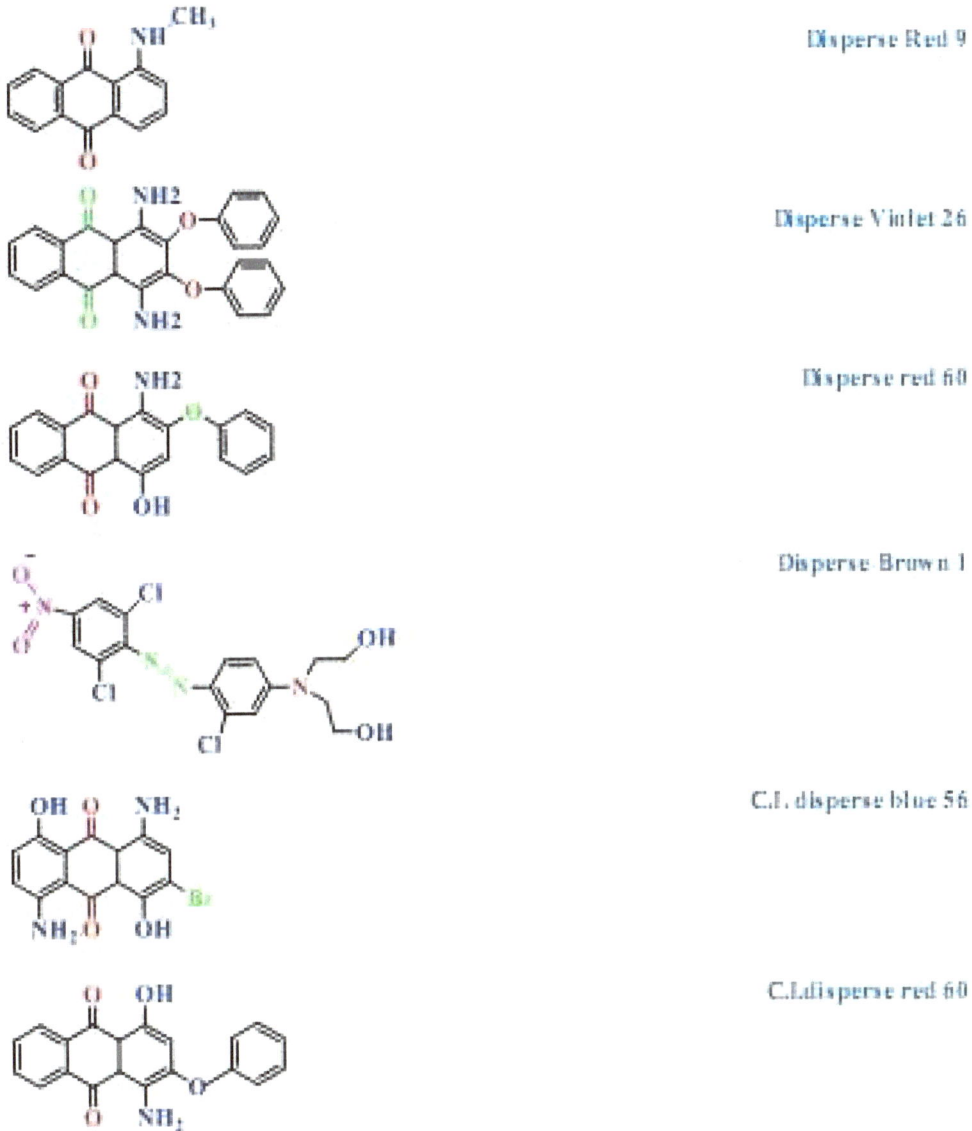

Disperse Red 9

Disperse Violet 26

Disperse red 60

Disperse Brown 1

C.I. disperse blue 56

C.I. disperse red 60

Fig. (4.2). Chemical structure of some disperse dyes.

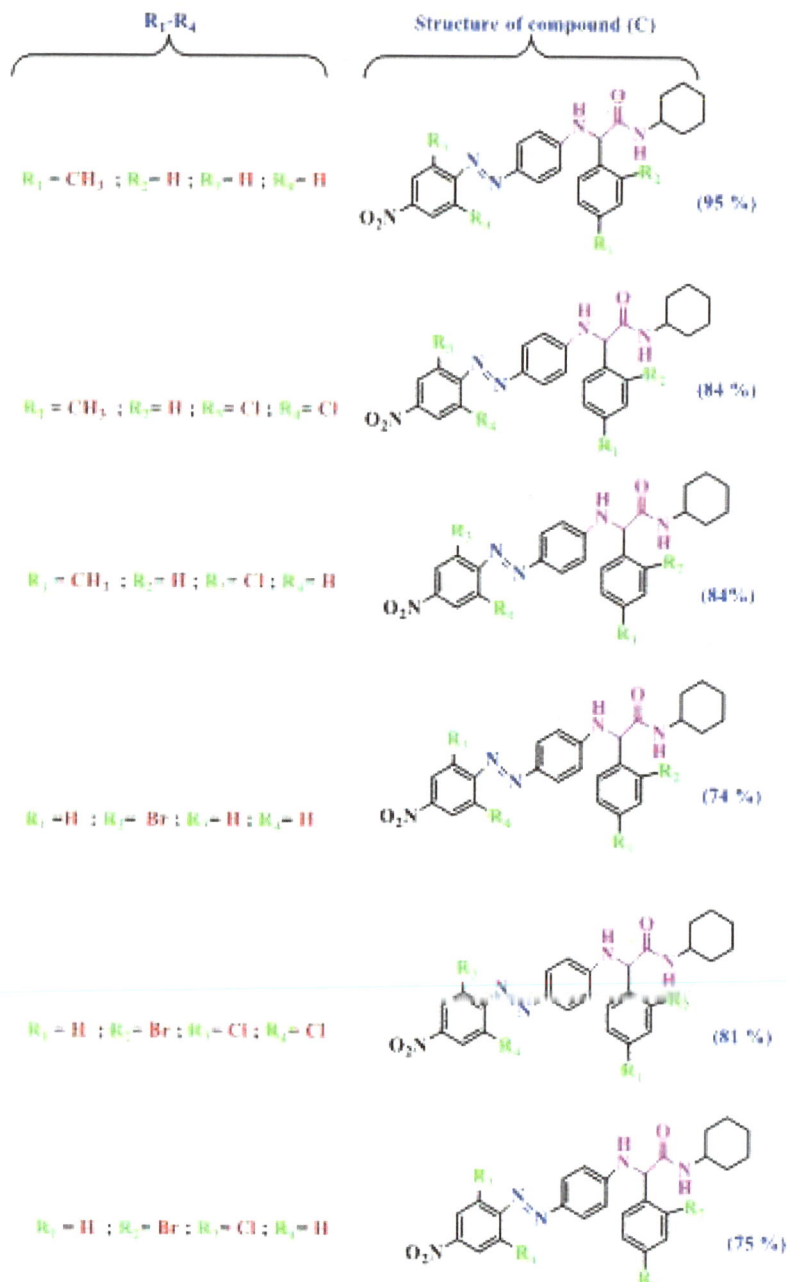

Fig. (4.3). Different structure of compound (Dispersive dye).

Direct Dyes

These dyes generally contain azo, oxazine, stilbene or phthalocyanine as chromophores and have a greater affinity towards cellulose fibers [19]. These kinds of dyes are mainly used in paper coloring and, when washed, take the appearance of solidity. Congo red is the first reported direct dye, and this class of dye is mainly further divided according to many parameters that include the presence of the chromophore, their fastness properties as well as application. It is to be noted that these dyes, though, have advantages in point of view of their easy application over a wide color gamut, but the wash fastness is not very good. This is the reason why these dyes are replaced by reactive dyes with much higher washing fastness. Fig. (**4.3**) shows the structures of a few direct dyes, whereas Table **4.2** summarizes a few azo chromophores based on direct dyes

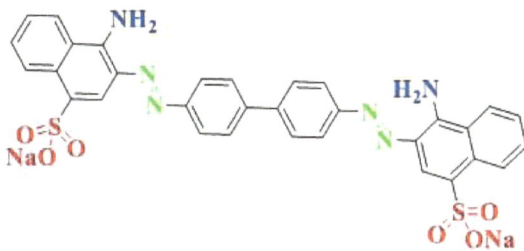

Direct Red 28

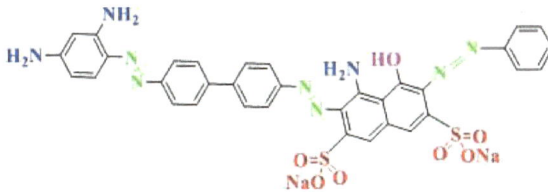

Direct Black 38

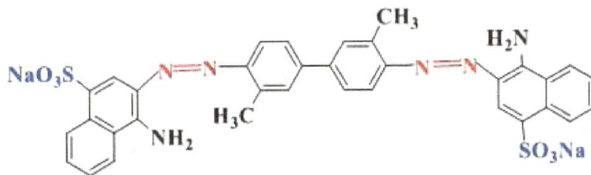

C.I. Direct Red 2

Fig. (4.4). Few direct dyes with their chemical structure.

Table 4.2. Examples of the certain direct dyes based anazo-chromophore and C.I. no, [16].

Chemical Structure	C.I. No.
	Direct Yellow 22 (a) and Direct Yellow (b).
	Direct Red 28 (c), Direct Yellow 12 (d), Direct Red 75 (e), Direct Orange 18 (f) , Direct Red 250 (g), Direct Orange 26 (h) and Direct Red 23 (i).

(Table 4.2) cont.....

Disazo	(f), (g), (h), (i)	
Trisazo	(j), (k), (l), (m)	Direct Black 166 (j), Direct green 33 (k), Direct Black 150 (l), Direct Blue78 (m)

(Table 4.2) cont.....

	Direct Black 151 (n) , Direct Black 22 (o) , Direct Orange 39 (p), Direct Black 19 (q) and Brown direct dye. C.I.3 5850 (r).
	Triphenodioxazine direct dyes

Reactive Dyes

These dyes are the second largest class of dyes and are mainly used in coloring cellulose fibres and sometimes silk or wool fibres. The wider uses of these particular dyes are mainly due to the fact that these dyes are associated with high wet fastness, brilliance and hue ranges. Also, dyes in this category have a higher attraction towards fibres, higher stability and more suitable processing condition [20, 21]. The better bond strength of these dyes comes from the fact that these dyes can form covalent bonds sulfhydryl groups of the protein structure of the textile dyes [22]. It is seen that reactive dyes with higher color gamut can be synthesized by reaction of dichloro-s-triazine dye with an amine at a temperature as low as 25–40° C [22]. Fig. (**4.5**) shows the basic structures of a few reactive dyes [23].

As mentioned before, this class of dye is mainly used in cellulose fibres however, for better fixation into cellulose fibres it needs a higher temperature of 80 °C and pH values (pH-11). One of the commonest and commercially available reactive dyes is Reactive Red-3 which gets lost more than 50 % during washing which is a matter of great worry [24].

Xiao *et al.* successfully synthesized three cationic dyes with the reaction schematics, as shown in Figs. (**4.6** and **4.7**). The dyes that are bound with copper, chromium or nickel are of greater concern as the decomposition of these dyes leads to the generation of toxic heavy metals that get ended in the food chain.

Vat Dyes

This class of dye mainly found its application in cellulose fabrics, more specifically in cotton, where they show excellent properties like exceptional fastness associated with a variety of agencies like washing, bleach, light, *etc.* This is mainly due to the insoluble nature of the dyes in water. Generally, Vat dyes are soluble in hot waters and sometimes, the presence of Na_2CO_3 helps to increase the solubility of the dye. However, the negative aspect of this type of dye is that it is not at all useful in the case of other types of dyes where it shows low fastness and also the pale depth of shades.

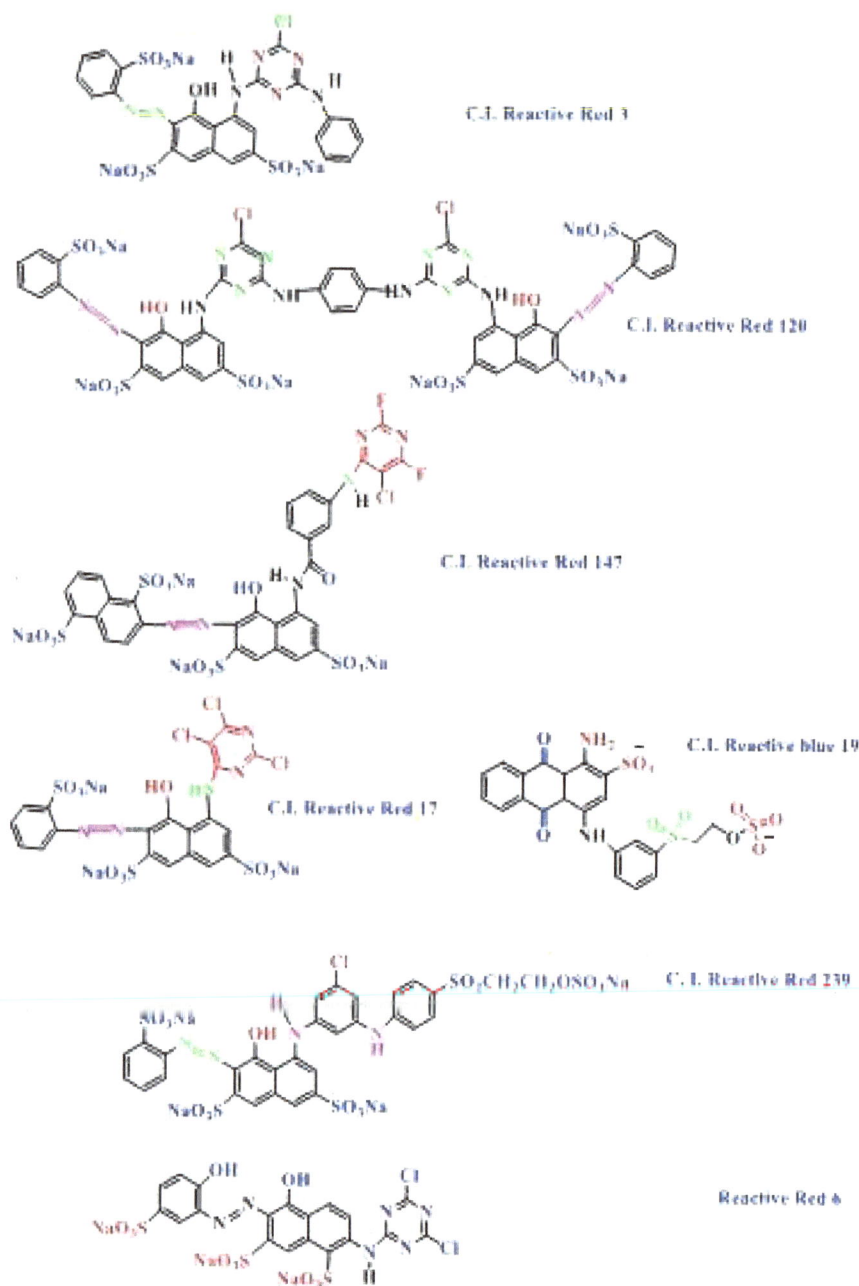

Fig. (4.5). Chemical structure of few reactive dyes.

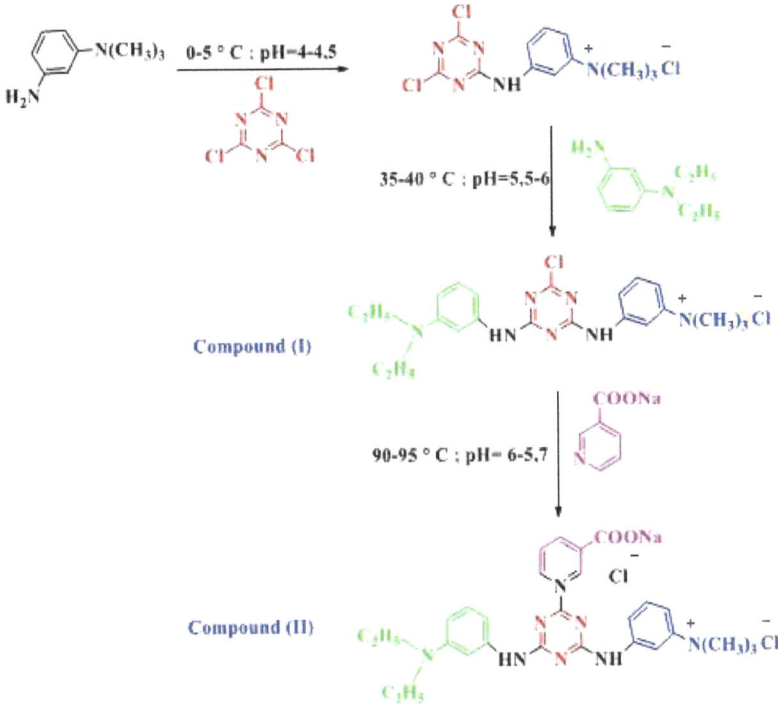

Fig. (4.6). Reaction of the coupling (I) and (II).

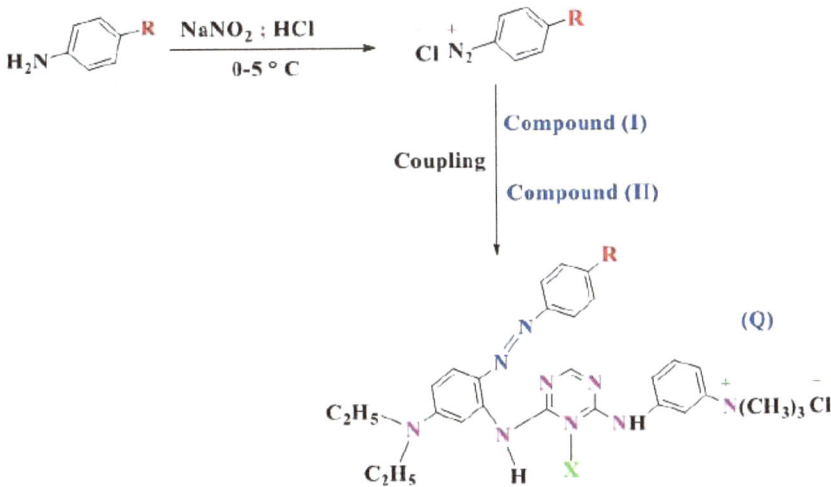

Fig. (4.7). Synthesis route for the cationic reactive dyes.

In separate works, Burkinshaw *et al.* and Muzamil Khatri *et al.* [25-27] proved that this class of dye possesses a few basic qualities of an ideal dye that include color fastness, wet fastness or characteristic light. However, in-depth study of the particular properties of the dyes are yet to be done. Indigo with color index C.I. Vat Blue 1 belongs to a generic family of VAT dyes. Indigo or Indigotin is one of the most important and widely used vat dyes that are generally found in indigofera, the indigo plant, as glucoside. Fig. (**4.8**) summarizes the structure of a few vat dyes.

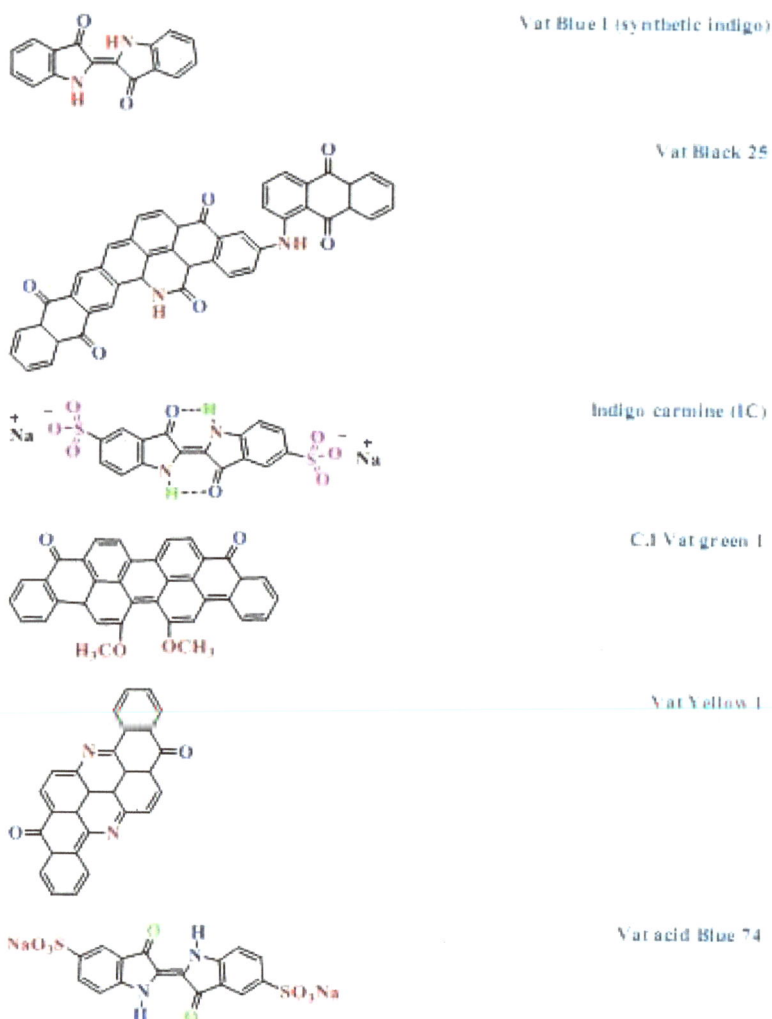

Fig. (4.8). Chemical structures of some vat dyes.

There are reports regarding the fact that halogenated derivatives of indigo (mainly bromine-induced derivatives) that give rise to other vat dyes, including indigoid, thioindigoid and anthraquinone [28]. The typical the structure of bromo derivative of anthraquinone has been shown in Fig. (**4.9**).

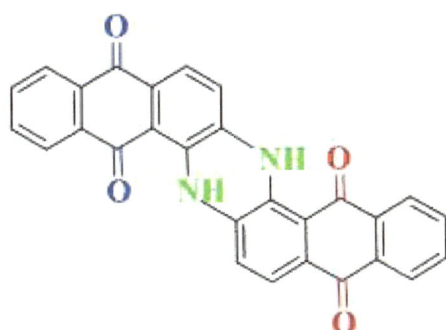

Anthraquinone vat dye

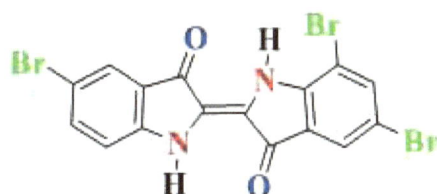

Vat dye (Bromo substituent)

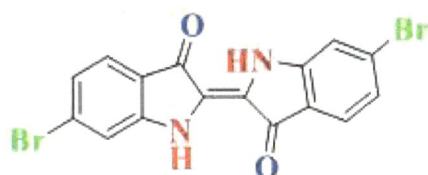

Tyrian purple

Fig. (4.9). Chemical structures of vat dyes anthraquinone, Bromo subtituent and Tyrian puple.

Basic Dyes

Much work has been done on this type of dye, mainly applied in acrylic, paper and nylon substrates and sometimes modified polyester substrates [29]. According to the work of Hunger and co-workers, the basic dyes contain a positive charge in the ammonium group or in different dyes like triarylmethane, xanthenes or acridine delocalized charges exist on dye cation [30]. The basic structures of a few basic dyes have been illustrated in Table **4.3**.

There is a simple explanation available regarding the working of basic dyes according to which basic dyes being water soluble, produce coloured cations when gets dissolved into the water. This positive color cation then get electrostatically attracted towards the fibres with a negative charge [29]. It is also observed that these classes of dyes contain a quaternary amine group which serves as an integral part of the dye. However, it is not mandatory, and there are exceptions.

Table 4.3. Few basic dyes with CI numbers and classes.

C.I. No.	Chemical Structure	Class
Basic Blue 26		Triarylmethane
Basic Red 1		Xanthene

(Table 4.3) cont.....

Basic Green 1		Triarylmethane
Basic Yellow 2		Diphenylmethane

Silkstone and co-workers report it that due to the poor migration properties, the basic dyes are generally accompanied by retarders, and it is also necessary to have good control while using retarders so that the anionic site of the fabrics does not get blocked; otherwise, dye uptake would be negatively affected. Apart from Table **4.3**, Fig. (**4.10**) also shows localized and delocalized charge basic dyes.

Acid Dyes

These dyes are mainly used in nylon, silk or wool and work in the low to moderate pH range of 3–7. These dyes are generally accompanied by one or more acid groups like SO_3H and $COOH$ [5]. The sulphonated group actually controls the solubility of the dyes and thus is of great importance. As these dyes are acidic in nature, they have a strong affinity towards basic fibres like polyamides [31]. Congo red is a typical example of acidic dye containing sulfonic acid groups and is generally used in various fields that include printing, textile, pharmaceutical, leather, papers and others due to its properties like high colour and greater solubility [32]. Table **4.4** summarizes the structures of a few acid dyes [33]. The pH value is maintained in the acidic regime by adding a suitable amount of formic or acetic acid. These dye molecules sometimes include metal complexes in their structures, and depending on the size of the molecule determines how the molecules interact with fibres hence the color fastness of the dye for a particular system. However, the light-fastness is generally in the blue scale range 5.0-6.0.

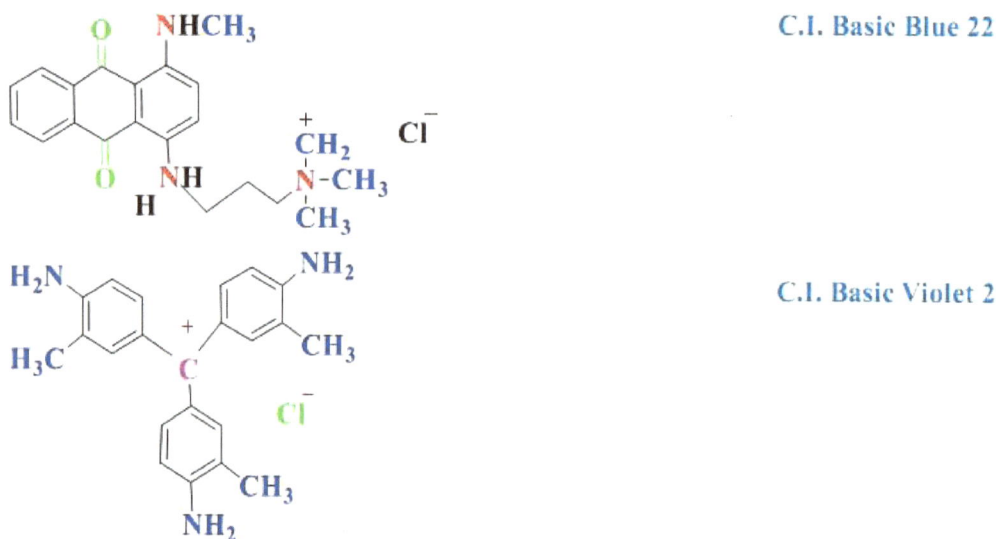

C.I. Basic Blue 22

C.I. Basic Violet 2

Fig. (4.10). Localized and delocalized charge basic dye.

Table 4.4. Example of a few acid dyes.

Structure of the Dye Molecule	C.I. No.
	Acid Red 249 Dye
	Acid Orange IV

(Table 4.4) cont.....

	Acid Blue 349
	Acid Blue 7
	Acid Blue 83

(Table 4.4) cont.....

	Acid Blue 25 (AB25)
	Acid Red 337
	Acid Yellow 36
	Acid Blue 147
	Acid Blue 25 (Anthraquinone)

(Table 4.4) cont.....

	Acid Violet 17
	Acid Orange 7
	Acid Black-234

Acid dyes generally contain azo chromophores or may be other like anthraquinone, triphenylmethane or copper phthalocyanine. As mentioned before, these dyes may be water soluble with the introduction of one or more sulphonate groups. A few examples shown in Acid dyes are azo chromophoric systems (the most important group), anthraquinone, triphenylmethane or copper phthalocyanine, which are soluble in water by the introduction of one to four sulphonate groups.

Sulphur Dyes

Sulfur dyes are one of the most extensively used dyes all over the globe. There are so many reports that suggest even 50 years ago, 9.1 % of the total dyes produced in the US are sulfur-based dyes that are most widely used for cellulose-based fabrics [34]. Approximately 110,000–120,000 tons of these dyes were produced every year, which happened to be the highest among any other groups of dyes.

In the year 2007, China produccd 8500-ton dyes, which is the third-highest in the world, and half of all the dyes used for coloring cellulose-based fibres are dyes, and 80 % of these are black sulfur.

In the case of sulfur dyes, the development of the product starts from the common initial material like benzene or naphthalene or diphenylamine, *etc.*, all having at least one nitro, nitro so, amino, or hydroxyl group. The detail synthesis process, then undergoes many intermediate processes like substitution, reduction, oxidation, ring formation, *etc.*

As mentioned before, sulfur dyes are used for both the pure cellulose fibres as well as blended fibres mixed with synthetic fibres [35]. However, these dyes are also used in silk as well as papers and even in the case of selected leathers. Sulfur dyes show good light fastness and washing, and these properties, along with their cheap cost, make them use most extensively. There are different classes of sulfur dyes that include:

Sulfur dyes, leuco- sulfur dyes, solubilized sulfur dyes, and condensed sulfur dyes. Fig. (**4.11**) represents a few sulfur dyes.

Azo Dye

Azo dyes consist of Azo bond linkage ($-N = N-$) may be more than that. Depending upon the numbers of azo bonds, they are named differently, like mono azo for single bond linkage and di or triazole respectively for double or triple bond linkage. Presently people are working with more than 2000 azo dyes. All these dyes collectively produce more than 1 million tons per year [36].

Apart from textile, these widely used dyes are also found in various other industries like textile, cosmetic, leather, pharmaceutical, paper, paint and food. As per the report given by Lucal and co-workers, approximately 70 % of the synthetic dyes are azo dye and thus, it is the largest class of synthetic dyes [37]. The increasing popularity of these dyes is due to several reasons that include its simple coupling reaction, possibilities of versatile applications, and, most importantly, adaptation to different structural variation to cope with different fields of needs. Sometimes very high molar extinction coefficients of azo dyes also offer them extra advantages over the other. One of the most alarming facts about these dyes is that, as per reports, 50,000 tons of such textile dyes come out as a waste product and are exposed to the environment [38].

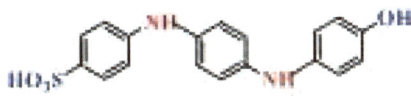

N-[p-(p-hydroxyanilino} phenyl]
sulphanilic acid

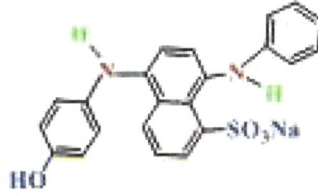

Sulfur brilliant green, CI 53570

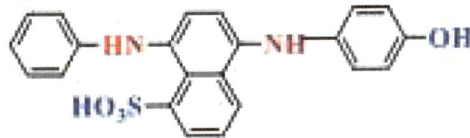

8-Anilino-5-[p-hydroxyanilino]-1-
naphthalenesulphonic acid.

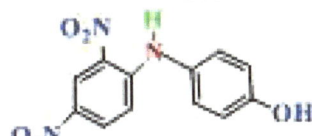

Sulfur blue dye, CI 53235

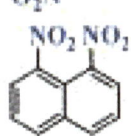

1,8-Dinitronaphthalene.

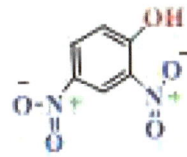

Sulfur black, CI 53185

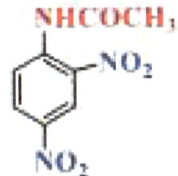

2',4'-Dinitroacetanilide

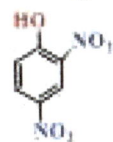

Leuco Sulfur black 1, C.I 53185

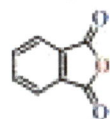

Phthalic anhydride

Fig. (4.11). Example of few sulfur dyes.

As mentioned before, there may be more than one azo linkage in such dyes which are connected on both sides of different aromatic rings like benzene, naphthalene and others. It is also possible that different aromatic heterocyclic units are attached to azo groups, and it is observed that these side groups are sometimes responsible for different shades associated with such dyes [39].

Rajguru and co-workers reported that sulfonated azo dyes (where sulfonate groups are present as substituents) and azo groups in conjugation with aromatic or enolizable groups constitute complex structures that are responsible for verities colors associated with such dyes.

Table **4.5** shows an example of a few azo dyes along with Figs. (**4.12** and **4.13**).

Table 4.5. Example of few azo dyes.

	Direct Black 22
	Acid Red 2

(Table 4.5) cont.....

	Disperse yellow 7
	Acid Orange 20
	Methy Red
	Trypan Blue
	Direct Blue 71

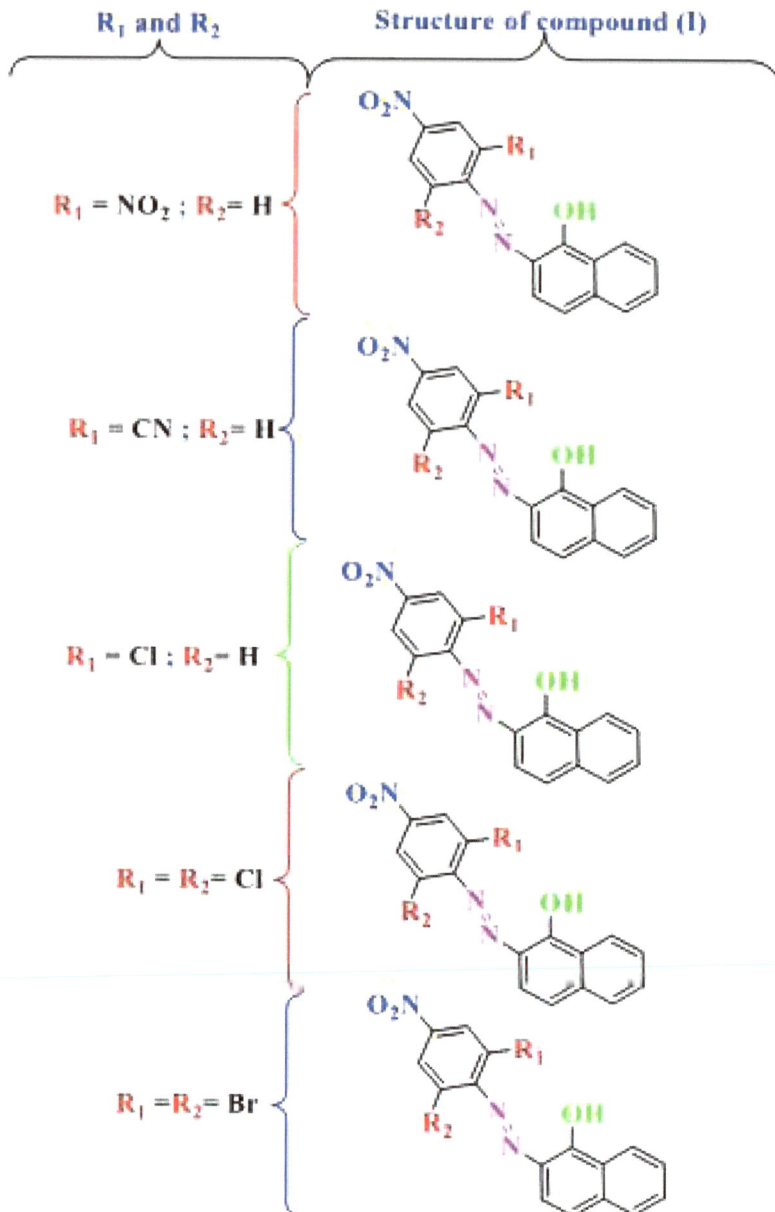

Fig. (4.12). Different structure of disperse azo dyes (I).

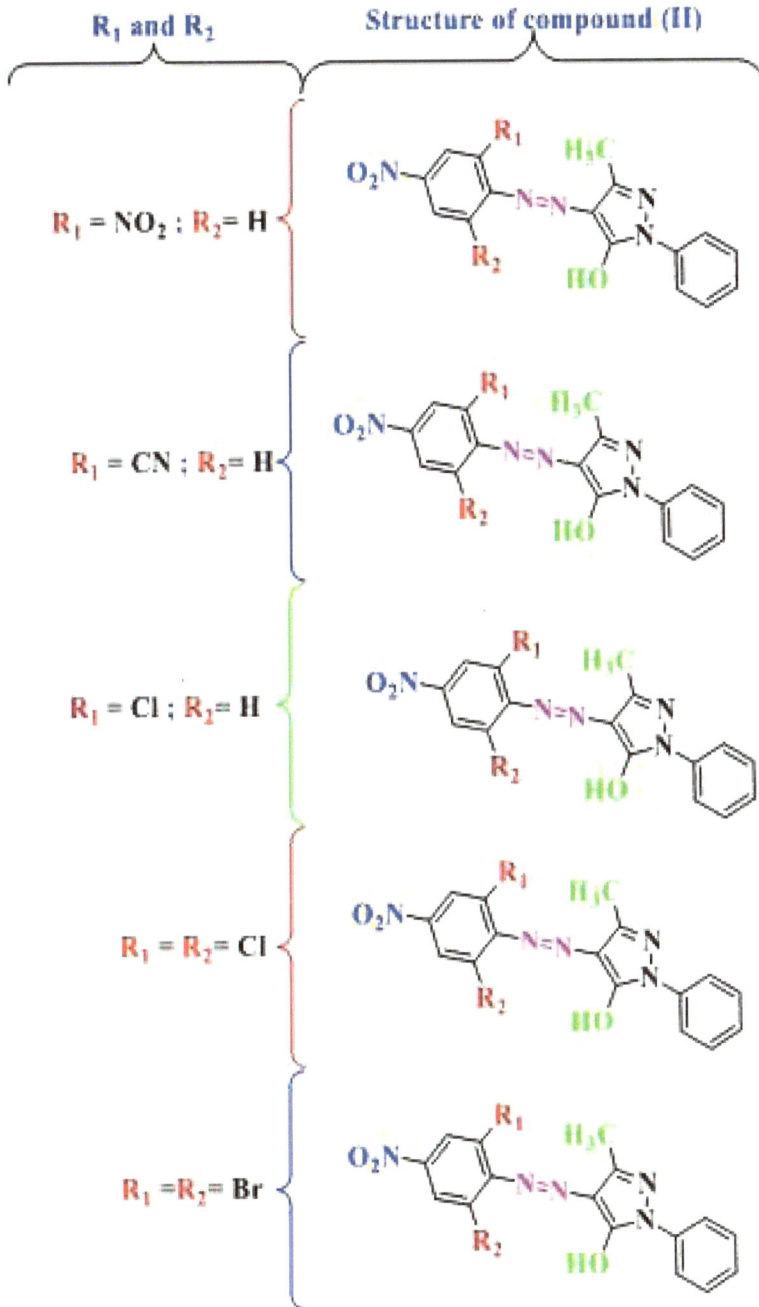

Fig. (4.13). Different structure of disperse azo dyes (I).

Impact of Textile Dyes on the Environment

The impact of textile dyes on the environment is huge and unfortunately, the impact is not very positive. It is reported that more than 10,000 tonnes of textile dyes are produced per year and of which, 100 tonnes of dyes are exposed to environment in the form of wastewater. Being the second-largest pollution sector nowadays, wastewater-induced textile dyes are responsible for several negative irreversible consequences like the disappearance of the Aral Sea [40]. This is mainly due to the overconsumption of water to irrigate cotton fields. The main reason that huge quantities of dyes are exposed to the environment is less affinity of the maximum surface towards various dyes.

The textile effluents mostly create water and air pollution because of the excess amount of use of dyes, mainly very colorful alkaline dyes characterized by salinity and eco-toxicity [41] mutagenic, teratogenic and carcinogenic, [42].

Maximum dyes are a very difficult process due to their extremely complex structure and thus are a serious threat to the environment.

CONCLUSION

In this chapter, mainly the classifications of dyes have been discussed. All kinds of dyes have been mentioned with their structures, and possible applications and many dyes like acid dyes, basic dyes, vat dyes, sulfur dyes or others. It has also been mentioned that a dye has different components like chromophore, auochrome and the matrix.

Chromophores basically determine the color response of the dyes. Few chromophores are anthraquinone, azo, phthalocyanines, triphenyl methanes, indigoids, xanthenes, polymethines, nitrosed, *etc.*

The chapter also summarizes the structures of dyes classified in terms of chromophores, colors, toxicity as well as applications. Basic synthesis processes of a few dyes have also been mentioned in this chapter.

REFERENCES

[1]   Ferreira, E.S.B.; Hulme, A.N.; McNab, H.; Quye, A. The natural constituents of historical textile dyes. *Chem. Soc. Rev.,* **2004**, *33*(6), 329-336.
        http://dx.doi.org/10.1039/b305697j PMID: 15280965

[2]   Abel, A. The history of dyes and pigments: from natural dyes to high performance pigments.*Colour Design*; Woodhead Publishing, 2012, pp. 557-587.
        http://dx.doi.org/10.1016/B978-0-08-101270-3.00024-2

[3]   Vázquez-Ortega, F.; Lagunes, I.; Trigos, Á. Cosmetic dyes as potential photosensitizers of singlet oxygen generation. *Dyes Pigments,* **2020**, *176*, 108248.
        http://dx.doi.org/10.1016/j.dyepig.2020.108248

[4]   Rehman, A.; Usman, M.; Bokhari, T.H.; Haq, A.; Saeed, M.; Rahman, H.M.A.U.; Siddiq, M.; Rasheed, A.; Nisa, M.U. The application of cationic-nonionic mixed micellar media for enhanced solubilization of Direct Brown 2 dye. *J. Mol. Liq.,* **2020**, *301*, 112408.
        http://dx.doi.org/10.1016/j.molliq.2019.112408

[5]   Nozet, H.; Majault, J. *Textiles chimiques: fibres modernes*; Eyrolles, 1976.

[6]   Silva, P.M.S.; Fiaschitello, T.R.; Queiroz, R.S.; Freeman, H.S.; Costa, S.A.; Leo, P.; Montemor, A.F.; Costa, S.M. Natural dye from Croton urucurana Baill. bark: Extraction, physicochemical characterization, textile dyeing and color fastness properties. *Dyes Pigments,* **2020**, *173*, 107953.
        http://dx.doi.org/10.1016/j.dyepig.2019.107953

[7]   Burkinshaw, S.M.; Salihu, G. The wash-off of dyeings using interstitial water. Part 4: Disperse and reactive dyes on polyester/cotton fabric. *Dyes Pigments,* **2013**, *99*(3), 548-560.
        http://dx.doi.org/10.1016/j.dyepig.2013.06.006

[8]   Laurent, A.D.; Wathelet, V.; Bouhy, M.; Jacquemin, D.; Perpète, E. Simulation de la perception des couleurs de colorants organiques. **2010**.

[9]   Temesgen, F.; Gabbiye, N.; Sahu, O. Biosorption of reactive red dye (RRD) on activated surface of banana and orange peels: Economical alternative for textile effluent. *Surf. Interfaces,* **2018**, *12*, 151-159.
        http://dx.doi.org/10.1016/j.surfin.2018.04.007

[10]   Zhenwang, L. **2000**The PT dye molecular structure and its chromophoric luminescences mechanism. *15th World Conference on Non-Destructive Testing,* , pp. 15-21.

[11]   Benkhaya, S.; M'rabet, S.; El Harfi, A. A review on classifications, recent synthesis and applications of textile dyes. *Inorg. Chem. Commun.,* **2020**, *115*, 107891.
        http://dx.doi.org/10.1016/j.inoche.2020.107891

[12]   Rauf, M.A.; Meetani, M.A.; Hisaindee, S. An overview on the photocatalytic degradation of azo dyes in the presence of TiO2 doped with selective transition metals. *Desalination,* **2011**, *276*(1-3), 13-27.
        http://dx.doi.org/10.1016/j.desal.2011.03.071

[13]   Christie, R. Colour Chemistry. Royal Society of Chemistry. *United Kindgom,* **2001**, *46*, 118-120.

[14]   Raman, C.D.; Kanmani, S. Textile dye degradation using nano zero valent iron: A review. *J. Environ. Manage.,* **2016**, *177*, 341-355.
       http://dx.doi.org/10.1016/j.jenvman.2016.04.034 PMID: 27115482

[15]   Shamey, R. Improving the colouration/dyeability of polyolefin fibres.*Polyolefin Fibres*; Woodhead Publishing, 2009, pp. 363-397.

[16]   Clark, M., Ed.; *Handbook of textile and industrial dyeing: principles, processes and types of dyes*; Elsevier, 2011.

[17]   Qin, Y.; Yuan, M.; Hu, Y.; Lu, Y.; Lin, W.; Ma, Y.; Lin, X.; Wang, T. Preparation and interaction mechanism of Nano disperse dye using hydroxypropyl sulfonated lignin. *Int. J. Biol. Macromol.,* **2020**, *152*, 280-287.
       http://dx.doi.org/10.1016/j.ijbiomac.2020.02.261 PMID: 32105691

[18]   Silkstone, K. The Influence of Polymer Morphology on the Dyeing Properties of Synthetic Fibres. *Rev. Prog. Color. Relat. Top.,* **1982**, *12*(1), 22-30.
       http://dx.doi.org/10.1111/j.1478-4408.1982.tb00221.x

[19]   Khatri, A.; Peerzada, M.H.; Mohsin, M.; White, M. A review on developments in dyeing cotton fabrics with reactive dyes for reducing effluent pollution. *J. Clean. Prod.,* **2015**, *87*, 50-57.
       http://dx.doi.org/10.1016/j.jclepro.2014.09.017

[20]   Zhang, S.; Ma, W.; Ju, B.; Dang, N.; Zhang, M.; Wu, S.; Yang, J. Continuous dyeing of cationised cotton with reactive dyes. *Color. Technol.,* **2005**, *121*(4), 183-186.
       http://dx.doi.org/10.1111/j.1478-4408.2005.tb00270.x

[21]   Mahmoodi, N.M.; Hayati, B.; Arami, M.; Mazaheri, F. Single and binary system dye removal from colored textile wastewater by a dendrimer as a polymeric nanoarchitecture: equilibrium and kinetics. *J. Chem. Eng. Data,* **2010**, *55*(11), 4660-4668.
       http://dx.doi.org/10.1021/je100248m

[22]   Benkhaya, S.; Cherkaoui, O.; Assouag, M.; Mrabet, S.; Rafik, M.; Harfi, A.E.L. Synthesis of a new asymmetric composite membrane with bi-component collodion: application in the ultra filtration of baths of reagent dyes of fabric rinsing/padding. *J. Mater. Environ. Sci.,* **2016**, *7*(12), 4556-4569.

[23]   Irfan, M.; Zhang, H.; Syed, U.; Hou, A. Low liquor dyeing of cotton fabric with reactive dye by an eco-friendly technique. *J. Clean. Prod.,* **2018**, *197*, 1480-1487.
       http://dx.doi.org/10.1016/j.jclepro.2018.06.300

[24]   Wantala, K.; Sriprom, P.; Pojananukij, N.; Neramittagapong, A.; Neramittagapong, S.; Kasemsiri, P. Optimal decolorization efficiency of reactive red 3 by Fe-RH-MCM-41 catalytic wet oxidation coupled with Box-Behnken design. *Key Eng. Mater.,* **2013**, *545*, 109-114.
       http://dx.doi.org/10.4028/www.scientific.net/KEM.545.109

[25] Burkinshaw, S.M. *Chemical principles of synthetic fibre dyeing*; Springer Science & Business Media, 1995.
http://dx.doi.org/10.1007/978-94-011-0593-4

[26] Hassan, M.M.; Carr, C.M. A critical review on recent advancements of the removal of reactive dyes from dyehouse effluent by ion-exchange adsorbents. *Chemosphere,* **2018**, *209*, 201-219.
http://dx.doi.org/10.1016/j.chemosphere.2018.06.043 PMID: 29933158

[27] Burkinshaw, S.M.; Son, Y.A. The dyeing of supermicrofibre nylon with acid and vat dyes. *Dyes Pigments,* **2010**, *87*(2), 132-138.
http://dx.doi.org/10.1016/j.dyepig.2010.03.009

[28] Preston, C. Dyeing of cellulosic fibres. *Distributed by the Society of Dyers and Colourists.,* **1986**.

[29] Broadbent, A.D. *Basic principles of textile coloration*; Society of Dyers and Colourists, 2001.

[30] Gupta, V.K.; Suhas, Application of low-cost adsorbents for dye removal – A review. *J. Environ. Manage.,* **2009**, *90*(8), 2313-2342.
http://dx.doi.org/10.1016/j.jenvman.2008.11.017 PMID: 19264388

[31] Benaissa, A. Etude de la faisabilité d'élimination de certains colorants textiles par certains matériaux déchets d'origine naturelle. **2012**.

[32] Wu, J.; Li, Q.; Li, W.; Li, Y.; Wang, G.; Li, A.; Li, H. Efficient removal of acid dyes using permanent magnetic resin and its preliminary investigation for advanced treatment of dyeing effluents. *J. Clean. Prod.,* **2020**, *251*, 119694.
http://dx.doi.org/10.1016/j.jclepro.2019.119694

[33] Cretescu, I.; Lupascu, T.; Buciscanu, I.; Balau-Mindru, T.; Soreanu, G. Low-cost sorbents for the removal of acid dyes from aqueous solutions. *Process Saf. Environ. Prot.,* **2017**, *108*, 57-66.
http://dx.doi.org/10.1016/j.psep.2016.05.016

[34] Nguyen, T.A.; Juang, R.S. Treatment of waters and wastewaters containing sulfur dyes: A review. *Chem. Eng. J.,* **2013**, *219*, 109-117.
http://dx.doi.org/10.1016/j.cej.2012.12.102

[35] Shore, J. Dyeing with reactive dyes. *Cellulosics dyeing,* **1995**, 189-245.

[36] Fatima, M.; Farooq, R.; Lindström, R.W.; Saeed, M. A review on biocatalytic decomposition of azo dyes and electrons recovery. *J. Mol. Liq.,* **2017**, *246*, 275-281.
http://dx.doi.org/10.1016/j.molliq.2017.09.063

[37] Lucas, M.S.; Dias, A.A.; Sampaio, A.; Amaral, C.; Peres, J.A. Degradation of a textile reactive Azo dye by a combined chemical–biological process: Fenton's reagent-yeast. *Water Res.,* **2007**, *41*(5), 1103-1109.
http://dx.doi.org/10.1016/j.watres.2006.12.013 PMID: 17261325

[38]   Chang, J.S.; Lin, C.Y. Decolorization kinetics of a recombinant Escherichia coli strain harboring azo-dye-decolorizing determinants from Rhodococcus sp. *Biotechnol. Lett.,* **2001,** *23*(8), 631-636.

   http://dx.doi.org/10.1023/A:1010306114286

[39]   McMullan, G.; Meehan, C.; Conneely, A.; Kirby, N.; Robinson, T.; Nigam, P.; Banat, I.M.; Marchant, R.; Smyth, W.F. Microbial decolourisation and degradation of textile dyes. *Appl. Microbiol. Biotechnol.,* **2001,** *56*(1-2), 81-87.

   http://dx.doi.org/10.1007/s002530000587 PMID: 11499950

[40]   Allouche, J. The governance of Central Asian waters: national interests *versus* regional cooperation. *Disarmament Forum,* **2007,** *4*(1), 45-55.

[41]   Benkhaya, S.; M'rabet, S.; El Harfi, A. Classifications, properties, recent synthesis and applications of azo dyes. *Heliyon,* **2020,** *6*(1), e03271.

   http://dx.doi.org/10.1016/j.heliyon.2020.e03271 PMID: 32042981

[42]   Wang, H.; Zhong, Y.; Yu, H.; Aprea, P.; Hao, S. High-efficiency adsorption for acid dyes over $CeO_2 \cdot xH_2O$ synthesized by a facile method. *J. Alloys Compd.,* **2019,** *776*, 96-104.

   http://dx.doi.org/10.1016/j.jallcom.2018.10.228

CHAPTER 5

Basic Structures and Properties of Few Potential Nanomaterials

Abstract: After the discussion of all the preceding sections/chapters, now we are in a position to review some typical nanomaterial systems that have established themselves as materials of immense potential in the field of water purification by successfully removing different textile dyes through different processes like catalysis adsorption or others. In this chapter, we will discuss a few such particular materials from different domains. The material will mainly include metal oxide nanostructures and related derivatives like zinc oxide, *etc.*, and also carbon nanostructures like carbon nanotube, graphene, *etc.*, and their hybrids. A detailed discussion regarding the dye removal ability of graphitic carbon nitride is also included here. In this regard, the results of other researchers will be accompanied by a few of our own findings, which will help the reader in better understanding the topic. Apart from these well-known materials that have higher dye removal effectiveness, a few other less studied and newly evolved systems, such as silicon nanowire and p-type conducting oxides like copper borate, have also been discussed with a few established results.

Keywords: Carbon, Carbon Nanotubes Zinc Oxide, Graphene, Nanostructures, Silicon nanowire.

CARBON NANOSTRUCTURES: BRIEF HISTORY AND PROPERTIES

Among all the materials, carbon nanostructures have attracted the attention of researchers as well as technologists from the last few decades due to their versatile applications in various fields of complete difference. Fullerene, carbon nanotube, nano fibers, nanocoils, *etc.*, have already established their importance in different fields, but still, day after day, new novel carbon nanostructures are being discovered and being used for different applications. The element carbon is unique in the sense that it has a wide variety of allotropes with a fascinating range of mechanical, optical and electrical properties. Among all other fields mentioned above, carbon fibers, tube, coil, diamond-like carbon, diamond, graphene, and even amorphous carbons have established themselves as field emitters as well as efficient coating elements and are used in different applications.

Diptonil Banerjee, Amit Kumar Sharma and Nirmalya Sankar Das

Carbon and its Different Hybridized States

Carbon is the chemical element with the symbol **C** and atomic number 6. As a member of group 14 on the periodic table, generally it is nonmetallic and tetravalent, making four electrons available to form covalent chemical bonds. There are three naturally occurring isotopes, of which ^{12}C and ^{13}C are stable, while ^{14}C is radioactive, decaying with a half-life of about 5730 years. There are several allotropes of carbon, of which the best known are graphite, diamond, and amorphous carbon. The physical properties of carbon vary widely with the allotropic form. For example, diamond is highly transparent, while graphite is opaque and black. Diamond is among the hardest materials known, while graphite is soft enough to form a streak on paper. Diamond has a very low electrical conductivity, while graphite is a very good conductor. Under normal conditions, diamond has the highest thermal conductivity of all known materials. All the allotropic forms are solids under normal conditions, but graphite is the most thermodynamically stable. A general list of different properties of carbon has been given in Table **5.1**.

Table 5.1. Some properties of carbon.

General Properties	
Name, symbol, atomic number	Carbon, C, 6
Group, period, block	14, 2, p
Standard atomic weight	12.0101gmol^{-1}
Electron configuration	$1s^2\, 2s^2\, 2p^2$
Physical Properties	
Phase	Solid
Density (at room temperature)	1.8–2.1 (amorphous carbon), 2.267 (graphite) 3.515 (diamond) g·cm^{-3}
Sublimation point	3915 K
Triple point	4600 K

(Table 5.1) cont.....

Specific heat capacity (at room temperature)	8.517 (graphite), 6.155 (diamond) $J \cdot mol^{-1} \cdot K^{-1}$
Atomic Properties	
Oxidation states	4, 3, 2, 1, 0, -1, -2, -3, -4
Electronegativity	2.55 (Pauli scale)
Ionization energies	1st: 1086.5 $kJ \cdot mol^{-1}$ 2nd: 2352.6 $kJ \cdot mol^{-1}$ 3rd: 4620.5 $kJ \cdot mol^{-1}$
Covalent radius	77(sp³), 73(sp²), 69(sp) pm
Vander Waals radius	170 pm

The element carbon is one of the most versatile elements on the periodic Table in terms of the number of compounds as it may form an infinite number of compounds virtually. This is largely due to the types of bonds it can form and the number of different elements it can bond. Carbon may form single, double, and triple bonds by means of different hybridization with the ground state configuration, as shown in below Schematic **1**.

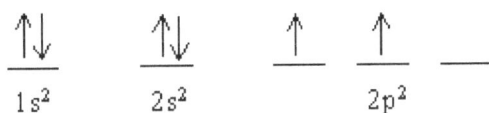

$$\uparrow\downarrow \qquad \uparrow\downarrow \qquad \uparrow \quad \uparrow \quad \_$$
$$1s^2 \qquad 2s^2 \qquad \quad 2p^2$$

From the ground state electron configuration, it can be seen that carbon has four valence electrons, two in the 2s subshell and two in the 2p subshell. The 1s electrons being the core electrons do not contribute in any kind of bonding. There are two unpaired electrons in the 2p subshell, so if carbon were to hybridize from this ground state, it would be able to form at most two bonds. It is known that any bond formation corresponds to the release of energy, so it would be to carbon's benefit to try to maximize the number of bonds it can form. For this reason, carbon will form an excited state by promoting one of its 2s electrons into its empty 2p orbital and hybridize from the excited state. By forming this excited state, carbon will be able to form four bonds. The excited state configuration is shown below in Schematic **2**:

↑↓   ↑   ↑   ↑   ↑
___   ___   ___ ___ ___
$1s^2$ $2s^1$ $2p^3$

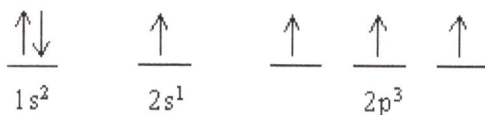

Since both the 2s and the 2p subshells are half-filled, the excited state is relatively stable. sp^3 hybridization occurs when one 's' orbital and all three 'p' orbitals of carbon atoms gets combined in the second energy level. As a result, four hybrid orbitals are formed and make four bonds with adjacent carbons and thus, the structure generated is called tetrahedral. The corresponding bonding arrangement is shown in Fig. (**5.1a**). The four sp^3 hybrid orbitals will arrange themselves in three-dimensional spaces to get as far apart as possible (to minimize repulsion). The geometry that achieves this is tetrahedral geometry, where any bond angle is 109.5°. For sp^2 hybridization, as the name suggests, two p orbital electros combine with one 's' electron in the second energy level, leaving one p orbital left alone. The situation will be as shown below in Fig. (**5.1b**). The three sp^2 hybrid orbitals will arrange themselves in three-dimensional spaces to get as far apart as possible. The geometry that achieves this is trigonal planar geometry, where the bond angle between the hybrid orbitals is 120°. The unmixed, pure p orbit will be perpendicular to this plane. Keep in mind, each carbon atom is sp^2, are trigonal planar. sp hybridized state is formed when only one p electron is combined with the s electron in the second energy state. The two sp hybrid orbitals arrange themselves in three-dimensional space to get as far apart as possible. The geometry, is a linear geometry with a bond angle of 180°. The two pure p orbitals, which were not mixed, are perpendicular to each other. The situation is shown in Fig. (**5.1c**).

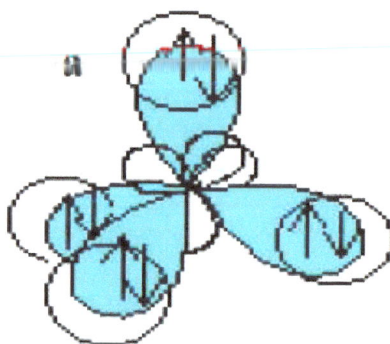

(Fig. 5.1) contd.....

b

pi bond (side by side overlap of pure p orbitals)

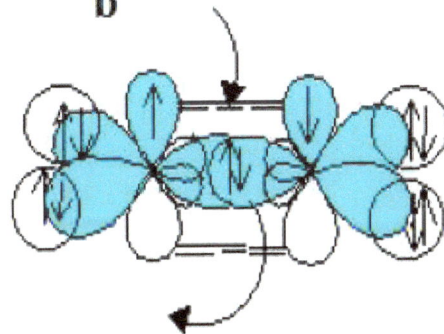

sigma bond (head on overlap of hybrid orbitals)

c

pi bond between two pure p orbitals

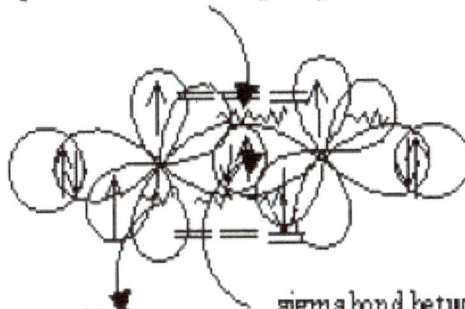

pi bond between two pure p orbitals

.. this pi bond is perpendicular to the other pi bond

sigma bond between two sp hybrid orbitals

three distinct ways in which a carbon sheet can be rolled into a tube, as shown in Fig. (5.2) below. The first two of these, known as "armchair" (b) and "zig-zag" (c) have a high degree of symmetry. The terms "armchair" and "zig-zag" refer to the arrangement of hexagons around the circumference. The third class of tube, which in practice is the most common, is known as chiral, meaning that it can exist in two mirror-related forms. An example of a chiral nanotube is shown at the bottom left (d).

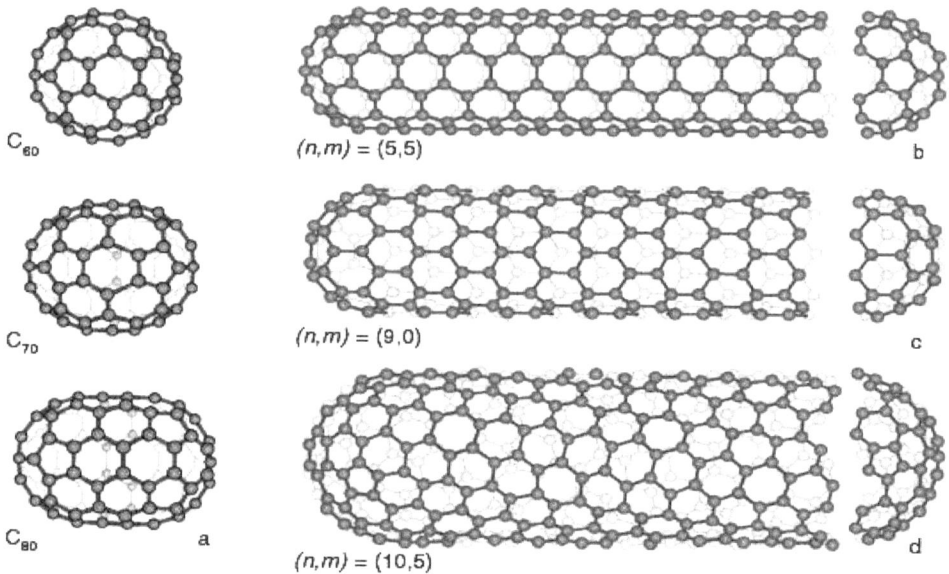

C_{60}

$(n,m) = (5,5)$ b

C_{70}

$(n,m) = (9,0)$ c

C_{80} a

$(n,m) = (10,5)$ d

Fig. (5.2). Types of caron nanotubes.

The structure of a nanotube can be specified by a vector, (n, m), which defines how the graphene sheet is rolled up. This can be understood with reference to Fig. (5.3). The chiral vector is defined on the hexagonal lattice as $C_h = n\hat{a}_1 + m\hat{a}_2$, where $\hat{a}_1$ and $\hat{a}_2$ are unit vectors, and n and m are integers. The chiral angle, θ, is measured relative to the direction defined by $\hat{a}_1$. Zigzag nanotubes correspond to (n, 0) or (0, m) and have a chiral angle of $0°$, armchair nanotubes have (n, n) and a chiral angle of $30°$, while chiral nanotubes have general (n, m) values and a chiral angle of between $0°$ and $30°$. According to the theory, nanotubes can either be metallic (green circles) or semiconducting (blue circles). To produce a nanotube with the indices (6, 3), say, the sheet is rolled up so that the atom labeled (0, 0) is superimposed on the one labeled (6, 3). It can be seen from the figure that m = 0 for all zig-zag tubes, while n = m for all armchair tubes [1]. As follows from elementary geometry and an assumption that the C-C bond has its normal length of 0.14 nanometers (nm), d being the tube diameter.

$$d = 0.078 \sqrt{n^2 + nm + m^2} \text{ nm}$$

$$\text{and } \theta = \tan^{-1} \left[\frac{\sqrt{3}m}{m + 2n} \right]$$

Fig. (**5.4**) has been constructed for (n, m) = (4, 2), and the unit cell of this nanotube is bounded by OAB'B. To form the nanotube, imagine that this cell is rolled up so that O meets A and B meets B', and the two ends are capped with half of a fullerene molecule. Different types of carbon nanotubes have different values of n and m.

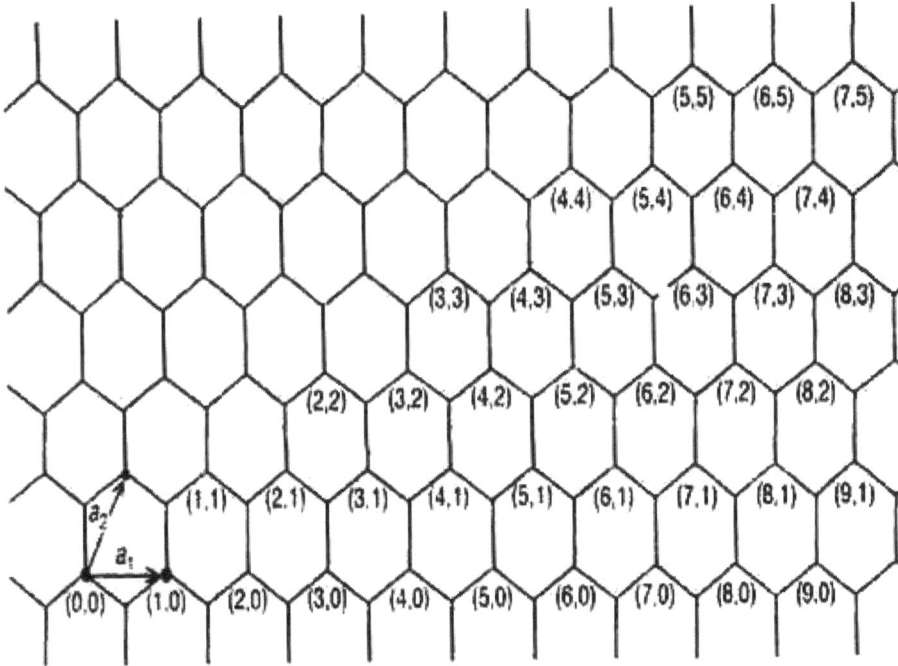

Fig. (5.3). Lattice structure of CNT explaining chirality.

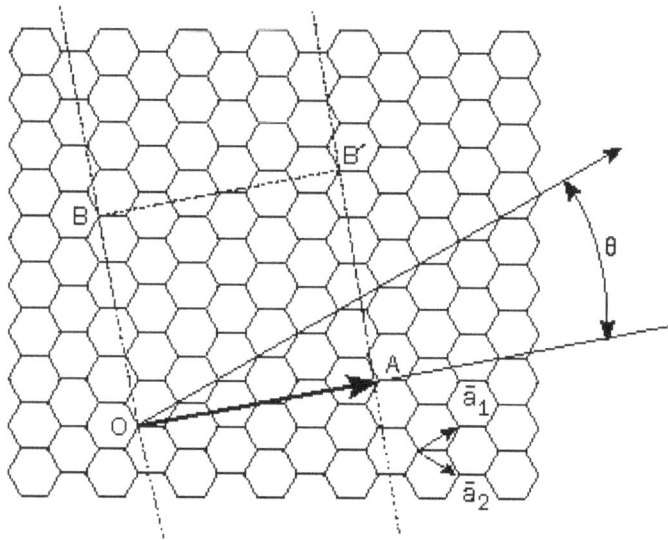

Fig. (5.4). Chiral vector and chiral angle in CNTs.

Carbon nanotubes have a set of fascinating properties. Firstly, depending upon the values of m, n mentioned before, CNT can act as metal as well as the semiconductors. When used as metal it has electrical conductivity comparable to or even better than copper. It has extremely good mechanical properties having Youngs modulus as high as 1200 GPa, Tensile Strength 150 GPa. It has a thermal conductivity of around 3000 W/mK. In addition to these, it has been seen that CNT are flexible and can be bent several times at 110°without undergoing structural changes. The structure is not easily changed with the effects of pressure. It has been demonstrated that CNT are only undergoing permanent structure changing at very high pressures (> 1.5 GPa) and that the deformations are totally elastic below that value. Some of the properties of carbon nanotube, which form a driving force for their wide range of applications, are shown in Table **5.2.** CNTs can be of two types; when a single carbon sheet gets rolled to form a CNT it is called a single-walled carbon nanotube (SWCNT), whereas when a number of carbon sheets roll, multi-walled (MWCNT) CNTs get produced. Obviously, these two structures have some fundamental differences regarding their properties which have been summarized in Table **5.3.**

There are lots of ways and means by which crystalline CNTs can be produced. These are mainly arc discharge method, laser ablation, thermal, chemical vapor deposition, plasma enhanced chemical vapor deposition, flame synthesis and chemical hydrothermal or solvothermal preparation. It is to be noted that all these are a high temperature process requiring suitable catalyst. CNTs can be very effectively used in various fields of applications like:

Table 5.2. Some fundamental properties of carbon nanotube.

Properties	Value
Average Diameter of SWNT's	1.2-1.4 nm
Density: (10, 10) Armchair	1.33 g/cm^3
(17, 0) Zigzag	1.34 g/cm^3
(12, 6) Chiral	1.40 g/cm^3
Interlayer Spacing: (n, n) Armchair	3.38 Å
(n, 0) Zigzag	3.41 Å
(2n, n) Chiral	3.39 Å
Young's Modulus: SWCNT	~ 1 TPa
MWCNT	0.3- 1 TPa
Maximum Tensile Strength	~ 63 GPa
Band Gap: For Metallic	0 eV
For Semi-Conducting	0.18 – 1.8 eV
Thermal Conductivity (Room Temp.)	3000 W/mK
Carrier mobility Semi-Conducting NT	10^5 cm^2/Vsec
Maximum Current Density	10^9 A/m2
Turn-on field	1.5 – 7.5 V/μm

(Table 5.2) cont.....

Threshold field for SWNTs and MWNTs (aligned and randomly aligned)	1.5 – 9.5 V/µm

Table 5.3. Comparison between SWNT and MWNT.

SWNT	MWNT
Single layer of graphene	Multiple layer of graphene
Catalyst is required for synthesis	Can be produced without catalyst
Bulk synthesis is difficult as it requires proper control over growth and atmospheric condition	Bulk synthesis is easy
Purity is poor	Purity is high
A chance of defect is more during functionalization	A chance of defect is less, but once occurred it's difficult to improve
Less accumulation in body	More accumulation in body
Characterization and evaluation is easy	It has a very complex structure
It can be easily twisted and are more pliable	It cannot be easily twisted

Branched Carbon Nanostructure is basically a special form of 1D carbon nanotubes or nanofibers and so can be discussed in the same section. Generally, these complex structures are synthesized via CVD or spray pyrolysis of different carbon sources like methane, acetylene, *etc.* This is also a high-temperature process. These kinds of structures have not been studied in much detail. So the detailed properties and growth mechanism are still now not very clear. However, it is shown that branched carbon nanostructure can be synthesized by PECVD using Ni as a catalyst with a gas mixture of acetylene and hydrogen. It was shown that the first vertical growth of CNT takes place by the usual tip growth mechanism taking catalyst particles on the tip of every tube. Now an excess hydrogen plasma treatment results in partial removal of catalyst, and branching begins to occur as schematically shown by them (here Fig. **5.5**).

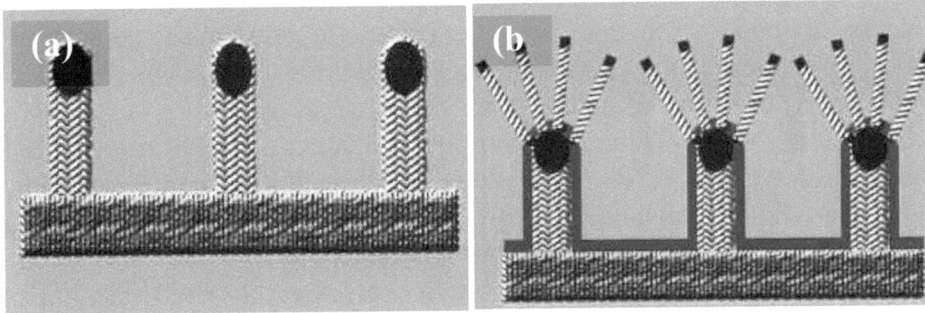

Fig. (5.5). Schematics of formation of branched CNT.

It is to be noted that there are a few other proposed mechanisms also, however all are a matter of debate as there is not any universally accepted mechanism. The structure has not been studied rigorously.

2D CARBON FLAKES AND GRAPHENE

Due to the ability of carbon to hybridize differently as stated earlier, it can take different shapes belonging to 0 D particles, 1D rod, 2D sheets or tetragonal 3D diamond depending upon the various external parameters as well as synthesis techniques. Among which, though from the very beginning of nanotechnology, carbon particles or diamond or fibers have gained intense interest, 2D carbon structures were not known fully until 2004 when Geim-Novoselov and his co-workers reported few atom thick graphene layers and studied the electric field effect on it [2]. Slowly it is seen that graphene may have many exceptional properties that are interesting from both application and theoretical points of view.

Specially, the exceptional electrical properties of graphene have attracted applications for future electronics such as field emitters, components of integrated circuits, transparent conducting electrodes and sensors. Graphene has a high electron (or hole) mobility as well as low Johnson noise (electronic noise generated by the thermal agitation of the charge carriers in an electrical conductor at equilibrium, which happens regardless of any applied voltage), and this helps in using it as the channel in a field effect transistor (FET).

It is noteworthy that 2D carbon structures are nothing but stacking graphite layers in some periodic ways. So it is expected that the properties of this structure would be highly dependent upon the number of graphene sheets that constitute the structures. Depending upon the number of graphene sheets, the entire carbon 2D family can be divided into several categories. Single layer graphene consists of a single two-dimensional hexagonal sheet of carbon atoms as shown in Fig. (**5.6a**) [3]. Bi-layer and few-layer graphenes are made of 2 to 10 layers of such two-dimensional sheets, respectively. Graphene structures consisting of more than 10 layers are considered a thick graphene sheet and termed differently, like carbon nanosheets, nanoflakes or nanowalls. They are obviously of less scientific interest and not much studied. In bi- and few-layer graphene, C atoms can be stacked in different ways, generating hexagonal or AA stacking, Bernal or AB stacking and rhombohedral or ABC stacking, shown in Fig. (**5.6b**) [3]. All these 2D carbons show three in-plane σ bonds/atom and π orbits perpendicular to the plane Fig. (**5.6c**) [3]. While the strong σ bonds work as the rigid backbone of the hexagonal structure, the out-of-plane π bonds control the interaction between different graphene layers. So when one wants to discuss the properties of the carbon 2D family, he has to remember that their properties are layered-number dependent and thus had to be discussed separately for single layer graphene (SLG) bi-layer graphene (BLG), few-layer graphene (FLG) and graphene sheets.

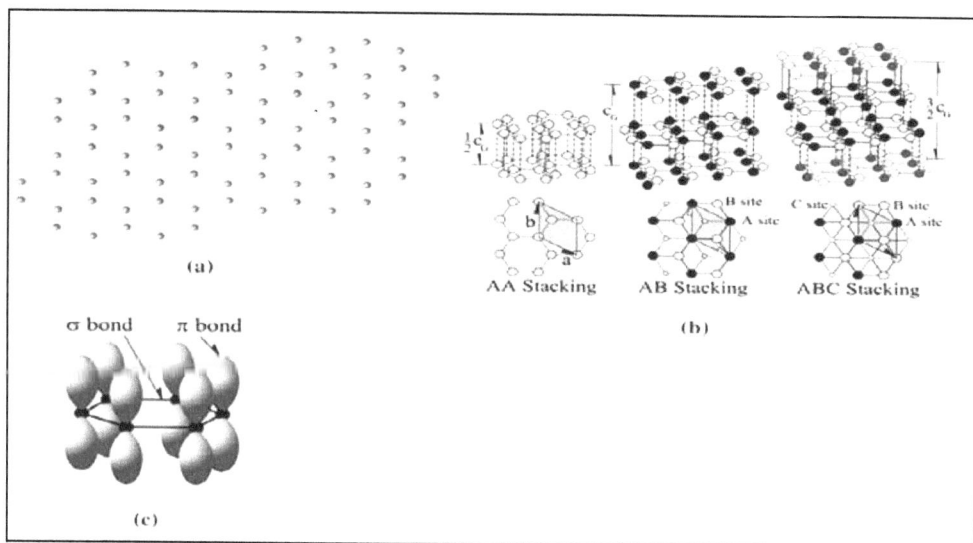

Fig. (**5.6**) (**a**) Graphene structure of single 2D hexagonal sheet of carbon atoms, (**b**) stacking sequences of graphene and (**c**) Schematic of the in-plane σ bonds and the π orbitals perpendicular to the plane of the graphene sheets.

The unique properties of SLG mainly stand on the fact that the charge carriers here, which can be described best by $(2 + 1)$-dimensional Dirac equations [4], known as mass-less Dirac fermions, and lose their rest mass, m_0. Room temperature anomalous Quantum Hall Effect (QHE) is reported from this kind of structure [5], which has made it suitable for being usedin electronics. Among different interesting properties, gas sensing ability of graphene is one of the most important from the application point of view.

Adsorption of certain molecules on graphene modifies the local carrier concentration and that, in turn, the resistance of the sample. These graphene-based sensors can be made so sensitive that they can detect adsorption or desorption of a single molecule of certain gases like CO, H_2O, NH_3 and NO_2. Molecular sensing capability could be achieved here, as graphene is electronically a very good low-noise material. Also, this structure has claimed to be one of the hardest materials, and both simulations and experiments show that its Young modules can take values as high as 1TPa. Apart from that, this material has proven to be bio-compatible also. These 2D carbon structures have many applications, such as fabricating high-performance transistors, transparent electrodes, ultra-capacitors and ultra-tough paper.

Unlike SLG, charge carriers in bi-layer graphene have finite mass and are called massive Dirac fermions. It also shows anomalous QHE but is different in nature from that of SLG. Though BLG is considered a hapless semiconductor, Zhou *et al.* have shown that an energy band gap of ~ 0.26 eV is produced in graphene, when it is epitaxially grown on an SiC substrate. Few layer graphene shows no gap, and the metallic character increases with the increasing graphene layer number. Like SLG, it also has a very high surface area and thus, it shows good gas adsorption properties. FLG can be functionalized by different covalent and non-covalent modifications in order to solubilize them in various solvents. Apart from covalent or non-covalent functionalization, graphene surface can be decorated with different metal particles like Pt, Ag, Au, *etc.* It is to be noted that graphene is a bio-compatible material and also it is very good material for being used in composite with different polymers [6]. It is rather difficult to enlighten the properties of thick carbon sheets as there has not been considerable work on this. In most of the cases, it has been synthesized as the bi-product of the carbon nanotube or graphene, and people have not shown much interest to study them further in order to explore their different properties. However, they have importance in deciphering the growth mechanism of the carbon 2 D family. There are a number of ways to synthesise 2D carbon structure of which exfoliation and cleavage, or thermal and plasma enhanced chemical vapor deposition has gained considerable interest.

DIAMOND-LIKE CARBON

Diamond-like carbon (DLC) is a metastable form of amorphous carbon, containing a significant fraction of sp^3 bonds. Diamond-like carbon is often discussed together with crystalline diamond films, even though their structure is significantly different. This is because the diamond-like solids exhibit unusual hardness and other properties resembling that of diamond, and they are sometimes made with a crystalline diamond in the same process. The diamond-like solids have very high number densities of atoms compared with other conventional hydrocarbons and hydrocarbon polymers. This high number density arises from their completely cross-linked structure and provides another point of similarity with crystalline diamond. The structure of DLC is still a matter of research. However, some overall features of this solid have become clear in the past several years. A proposed structure of DLC is shown in Fig. (**5.7**). The shaded circles are sp^3 coordinated carbon, tetrahedrally bonded to four other atoms. The fully black circles represent sp^2 carbon atoms, trigonally bonded to three other atoms. The open circles are hydrogen atoms. One major feature of DLC is the type of clustering that the sp^2 carbon atoms undergo. It is proved, on the basis of optical spectroscopic evidences that, many of the sp^2 carbon atoms are present as large graphite-like clusters within DLC. These clusters appear to be spatially isolated from one another, because the films have extremely low conductivity. In DLC, it is believed that the structure resembles an atomic network of hydrogenated silicon (a-Si: H) and is formed based on the configuration of the local equilibrium.

As mentioned before, DLC is a mixed network of amorphous hydrogenated carbon (a-C:H), where both sp^3 and sp^2 hybridized carbon are present, and their relative abundance depends upon the minimization of free energy. The corresponding model is called the "cluster model" where sp^2 sites are supposed to make pairs resulting in C=C bonds as one can see in ethylene, the double bonds further rearrange themselves to form 6 fold aromatic benzene ring. Finally, all of these form a mixed network consisting of aromatic ring and a graphitic cluster within sp^3 bonded matrix.

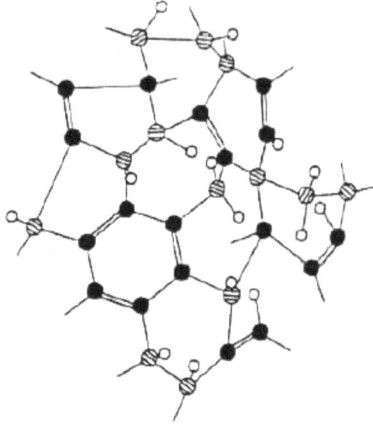

Fig. (5.7). A two-dimensional representation of a proposed structure for diamond like hydrocarbon (a-C:H).

Here the σ-σ* pair and π-π* pair respectively form the deep band states (both valence and conduction band) and band edge; the latter, in turn, determines the optical gap of the system as schematically shown in Fig. (**5.8**). The optical gap of the system is associated with the relative abundance of the graphitic cluster [7, 8] and the variation may be assumed to follow the simple rules like $E_g = 6/M^{1/2}$ eV, where M is the number of six-fold rings in the cluster. The sp^3 bonding in DLC governs its mechanical properties, whereas the collective behaviour of sp^2 bonding is responsible for electrical and optical properties.

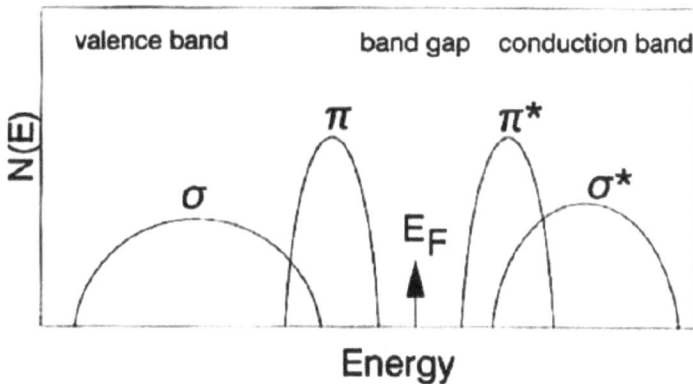

Fig. (5.8). Schematic diagram of σ and π states in amorphous carbons (**a-C**).

The DLC has some extreme properties like diamond, such as high mechanical hardness, chemical inertness, dielectric strength, low wear and friction, optical transparency in the visible and infrared region, high electrical resistivity, high thermal conductivity and low electron affinity. Also, there are some practical advantages of DLC over the diamond. DLC can be smooth on the nanometer scale, with its hardness approaching that of diamond. It can be deposited at low temperature on a large area substrate; hence the range of materials it can coat is greatly increased and low-cost, large-area electronic devices can be easily realized. DLC films have widespread applications, as protective coatings in areas, such as optical windows, magnetic storage disks, car parts, biomedical applications and as micromechanical devices (MEMs). Typical properties of the various forms of DLC are compared to diamond and graphite in Table **5.4** [9] . DLC consists of not only amorphous carbon (a-C), but also amorphous hydrogenated carbon (a-C:H). The composition of DLC is most conveniently displayed on a ternary phase diagram as in Fig. (**5.9**), first used by Jacob and Moller [10]. If the fraction of sp^3 bonding reaches a high degree, McKenzie [11] suggested renaming the a-C as tetrahedral amorphous carbon (ta-C), to distinguish it from sp^2 a-C. Koidl *et al.* [12] produced DLC by plasma enhanced chemical vapour deposition (PECVD) method, whose composition reached the interior of the triangle, this was named as a-C:H. A more sp^3 bonded material with less hydrogen content, which can be produced by high plasma density PECVD reactors, was denoted as hydrogenated tetrahedral amorphous carbon (ta-C:H).

There have been major developments in the synthesis processes and characterization techniques of diamond-like materials. Various n-and p-type doping with nitrogen, phosphorous, boron, *etc.*, in the DLC matrix, has been achieved to increase the conductivity of the material. Nitrogen has also been proved to decrease the intrinsic stress generated in the DLC. Nowadays, diamond and DLC films have attracted great interest due to their negative or low electron affinity. As electron affinity depends fundamentally on the type of bonding, not on crystallinity, amorphous DLC also has a low electron affinity. So, they have great potential in the application as electron emitters in vacuum microelectronics, such as field emission displays. For the past several years, metal-incorporated nanocomposite DLC (I-C: H) films have aroused immense interest among workers due to their interesting tribological, electrical and optical properties. Nanometer-sized metal clusters incorporated into the DLC matrix can increase drastically the electrical conductivity and increase the adhesive property of the film on the material, resist peeling off the material from the substrate and also optimum metal incorporation increases the hardness of the film. With all these fascinating properties mentioned above, DLC can very effectively be used as:

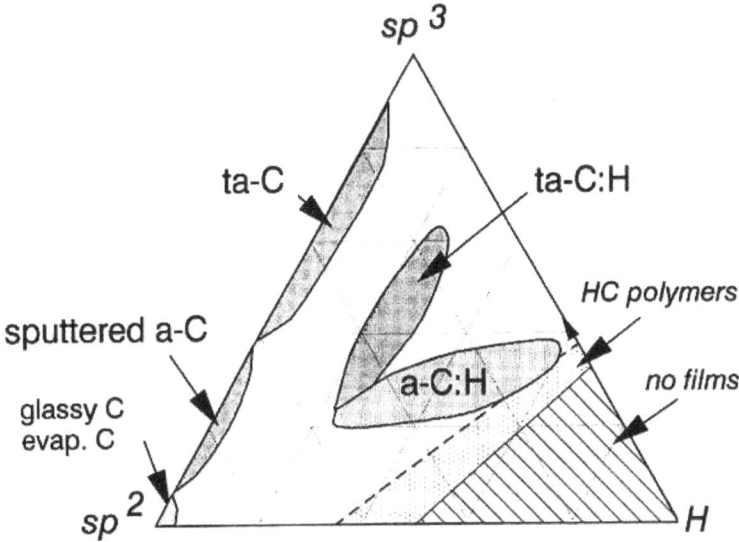

Fig. (5.9). Ternary phase diagram of various forms of diamond like carbon.

Table 5.4. Comparison of major properties of amorphous carbon with those of other materials, like diamond, graphite, C$_{60}$ and polyethylene.

Sample	sp^3 (%)	H (%)	Density (gm.cm^{-3})	Optical Gap (eV)	Hardness (Gpa)
Diamond	100	0	3.515	5.5	100
Graphite	0	0	2.267	0	-
C60	0	0	-	1.6	-
Glassy C	0	0	1.3-1.55	0.01	3
Evaporated C	0	0	1.9	0.4-0.7	3
Sputtered C	5	0	2.2	0.5	-

(Table 5.4) cont.....

Ta-C	80-88	0	3.1	2.5	80
a-C:H hard	40	30-40	1.6-2.2	1.1-1.7	10-20
a-C:H soft	60	40-50	1.2-1.6	1.7-4.0	<10
ta-C:H	70	30	2.4	2.0-2.5	50

INTRODUCTION TO GRAPHITIC CARBON NITRIDE

Basic Structure and Synthesis Techniques of Graphitic Carbon Nitride

The history of graphitic carbon nitride dates back to 1834, when a polymeric derivative was synthesized by Berzelius and named Liebig as melon'', which is regarded as one of the oldest synthetic polymers [13]. Early in 1922, closer insights into the structure of these compounds were described by Franklin. He found that the empirical composition of melon derivatives derived from mercuric thiocyanate varies with the method of preparation, and hydrogen content varied from 1.1 to 2.0 wt% [14]. Based on these findings, it was indicated that probably not one single structure should be assigned to melon. It is most likely a mixture of polymers of different sizes and structures.

After a long span of 170 years, various carbon and nitrogen-rich starting compounds or precursors have been employed for the diverse syntheses of g-C_3N_4-like materials. For example, Kouvetakis *et al.* a decomposed derivative of melamine precursors at 400–500°C and obtained an amorphous carbon nitride with more or less the correct elemental composition and a broad graphitic stacking peak [15].

Condensation performed at 250°C and 140MPa revealed a weakly condensed product with low order and many pendant amino groups acting as defects. Thermal treatment of melamine in the presence of hydrazine at 800°C and 2.5GPa resulted in C_3N_4 with almost perfect graphitic stacking, but the high temperature resulted in a high depletion of nitrogen, and there was a very scarce amount of remaining nitrogen to show a regular distribution. This work was resumed by Qian *et al.*, who synthesized C_3N_4 by condensing cyanurchloride and calcium cyanamide [16] and identified a temperature range between 500 and 600°C to optimize the order and composition of the potentially accurate g-C_3N_4 phase.

Wolf *et al.* used 2-amino-4,6-dichlorotriazine as a precursor and obtained a close-to crystallinegraphitic C_3N_4 derivative in a high pressure–high-temperature approach, where the generated HCl played the role of a template to fill the nitridic in-plane pores of a triazine based condensation pattern [17]. The structure synthesized by Wolf was quite perfect, as the system exhibited both regular in-planes packing of the nitridic triazine-based pores and X-ray patterns which could be quantitatively explained.

Xie *et al.* were able to solvothermally produce g-C_3N_4 from cyanurchloride and sodiumamide in benzene by heating to about 200°C for 8–12 hours [18]. The same group reported on the production of carbon nitride nanotubes by a similar reaction between cyanurchloride and sodium azide [19]. Graphitic packing in those samples was less pronounced, which, however, may also be due to the modified nanostructure.

In 2001 Komatsu *et al.* [20] reported a highly ordered model of carbon nitride, $C_{91}H_{14}N_{124}$. The same group also reported a highly crystalline species which they termed the "high molecular weight" melon. The X-ray diffractograms of both observed species for polymeric materials were highly ordered.

Another landmark step towards better-defined and organized graphitic C_3N_4 systems was published by Schnick *et al.* [21], who were able to isolate and solve the crystal structure of another intermediate, 2,5,8-triaminotri-s-triazine, or melem ($C_6N_{10}H_6$). Melem was found to be a very stable intermediate, but further heating, in contrast to the Komatsu experiments, resulted in poorly defined amorphous graphitic C_3N_4, as reported by the authors. In 2007 the same group was able to explain the structure of a highly defined melon polymer, thus providing further evidence that the polymeric species can exhibit high local crystal packing. Physical and chemical vapour deposition methods to synthesize g-C_3N_4 thin films are described quite often, but all of them are associated with the elimination of the very stable nitrogen, which results in disordered, carbon-rich materials hampering the C: N ratio. As g-C_3N_4 is a potential candidate for the useful substitution for amorphous and graphitic carbon in a variety of materials science and technology applications, *e.g.*, as a catalyst or active catalytic support, as a membrane material or for gas storage, novel procedures to define carbon nitride materials and a better understanding of the reaction sequence and evolution of different morphologies are on-going lucrative topics of research. Both triazine and tri-s-triazine are considered tectonic units to constitute potential allotropes of g-C$_3$N$_4$, which differ in their stability due to the different electronic environments of the N atom and also due to the sizes of the nitride pores. The two structures are given below in Fig. (**5.10**).

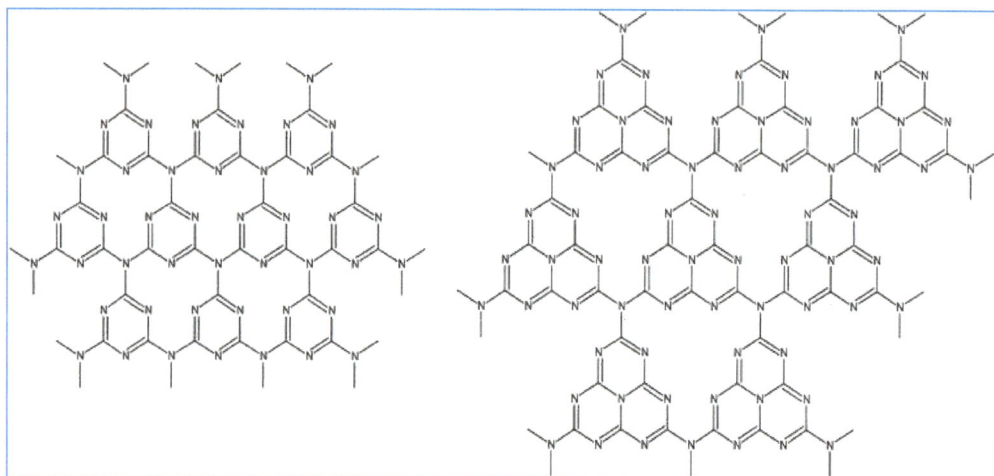

Fig. (**5.10**). (**a**) Triazine and (**b**) tri-s-triazine based structures of g-C$_3$N$_4$.

Kroke *et al.* [22] calculated that the tri-striazine-based structure is 30 kJ mol^{-1} more stable on the basis of density functional theory (DFT).

A reaction that results in formation of the ideal graphitic carbon nitride is a combination polyaddition with a polycondensation process where the first intermediate product that gets formed is melamine. In the second, *i.e.*, condensation steps, the ammonia is eliminated, and subsequently, different products get formed under the performance of the process in closed and open reaction flasks. The formation is highly dependent upon the temperature, and it is seen that melamine, which is an intermediate product as stated before, exists in the highly crystalline form up to 350 °C. This is then further rearranged at around 390 °C, and the tri-s-triazine unit begins to form. Condensation of this unit, then begins to take place, and the final product may be obtained ideally at 520 °C. However, a further increase in temperature makes the product slightly unstable at around 600 °C, and when the temperature further increases over 700 °C, the product simply disappears by the formation of nitrogen and other cyano-products. The main problem associated with this problem is that the melamine gets sublimated at a higher temperature. Also, the proper control of nitrogen gas flow is another serious issue as this is one of the main controlling parameters. This problem may be partially overcome by making an arrangement so that melamine transiently formed and coexisted with other species retarding the process mentioned before. Thus to increase the mass efficiency, which is very crucial in the process, it is recommended to use urea or dicyandiamide as source material.

At a temperature around 400 °C, rearrangement of cyameluric nucleus results in the generation of melamine, the unit cell of melam. This, being a stable intermediate product, may be isolated in a closed system under the excess pressure of ammonia. When the reaction is made in an open atmosphere, one can see a higher carbon: nitrogen ratio from elemental analysis suggesting a higher mass loss which is highly discouraged. When the system is further heated to 500 °C, there is further loss of ammonia, and the loss is more significant as it is associated with condensed C_3N_4 polymer. These steps of the reaction may also further be studied from a theoretical calculation using density functional theory (DFT). It is to be noted that melamine-like materials are molecular crystals where the force of interaction mainly comes from the covalent bonds, whereas the crystals are found to be held together by weak forces like hydrogen bond formation or van der Waals interactions.

Fig. (**5.11**) shows that the cohesive energy of molecules has increasing trends with the polyaddition as further confirmed by theoretical calculation as well. This further confirms the formation of melamine from cyanamide when heated.

However, there exists still confusion regarding the fact that if triazine units of C_3N_4 sheets get formed by further condensation or if melam molecules first get produced *via* condensation of melamine, then the as-produced melam gets polymerized into the required C_6N_8 sheet structure based on tri-*s*-triazine units.

Experiments indicated that melam(melamine dimers) is the probable product after polymerization of melamine and are metastable intermediates. This is in agreement with DFT calculations that show that the cohesive energy increases from melamine *via* melam to melem. This further suggests that the polymerization continues *via* the tri-*s*-triazine path, and the penultimate step is the connection between adjacent molecules in the melem crystal to form dimelem molecules. These dimelem molecules can then undergo further condensation by the further release of NH_3 molecules to produce melon which is a linear polymer of tri-*s*-triazine units.

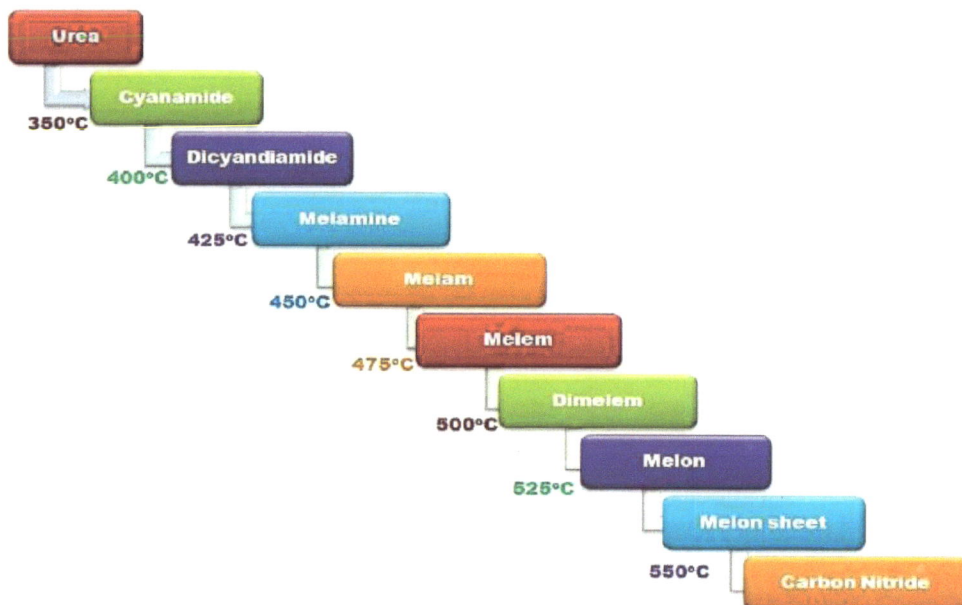

Fig. (5.11). Calculated energy diagram for the synthesis of carbon nitride. The starting precursor cyanamide is condensing into melamine. Further condensation can then proceed to C_3N_4.

The calculations indicate that the shape of the melon chain is determined by a competition between the energy gain by extending the π electron system into a linear chain and the repulsion between the lone pair electrons in the N atom on the edge of the tri-*s*-triazine unit. The linear chain is therefore strongly buckled, since the tri-*s*-triazine molecules try to maximize the distance between the N atoms on the edge of adjacent molecules due to the N-N repulsion.

It was proposed by Sehnart *et al.* that the melon chain is instead constructed from dimmers where the top corner is connected to a side corner, giving the melon chain a staircase shape [23]. DFT calculations show that the melon in this dimerized form is more stable by 20 kJ per mol tri-*s*-triazine molecules than a linear chain. The final step of the polymerization is the formation of carbon nitride sheets by fusion of the melon chains.

There are a number of ways to connect the melon chains into a sheet-like structure, and the actual form depends on the degree of condensation. The first possibility is connecting the melon chain two by two, where every second melon chain is inverted [24]. This forms a structure where the triangular tri-*s*-triazine molecules fit like a lock and key into each other, and the structure is held together by hydrogen bonds between the amino groups and the edge N atoms. This melon sheet structure is indeed more stable than the isolated melon chains by 63 kJ monomol^{-1}.

The formation energy of a melon sheet with respect to graphite and N_2 is slightly negative ΔE melon sheet ($T = 0$ K) $= 16$ kJ mol-1tri-*s*-triazine units indicating that it is a stable structure and it may be considered as the condensed structure of melon. Releasing another NH_3 molecule per tri-*s*-triazine unit would remove the remaining hydrogen to form a pure carbon nitride material with C_6N_8 stoichiometry.

The cohesive energy increases significantly upon removing the remaining hydrogen, and eventually indicates that C_6N_8 would theoretically be the most stable form of carbon nitride. However, there may be kinetic hindrance for the connection of the tri-s-triazine units into the ordered pattern of the C_6N_8structure, starting from the melon sheet structure, since every second melon chain is connected in the opposite direction. In addition, the N repulsion becomes much more severe due to the closer proximity of the tri-*s*-triazine units in the sheet structure, such that the C_6N_8sheets are heavily buckled.

Properties of Graphitic Carbon Nitride

Intrinsic features and properties of graphitic carbon nitride are as follows:

a. Being Metal Free: Its property of being metal free has always been an interesting subject of research on graphitic carbon nitride from the very beginning. The compositional elements for g-C_3N_4 are just C, N, and residual hydrogen in defects and surface edges. It is to be noted that the fabrication of an ideal graphitic carbon nitride with a C/N stoichiometric ratio of 0.75 is difficult to achieve. Elemental analysis could provide evidence for incomplete condensation and the presence of excess nitrogen. An average C: N atomic ratio of 0.72 is usually obtained with trace, but not significant amounts of hydrogen (2%) from uncondensed amino groups on the edges. The simple constitution of g-C_3N_4 accounts for the intoxicate nature of the materials and thus contributes to the biocompatibility for some potential biological applications. The non-toxic and biocompatible nature of g-C_3N_4 was investigated by Zhang *et al.* in 2012 by incubating the HeLa cells with g-C_3N_4 [25]. They reported that the HeLa cells could maintain their activity after incubation with g-C_3N_4 nanosheets with concentrations up to 600 µg mL^{-1}in aqueous solution.

b. Optical Properties: Optical properties of g-C$_3$N$_4$ have been studied in details by employing UV-Vis diffuse reflectance spectroscopy and photoluminescence for various applications. As reported from the diffuse reflectance spectrum, graphitic carbon nitride shows the typical absorption pattern of an organic semiconductor. It has a wide band gap of 2.7 eV and an absorption edge around 420 nm. Graphitic carbon nitride usually exhibits strong blue photoluminescence at room temperature. The luminescence is found to cover a wide visible range (430–550 nm) and has a maximum emission around 470 nm. In 2013 Zhang *et al.* demonstrated that photoluminescence of g-C$_3$N$_4$ is dependent on condensation temperature and synthesis conditions [26]. Electrically generated chemiluminescence of g-C$_3$N$_4$ was observed by Cheng *et al.* in 2013 [27]. A unique phenomenon of two photon absorption of graphitic carbon nitride was discovered by Zhang *et al.* in 2014 [28]. Graphitic carbon nitride has the property to simultaneously absorb two near-infrared photons and emit bright fluorescence in the visible light region. Zhang *et al.* showed that when g-C$_3$N$_4$ was excited by 780 nm red laser, it emitted strong green light [28]. But in case of two-photon excitation a large red shift of the emission spectrum takes place with respect to one-photon excitation. However, in case of other semiconductors, the emission spectra of two-photon and one-photon excitation are located at approximately the same position. Fig. (**5.12**) shows the basic structural unit of the material responsible for optoelectronic properties of the material.

c. A Polymeric Semiconductor: g-C$_3$N$_4$ possesses a significantly wide band gap of 2.7 eV. The conduction band (CB) is at -1.1 eV and the valence band (VB) positions at +1.6 eV, *vs.* normal hydrogen electrodes (NHEs) [29]. This signifies that it can act as a visible-light-active photocatalyst for overall water splitting. The suitable electronic band structure also makes graphitic carbon nitride a potential candidate for solar energy converting systems, like photo-electrochemical cells.

d. Stability: A Thermogravimetric analysis of g-C$_3$N$_4$ showed that this material can remain stable thermally up to 600 °C. A strong endothermal peak appears at 630 °C in Differential Thermal analysis, which is consistent with the consecutive, complete weight loss, thereby indicating that the thermal decomposition and vaporization of the material have started at this temperature. The complete decomposition of graphitic carbon nitrides occurs at 700 °C, and no residue of the material is left if the temperature is further increased. This thermal stability is one of the highest for an organic material, and thus it has promising high-temperature applications. The thermal stability of carbon nitride depends upon synthesis procedures and the different degrees of condensation. Moreover, the decomposition temperature is lower than the graphitization temperature of carbon, which suggests that the material can act as template for synthesizing a refined carbon nanostructure. The excellent chemical stability against acid corrosion was also revealed in the study of other properties.

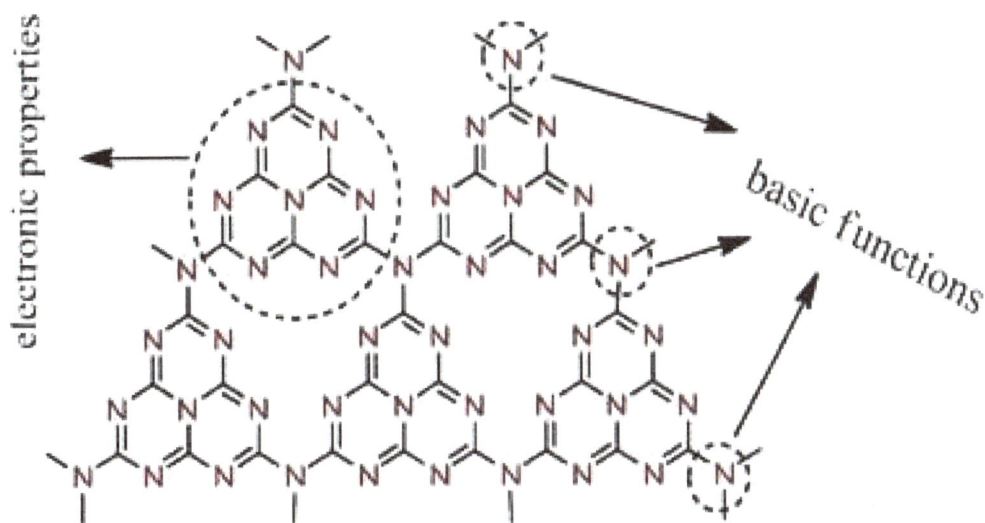

Fig. (5.12). Structural units contributing to electronic and basic functions

Zhao *et al.* in 2015 found out that the dispersion of g-C₃N₄ could form a true solution [30]. The recovered g-C₃N₄ from acid solution still possessed a predominant (002) peak in the X-ray diffraction spectrum, indicating that g-C₃N₄ could easily restack even after acid treatment. In addition to these, Bai *et al.* in 2014 and Zhu *et al.* in 2015 reported that g-C₃N₄ nanosheets or single-layered quantum dots in neutral water are generally negatively charged with a measured zeta potential of about-40 mV [31, 32]. The highly negative surface charge ensures the stable suspension of g-C₃N₄ nanostructures in an aqueous solution and could remain stable for several weeks without aggregation or precipitation. Like graphite, the stacking of g-C₃N₄ with optimized van der Waals interactions between the single layers makes it insoluble in most solvents. No detectable solubility or reactivity of carbon nitride has been observed in conventional solvents, including water, alcohols, DMF, THF, diethyl ether, and toluene.

e. Crystal Structure of Idealized g-C₃N₄: It is a well-known fact that carbon nitrides can exist in several allotropes with diverse properties and features. The graphitic phase is regarded as the most stable allotrope under ambient conditions. Triazine units (C_3N_3) had been proposed as one of the basic building blocks of g-C_3N_4. The other structural models, tri-s-triazine (heptazine, C_6N_7) rings, which are structurally close to the Melon structure, are more energetically favoured building blocks of g-C_3N_4 than triazine, as reported by the same group. The triazine or tri-s-triazine rings are cross-linked by trigonal nitrogen atoms to form extended networks.

Applications of Graphitic Carbon Nitride

Graphitic carbon nitride has interesting properties, such as good accessibility at a low price, a high thermal and chemical stability, and amenability to chemical modification, which makes it suitable for a range of applications. Even in this early stage, carbon nitrides together with their modifications, have already found numerous applications in relevant fields of chemistry. Due to the special semiconductor properties of carbon nitrides, they show unexpected catalytic activity for a variety of reactions, such as for the activation of benzene, trimerization reactions, and also the activation of carbon dioxide (artificial photosynthesis).Some of the important applications are stated as follows:

1. Carbon Nitride can act as a Photocatalyst for water splitting.

2. Carbon Nitride can act as a catalyst for oxidation reactions:-

a) Oxidation of alkanes.

b) Oxidation of olefins.

c) Oxidation of alcohols.

d) Oxidation of heteroatoms.

e) Photo-degradation of pollutants.

3. Carbon Nitride has applications in Hydrogenation reactions.

4. Carbon Nitride can act as a basic catalyst.

5. Carbon Nitride has applications in NO Decomposition.

6. Carbon Nitride can activate π- bonds and Aromatic systems.

Such materials are important in high-performance engineering applications for high-hardness, high-temperature, high-power, or high-frequency devices ranging from microelectronic to space flight applications.

The suitable electronic band structure also makes graphitic carbon nitride a promising candidate for solar energy converting systems, such as photo electrochemical cells. Indeed, photocurrent was observed even in the case of bulk g-C_3N_4 under the illumination of visible light (λ>420).The high chemical and thermal stability of carbon nitride makes photo electrochemical cells stable under an oxygen atmosphere. Furthermore, the electronic band structures of g-C_3N_4 could be tuned by modification of the nano morphology or doping, which makes the improvement in photocurrent possible. For example, mesoporous carbon nitride (mpg-C_3N_4) can in principle, enhance the light harvesting ability owing to its large surface area and multiple scattering effects, and therefore showed an increase in photocurrent.

Other modifications, including doping and protonation, can also increase the photocurrent.

Introduction to Zinc Oxide

The Basic Structure and Synthesis Techniques of Zinc Oxide Nanostructures

Zinc oxide is an inorganic compound with the formula ZnO. ZnO is a white powder that is insoluble in water, and it is widely used as an additive in numerous materials and products, including rubbers, plastics, ceramics, glass, cement, lubricants, paints, ointments, adhesives, sealants, pigments, foods, batteries, ferrites, fire retardants, and first-aid tapes. Although it occurs naturally as the mineral zincite, most zinc oxide is produced synthetically. Crystalline zinc oxide changes its color from white to yellow when heated and gets reverted back to white again upon cooling thus it is thermochromic in nature. This is due to loss of oxygen at high temperature and thus formation of non-stoichiometric chemical structure like $Zn_{1+x}O$ where x = 0.00007 at 800 °C. ZnO is a wide-bandgap semiconductor of the II-VI semiconductor group. The native doping of the semiconductor due to oxygen vacancies or zinc interstitials is n-type. This semiconductor has several favorable properties, including good transparency, high electron mobility, wide band-gap, and strong room-temperature luminescence. Those properties are valuable in emerging applications for: transparent electrodes in liquid crystal displays, energy-saving or heat-protecting windows, and electronics as thin-film transistors and light-emitting diodes. Zinc oxide crystallizes in two main forms, hexagonal wurtzite and cubic zinc-blende. The wurtzite structure is most stable at ambient conditions and thus most common. The zinc blende form can be stabilized by growing ZnO on substrates with cubic lattice structure. In both cases, the zinc and oxide centers are tetrahedral, the most characteristic geometry for Zn(II). ZnO converts to the rock salt motif at relatively high pressures about 10 GPa. Hexagonal and zinc blende polymorphs have no inversion symmetry (reflection of a crystal relative to any given point does not transform it into itself). This and other lattice symmetry properties result in piezoelectricity of the hexagonal and zinc blende ZnO, and pyro-electricity of hexagonal ZnO.

ZnO belongs to a point group 6 mm (Hermann-Mauguin notation) or C_{6v} (Schoenflies notation), and the associated space group is $P6_3mc$ or C_{6v}. The lattice parameter values are $a = 3.25$ Å and $c = 5.2$ Å; and thus their ratio $c/a \sim 1.60$ is close to the ideal value of $c/a = 1.633$. Like other II-VI semiconductor, bonding in ZnO has an ionic character with the corresponding radii values of Zn^{2+} and O^{2-} are respectively 0.074 and 0.140 nm. Fig. **(5.13)** shows the basic crystal structure of ZnO.

Because of the polar Zn-O bonds, zinc and oxygen planes are electrically charged. To maintain electrical neutrality, those planes reconstruct at the atomic level in most relative materials, but not in ZnO – its surfaces are atomically flat, stable and exhibit no reconstruction. This anomaly of ZnO is not fully explained.

ZnO may be synthesized in various shapes and sizes from bulk to nanoregime with various shapes including zero (quantum dot), one (nanowire), two (nanoflakes) and three dimensions (any other structure like tetrapods, flowers, *etc*). The specialty of the ZnO preparation by different technique is that most of the processes involve low temperature technique and the synthesis process would not exceed 90 °C.

Few effective reagents/precursors are zinc nitrate, zinc acetate, hexamine, *etc*. The latter creates a basic environment which favors the reaction kinetics. Polyethylene glycol or polyethylenimine like additive helps the growth of certain ZnO nanostructure. Also seeding technique like $KMnO_4$ activation drastically helps the growth of one dimensional ZnO nanostructures. When one seeks for doping it is customary to add nitrate precursors of the associated metals. The morphology of the system may be changed either by changing precursor ratio or giving suitable thermal treatment like changing the reaction temperature or the heating rate.

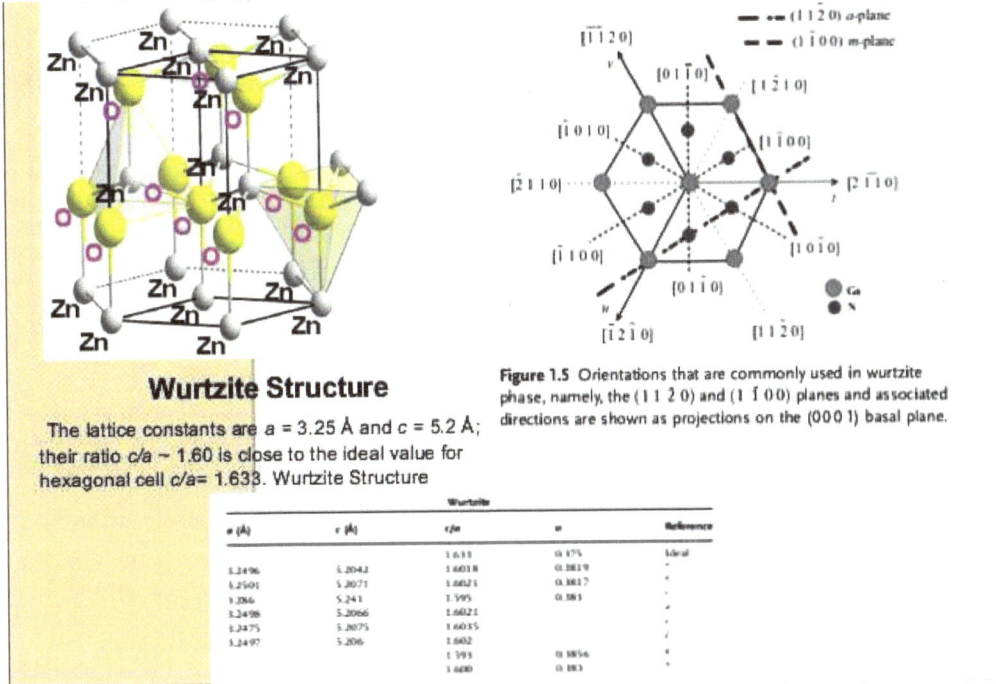

Wurtzite Structure

The lattice constants are $a = 3.25$ Å and $c = 5.2$ Å; their ratio $c/a \sim 1.60$ is close to the ideal value for hexagonal cell $c/a = 1.633$. Wurtzite Structure

Figure 1.5 Orientations that are commonly used in wurtzite phase, namely, the $(1\,1\,\bar{2}\,0)$ and $(1\,\bar{1}\,00)$ planes and associated directions are shown as projections on the $(000\,1)$ basal plane.

Fig. (5.13). Wurtzite Structure of Zinc Oxide.

As mentioned before vertically aligned one dimensional ZnO nanostructures have been developed by simple sol-gel technique over $KMnO_4$ seeded substrates like silicon, glass using zinc salts like zinc nitrate or acetate. The seeding technique facilitates uniform nucleation throughout the surface thus supporting the uniform development of the nanostructures. Few suitable or common seeding techniques like sol-gel, spin coating, lithography, physical vapor deposition and other. ZnO nanostructures has versatile applications like anti-bacterial activity, cold emitter, solar-cell, hydrophobic coating, catalyst to be discussed later.

Properties of Zinc Oxide

Mechanical Properties

Hardness of ZnO does not exceed 4.5 (Mohs scale) and thus may be considered to be a relatively soft material having elastic constant smaller than most of the III-V semiconductors. Different properties like high heat capacity or thermal conductivity, high melting point or low thermal expansion coefficients make it a good candidate for being used as ceramics.

This is one of the most important materials in view of piezoelectric applications that includes very high electro-mechanical coupling as it has the highest piezoelectric tensor comparable to the value of that of GaN and AlN.

Electrical Properties

The electrical, as well as the optical properties of a material, are directly related to its optical band gap, which generally comes out to 3.3 eV for ZnO an can be further tuned between 3–4 eV by suitable doping with foreign materials as well as by changing the cluster size and bringing the quantum confinement effect. The value 3.3 eV is rather large and this in turn, makes it suitable to be used as high temperature-high power device operation. It further equips ZnO to withstand high breakdown voltage and have lesser electrical noise.

Doping is always a crucial factor in such kinds of semiconductors to get a tuning over the conductivity. ZnO is by nature an n-type semiconductor even without any external effort, which is though believed to be due the improper stoichiometry, but the matter is still under debate. In some theoretical calculations, it is predicted that unintentional hydrogen substitution is the key reason for such inherent n-type conductivity. Apart from that, intentional n-type doping is easily done by using group III elements like Al, Ga, In as an oxygen substitute in ZnO.

In comparison with the previous discussion, it is much more complicated to get a p type conductivity from ZnO. The main two reasons are the lesser solubility of the p type dopant and the compensation by excess n type impurities that are inherently present in the host matrix. The measurement of p type material gets further complicated by the inhomogeneity of the material [33].

The p-type doping in ZnO still remains difficult, but so far, little success has been made with selected groups I materials like Li, Na, K; or group-V elements including N, P, As or even with Cu and Ag. However, most of the time, these resulting formation of deep acceptor levels without contributing significantly, in the p type doping. It is this reason due to which even after having so many applications use of p type ZnO is not so common. The mobility of the system is highly dependent on temperature and the maximum value of electron and hole mobility are respectively ~ 2000 cm^2/(V·s) and 5–30 cm^2 at 80 K.

Various Synthesis Procedures of Zinc Oxide Nanostructures

Indirect Process

In the indirect or French process, metallic zinc is melted in a graphite crucible and vaporized at temperatures above 907 °C (typically around 1000 °C). Zinc vapor reacts with the oxygen in the air to give ZnO, accompanied by a drop in its temperature and bright luminescence. Zinc oxide particles are transported into a cooling duct and collected in a bag house. This indirect method was popularized by LeClaire (France) in 1844 and therefore is commonly known as the French process. Its product normally consists of agglomerated zinc oxide particles with an average size of 0.1 to a few micrometers. By weight, most of the world's zinc oxide is manufactured *via* French process.

Direct Process

The direct or American process starts with diverse contaminated zinc composites, such as zinc ores or smelter by-products. The zinc precursors are reduced (carbothermal reduction) by heating with a source of carbon such as anthracite to produce zinc vapor, which is then oxidized as in the indirect process. Because of the lower purity of the source material, the final product is also of lower quality in the direct process as compared to the indirect one.

Wet Chemical Process

A small amount of industrial production involves wet chemical processes, which start with aqueous solutions of zinc salts, from which zinc carbonate or zinc hydroxide is precipitated. The solid precipitate is then calcined at temperatures around 800 °C.

Laboratory Synthesis

Numerous specialised methods exist for producing ZnO for scientific studies and niche applications. These methods can be classified by the resulting ZnO form (bulk, thin film, nanowire), temperature ("low", that is close to room temperature or "high", that is T ~ 1000 °C), process type (vapor deposition or growth from solution) and other parameters.

Large single crystals (many cubic centimeters) can be grown by the gas transport (vapor-phase deposition), hydrothermal synthesis, or melt growth [34]. However, because of high vapor pressure of ZnO, growth from the melt is problematic. Growth by gas transport is difficult to control, leaving the hydrothermal method as a preference [34]. Thin films can be prepared by chemical vapour deposition, metalorganic vapour phase epitaxy, electro-deposition, pulsed laser deposition, sputtering, sol-gel synthesis, atomic layer deposition, spray pyrolysis, *etc.*

Ordinary white powdered zinc oxide can be produced in the laboratory by electrolyzing a solution of sodium bicarbonate with a zinc anode. Zinc hydroxide and hydrogen gas are produced. The zinc hydroxide upon heating decomposes to zinc oxide.

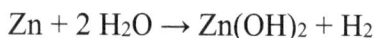

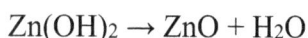

$$Zn + 2\ H_2O \rightarrow Zn(OH)_2 + H_2$$

$$Zn(OH)_2 \rightarrow ZnO + H_2O$$

Potential Applications of Zinc Oxide

Electronics

The optical gap of ZnO corresponds to a wavelength around 375 nm which corresponds to a practical value of 3.37 eV. This value has made it suitable for using it in different optoelectronic applications that include LASER diode or light emitting diode (LED). There are a good portion of the optoelectronic application of ZnO that gets overlapped with that of another opto-electronic material GaN. However, the exciton binding energy of ZnO (~60 meV) is much higher compared to that of GaN and it is almost 2.4 times higher than the room temperature thermal energy. The latter property successfully predicts ZnO can emit light at room temperature and thus it is possible to get fascinating Opto-electronic properties/devices when it is coupled with GaN for instance, in LED applications where a transparent oxide coating of ZnO increases the light out-coupling. Also, as ZnO has radiation resistance properties it is thus one of the best candidates for being used in space applications. Also it is found application in the development of LASER in UV region that gets pumped electronically. ZnO has a tendency to grow along a particular direction, thus favouring the growth of one-dimensional structure. The sharp tip of the as synthesized structure attracts electric field offering huge field enhancement and thus work as efficient cold cathode emitter.

Another use of ZnO comes when it is doped with Aluminum and Al doped ZnO layer may be used in transparent electrodes and the system is much lesser toxic compared to the conventional systems like indium tin oxide (ITO) system. ZnO found its application in the development of transparent thin film transistor also. The reason is that conventional FET does not necessarily need a p-n junction and thus one can avoid the problem of p type doping in ZnO. ZnO nanorods may also be used in FET as conducting channel.

Zinc Oxide Nanorod Sensor

ZnO may behave as an effective sensor that works on the principle of changing of electric current passing through ZnO nanorods due to adsorption of different gas molecules. One of the most prominent examples is related to hydrogen sensing. ZnO can sense the presence of hydrogen and the sensitivity even increases significantly when coated with a small layer of Pd may be developed by sputtering. The presence of Pd helps dissociation of hydrogen into atomic hydrogen through the process of catalysis thus increasing the device sensitivity. The system surprisingly has the sensitivity to detect hydrogen as less as 10 parts per million and that too at room temperature but it has no impact on oxygen [35].

Spintronics

ZnO has been considered to be one of the most promising candidates for spintronic application especially when doped with magnetic materials like (Mn, Fe, Co, V, *etc.*) and then we call the product formed as diluted magnetic semiconductors (DMS). The materials are extremely useful for storing memories and thus can be used as shape memory alloys. It has been long since the room temperature ferromagnetic properties have been discovered in Mn doped ZnO. However, there is still debate regarding the fact that this magnetism comes from the host matrix or any intermediate oxide phase formed due to doping.

Piezoelectricity

ZnO has established itself to be potential piezo electric material. The can generate power from slight mechanical stress. Experimentally it has been proved that cloths coated with ZnO can generate power from the mechanical stresses that gets generated from wind or from the movement of the body.

In the year of 2008 "*Center for Nanostructure Characterization* at the Georgia Institute of Technology" reported the development of a device that is capable of generating alternating current by giving mechanical stresses to ZnO nanowires. The associated voltage may be up to 45 millivolts with conversion efficiency (mechanical to electrical energy) around seven percent. In that worker, the nanostructure used by the researchers has length and diameters, respectively 0.2–0.3 mm and 3-5 µm micrometers, however the device may be equally functioning even it is made much smaller than this.

Li-Ion Battery

ZnO has another important field of applications to be used and, *i.e.*, as anode in Li-ion battery. The material has a very high capacity of over 978 mAh $g^{-1,}$ higher than the other similar materials like different transition metal oxide CoO (715 mAh g^{-1}), NiO (718 mAh g^{-1}) and CuO (674 mAh g^{-1}) with further advantages over the later regarding simplicity, cost-effectiveness and environmental friendliness.

It is the varistor where the material ZnO first found its application in the field of electronics. It is believed that without this first step, it would have been possible to use ZnO-based electronics in household applications.

In the year of 1954 the Hall Effect has been first discovered, claiming the inherent n type nature of the material. Almost at the same time, emission properties of ZnO came under the attention. Since then the work on the ZnO and GaN has been done with much more focus. Even statistics said that the most of the articles in the related fields are mainly on the two materials *i.e.* ZnO and GaN. However, it is seen that research related to GaN is somewhat being faded whereas that on ZnO is still climbing up even after almost 90 years of its initial boom. Apart from the applications mentioned before ZnO is being UV protector found applications in most of sunscreens. ZnO is an active carrier in drug delivery found application in medical fields. ZnO is also used as coating with the suitable conditions it may be made super-hydrophobic with an extremely less rolling angle at the same time it can be tuned to the super hydrophilic region by simple UV light treatment.

After all these discussions, now is the time to discuss the present trend of research related to this material. Currently, the work has mainly been focused on several topics like the development of stable p-type ZnO, and in this context, significant progress has been made in the development of UV laser diodes based on ZnO p-n junctions. However, such work mainly lacks from the point of view yield and stability in large scale and p-type ZnO is still under a great challenge. Another topic of interest is the transparent electronic that will result in transparent thin film transistors for flexible electronics on plastics and polymer substrates. The tuning of electrical conductivity of ZnO by suitable doping like Al, B, Ga or other impurities is another topic of current interest and the addition of transparency from the point of view of the optics can make the material suitable for solar cell and flat panel display. This is very important from the fact that currently, indium tin oxide (ITO) is the material that serves as industry-standard transparent conducting material but lacking from the cost point of view. The reason that ITO is costly rests on the fact that indium is a very less abundant material. Thus, doped ZnO really has the potential to replace this, provided the other issues may properly be taken care of.

Silicon Nanowires (SiNWs)

Silicon is no doubt the basic material in the industry of the semiconductor and thus, silicon-based different nanostructures especially one-dimensional nanowires or nanorods have attracted the central attraction of the researchers. The silicon-based materials may include pure silicon, different silicate or silicide, all of which comes out to be almost equally important due to the central role of the parent material, *i.e.*, silicon. The systems show various potential applications to be discussed after words few of which may be mentioned here, and they are FET, cathode emitter, anti-reflection coating, sensors different MEMS or NEMS and many others. There is an increase in the growing interest in the research work on silicon nanowires because of its application in many fields mentioned above. Their one-dimensional geometry on the nanometer scale provides an extremely high surface area with a nanoscale radius of curvature and great mechanical flexibility. These properties are advantageous in many chemical and mechanical applications. As one goes on to reduce the size of silicon (Si) bulk material, the number of surface atom increases.

As the number of surface atoms increases, its optical absorbance also increases. That means optical absorbance of SiNWs is more than Si bulk material. In UV and IR spectrum, the reflectance of the SiNW is less than 5%. As SiNWs have a greater surface-to-volume ratio, its surface reactivity is more. The surface of SiNWs is easily oxidized in air due to a high density of dangling bonds at the surface. If SiNWs can be used properly in photovoltaic cells, then it will be more efficient and can help in solving the energy crisis.

For SiNWs, the carrier type and concentration can be controlled by doping, as in bulk silicon. Furthermore silicon turns into a direct-band-gap semiconductor at nanometer size, due to quantum confinement, opening an entirely new field of optoelectronic applications for SiNWs. But SiNWs having extremely small diameters show indirect band gaps. This suggests using medium-sized nanowires for optical applications. One of the most interesting facts about the SiNWs is the dependence of nanowire diameter on the growth direction and systematics studies has shown that the nanowires have diameters in the range between 3-10 nm, 10-20 nm and 20-30 nm respectively grow along (110), (112) and (111) direction. SiNWs have shown promises regarding the applications associated with photovoltaic in room temperature, which is mainly due to the quantum confinement effect in nanostructured silicon. Though significant works have been done on the different aspects of SiNW properties and applications still there are ample of scopes to study other aspects of the materials.

Why Silicon Nanowires?

Silicon is by far the major player in today's electronics market, dominating the microelectronics industry, with about 90% of all semiconductor devices sold worldwide being silicon based. Silicon is a semiconductor material with the band gap of 1.12eV. Silicon possesses two of the most outstanding natural dielectrics, silicon dioxide (SiO_2) and silicon nitride (Si_3N_4), which are essential for device formation. In particular, SiO_2, which is the basis of the metal–oxide – semiconductor devices (MOS) can be grown thermally on a silicon wafer. It is chemically very stable and can achieve a very high breakdown voltage. Silicon is non-toxic, relatively inexpensive (silicon comprises about 26% of the earth's crust, which makes it second in abundance only to oxygen), easy to process (a very well established industrial infrastructure in silicon processing exists around the world), and has quite good mechanical properties (strength, hardness, thermal conductivity, *etc.*). Silicon materials, which exhibit a wide optical adsorption range, high optical absorption efficiency, and high electron mobility, can become a great potential photoelectric conversion material for its important applications in the field of photovoltaics and photocatalysis.

Synthesis of Silicon Nanowires (SiNWs)

There are several methods to synthesize SiNWs. These include CVD, Laser ablation, thermal evaporation or, most commonly, metal-assisted chemical etching. Metal-assisted chemical etching has gained special attention due to its simplicity, low cost and higher yield of the final product. The detail of the process is given as follows.

Metal Assisted Chemical Etching Process

Vertically aligned SiNW arrays can be achieved by using top-down metal-assisted chemical etching of single crystalline silicon. In the etching process, the elements that are usually used for deposition are silver (Ag), gold (Au), and copper (Cu) on silicon substrate. These metals attract electron from the silicon substrate. Metal-assisted chemical etching of SiNWs is based on the galvanic displacement reaction where simultaneously silver nitrate ($AgNO_3$) is reduced to silver (Ag) nanoparticles and silicon gets oxidized. In this process, dissolution of silicon adjacent to Ag particles occurs. The nucleation energy of the silver particles near the defective sites, such as dopants and dislocations, is much lower compared to the other sites. The presence of point defects reduces the corrosion potential of silicon and results in localized pit corrosion. The reactions that take place at the electrodes are as follows:

$$H_2O_2 + 2H^+ + 2e^- \longrightarrow 2H_2O \text{ (Cathode)}$$

$$Si + 6F^- \longrightarrow [SiF_6]^{2-} + 4e^- \text{ (Anode)}$$

The whole experimental process is schematically shown in Fig. (**5.14**).

In the process described before, the pit environment gets depleted into the reactants of the cathode side. This shifts the majority of the cathodic reaction to the exposed surface. In this entire process, the smaller size of the fluorine (F^-) ions and high diffusivity plays the main trick here. It is, in this context, mentioned worthy that when with the growth of pits F^- ions get migrated into the pit and continuously goes on dissolving the silicon oxide favouring the nanowire growth. Here silver ions work as a catalyst in nanowire formation. Here the catalyst particle determines the nanowire diameter and thus plays an important role. Apart from Ag there are other catalysts like copper or gold or others that can serve the purpose with equal efficiency and it is the equilibrium phase diagram that may be used to choose a particular catalyst as per the requirement and growth condition.

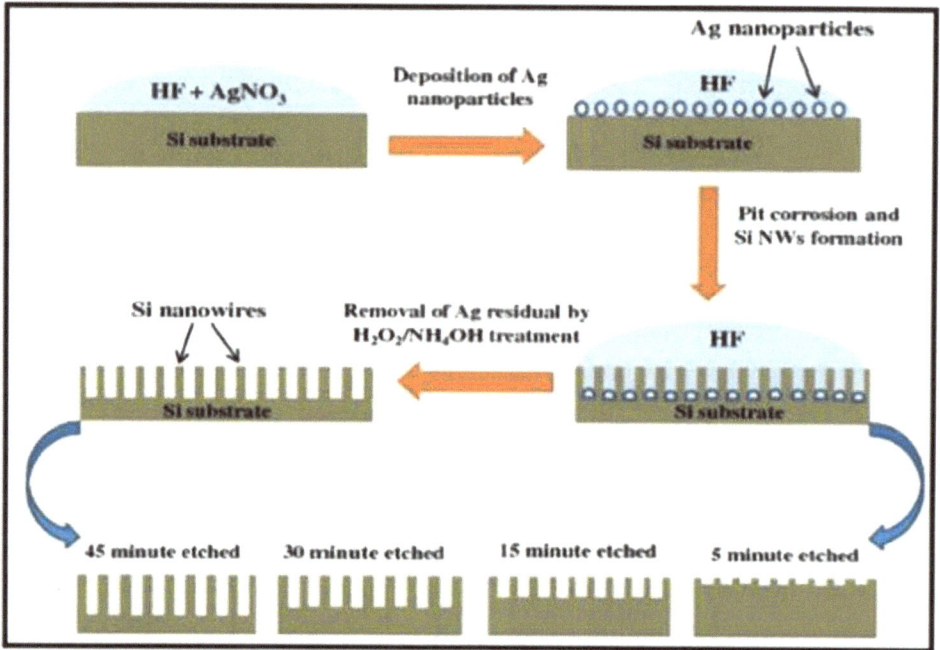

Fig. (5.14). Schematic, showing the growth mechanism of SiNWs by metal assisted chemical etching.

Properties of Nanowires

SiNWs is one-dimensional material and thus here, the electron can flow in only one direction and here, the density of state (N(E)) varies with energy (E) in a pattern like $N(E) \propto E^{-1/2}$. Thus the properties of the SiNWs differ significantly from their bulk form due to the very strong quantum confinement effect. Due to the higher surface area, there exist number of dangling bonds in the materials, which makes SiNWs easily vulnerable to oxidation due to atmospheric oxygen present. Few particular properties are as follows:

Electronic and Optical Properties

SiNWs show the strong confinement effect and thus, there is a change in the optical band gap that varies between 1.1 to 3.95 eV with wire diameter between 7 to 1.3 nm. Other related properties like defects, valley splitting or effective mass all depends upon the value of the diameter. When SiNWs are externally treated with oxygen or hydrogen what one gets is oxygen or hydrogen terminated SiNWs that has properties significantly differ from pure SiNWs and thus are one of the important topics of research. It is seen that due to the strong quantum confinement effect, the adsorption edge gets blue-shifted from the bulk indirect band gap of 1.1 eV. More the silicon nanowire gets entered into the nano-regime more deviation from the bulk band gap gets occurred. It is mentioned worthy here that the growth orientation and the surface of the substrate have a marked influence on the opto-electronic properties of the material. The electrical conductivity of the system may be increased by many orders by suitable doping like boron or phosphorus.

Electronic and Optical Properties

Magnetic Properties

Magnetic nanowire arrays consisting of isolated needle-like magnetic nanowires have recently aroused considerable interest from the viewpoint of perpendicular magnetic recording. When the magnetic field is applied parallel to the long axis of the magnetic nanowire, it exhibits a coercive field, which is inversely proportional to the pore diameter. It is also important to note that changes in the diameter of the nanowires may influence the surface energies of the different crystallographic planes. Therefore, it is apparent that one can tune the magnetic properties of the nanowires by simply changing their diameter, which will change coercivity, magnetization and the shape of the hysteresis loop. The high aspect ratio of the nanowires results in enhanced coercivity and suppresses the onset of the super-paramagnetic limit', which is considered to be very important for preventing the loss of magnetically recorded information between the nanowires. The suitable separation between the nanowires is maintained to avoid the inter-wire interaction and magnetic dipole coupling.

Thermal and Mechanical Properties

SiNWs have established themselves to be one of the promising thermo-electric materials and thus as a thermo-electric power generator. This makes it extremely important to have an idea about the thermal conductivity of the material. When the diameter of the SiNWs gets smaller and smaller, the surface-to-volume ratio gets increased, which in turn increases the surface scattering effects resulting decrease in the thermal conductivity values. When the diameter further decreases (<.5nm), it shows an increasing trend. The values of Young' Modulus of the SiNWs do not show any marked difference from that of bulk silicon and are estimated to be around 195 GPa, which is closed to Young's modulus of bulk Si 169 GPa for Si (111) plane.

Applications of SiNWs

Semiconductor nanowires have recently been used as building blocks for assembling a range of nanodevices, including FETs, p-n diodes, BJTs and complementary inverters. The nanotube and nanowires with sharp tips are promising materials for applications as cold-cathode field emission devices. Optical communications industrial growth has generated a high demand for efficient and low-cost materials to be used for properties and functions such as light emission, detection and modulation. Also, silicon-based materials with enhanced optical properties have applications in accelerating the efficiency of photovoltaic solar cells, which is a market also dominated by silicon, and which is expected to experience tremendous growth in the near future. The importance of developing a technology that would allow optical and electronic devices to be easily integrated on a silicon wafer has long been recognized. Over the past 15 years, considerable efforts have been carried out within the research community to achieve this goal. Several materials and methods have emerged out to be as possible contenders for silicon-based optoelectronic devices and applications, which includes silicon-based superlattices and quantum dots facilitating quantum confinement in silicon nanocrystals. SiGe and SiGeC devices doped with optically efficient rare earth impurities such as erbium direct integration of III-V materials on silicon; porous silicon; silicon and carbon clusters embedded in oxide or nitride matrices; superlattices of epitaxially grown silicon with adsorbed oxygen.

SiNWs in Device Application

SiNWs are very good solar cell materials with several advantages that include improved charge transport, higher efficiency and lower cost. It also offers reflectance a couple of orders less than the other conventional materials within the wavelength range between 300 – 1100 nm. SiNWs may also be efficiently used as electrochemical switches that found their application in different logic devices. The advantages of these switches rest in the fact of zero power consumption in off states and negligible power consumption in on states and temperature-independent operation over a wide temperature range. Traditional FET has p-n type material with different doping. FET with the enhanced mode operation may be developed from SiNWs with Schottky contacts providing numbers of advantages. This offers a very high on/off current ratio of up to $\approx 10^7$. Not only that, it also offers good thermal emission leakage with 3-4 times improved on/off current ratio compared to traditional FET. Overall performance also gets enhanced equal extend.

SiNWs in Data Storage Technology

As mentioned before, magnetic material doped silicon nanowire called DMS has the ability to store memories and thus may be used as a memory device and the advantage of this is that it has better stability at a higher temperature. There are reports of the development of SiNW-based non-volatile memory device. The system has further advantages regarding the cost-effectiveness as well as environmental friendliness.

SiNWs in Rechargeable Lithium-Ion Batteries

Silicon is also a material of potential to be used as the anode material in lithium-ion battery, especially due to the low potential associated with the discharge phenomenon. The highest charge capacity, as has been calculated by different workers, has reached a value as high as 4200 mAh g-1. The mechanism is rather simple and thus clear. Here Li positive ion reacts with SiNWs, resulting formation of lithium silicide having much greater capacity. In the reverse cycle, when energy is exhausted, lithium silicide gets converted back to silicon, giving a rechargeable mechanism. SiNWs found applications as electrodes in lithium-ion micro batteries. It offers a boost in the storage capacity with the increasing temperature almost three times as the temperature changes from 525 to 575 °C.

SiNWs as Ultra-Violet (UV) LEDs and Lasers

As Si is an indirect band gap material, it has less efficiency in the emission of light. Still single crystal SiNWs have found applications in UV-LED and lasers. The SiNWs have a wide band gap, giving a wide band gap. This wide band gap is associated with a direct electronic transition that corresponds to short radioactive lifetime emission. Various workers have shown that the system has various properties like stimulated emission, optical gain, and population inversion required for efficient laser action.

SiNWs as Efficient Thermoelectric Materials

When there is a temperature gradient, SiNWs generate considerable electricity and thus can be considered a very efficient thermo-electric material. However, there is no marked change between SiNWs and bulk silicon in terms of Seebeck coefficient and electrical resistivity. The SiNWs can give a figure of merit ZT = 0.60 at room temperature. This value may be changed and goes beyond unity at room temperature. It may be mentioned here that bulk Si has poor thermo-electric material, while SiNWs, when its' thermal conductivity is made less, may be useful thermo-electric materials.

SiNWs as Chemical and Biomedical Sensors

Low dimensional material, especially nanowire/tube based dimension material, has a very high surface-to-volume ratio due to which they have very attractive sensing properties. This sensing mechanism is based on the change of the conductance of the narrower due to adsorbed atoms/molecules. SiNWs are considered a very good sensing material because of their very high numbers of dangling or free bonds that may be passivated by suitable chemical treatment. SiNWs have proved their potential to be used as a bio or chemo sensors. Not only this, SiNWs sensor based FET can provide a very high on/off ratio. When the material is exposed to a suitable gas like ammonia or cyclohexane the system shows an acute charge transfer between the analytes and the nanowire. It is a known fact that the SiNWs prepared by 20% H_2O_2 etching solution exhibit the best activity in the decomposition of the target organic pollutant, Rhodamine B (RhB), under Xenon (Xe) arc lamp irradiation. To enhance the photocatalytic activity of the SiNWs, its surfaces were doped with HF, Pt, *etc.* When SiNWs get doped with HF, hydrogen-terminated SiNWs (H-SiNWs) were obtained. The H-SiNWs enhance the performance of SiNW arrays by increasing their photocatalytic activity. This was mainly due to the hydrogen atoms on the H-SiNWs surface, which produce many more •OH radicals and have high reductive activity. Also, they have high stability and catalytic activity with at least 10 recycle times. Also, H-SiNWs were ideal photocatalysts for the selective hydroxylation of benzene into phenol and for methyl red decomposition. When the SiNWs was doped with the platinum nanoparticle, they were used as effective photo-catalysts for photocatalytic degradation of organic dyes and toxic pollutants for organic waste treatment and environmental remediation. When the SiNWs were decorated with copper nanoparticles (SiNWs–Cu), they exhibited excellent and stable activity for the catalytic reduction of 4-nitrophenol to 4-aminophenol by sodium borohydride(NaBH4) in an aqueous solution. This novel catalyst also shows excellent catalytic performance for the degradation of organic dyes, such as methylene blue (MB) and rhodamine B (RhB). The degree of degradation for SiNWs–Cu was about 64% in comparison to 30% for H–SiNWs after 20 min. Thus, SiNWs–Cu acted as an attractive catalyst, because it was highly efficient, cost-effective, eco-friendly and could replace noble metals for certain catalytic applications. When SiNWs were modified with Platinum and Gold, it degraded about 95% of the methylene blue dye under UV light irradiation, whereas for H-SiNWs, it was only 91% degradation. Both SiNWs and SiNWs doped with silver (SiNWs–Ag) photocatalysts displayed a comparable activity, but remain less efficient than the H–SiNWs substrate.

CONCLUSION

This chapter discusses the basic structures and properties of a few materials that have established themselves as materials of potential in the field of water purification, especially in dye removal. In this consequence, few common materials that include carbon nanotube, graphene, graphitic carbon nitride, zinc oxide, or silicon nanowires, *etc.*, have been discussed in depth.

It has been shown that carbon is one of the most important materials in the field of Nanoscience as it has different allotropes of complete, versatile properties, and it can be found in different morphologies. It is mentioned that a carbon nanotube is a rolled graphite sheet, and depending upon the numbers of parallel sheets, it is called single walled, double walled, multi-walled and so on. Depending upon the chirality of CNT, it can be classified as armchair, zigzag and chiral. Synthesis of high-quality CNT needs a suitable catalyst for which transition metals like nickel, iron, cobalt and so on are ideal. On the other hand, graphene, a zero band gap material, is nothing but a graphite sheet, and depending upon the number of sheets, it is called single-layered, double-layered or few-layered graphenes. Mixed phased amorphous DLC contains both sp^2 as well as sp^3 hybridized carbon, and thus, its properties can be very easily tuned if the relative presence of the specific hybridized carbon gets changed. Graphitic carbon nitride is another planar efficient optoelectronic material and one of the most promising dye removers from water. Zinc oxide is another most important material in the nano regime as due to its favourable band gap and structure formation, it can be used in various optoelectronic applications. The detail structures, properties, applications and synthesis processes of SiNWs have also been discussed in the chapter.

REFERENCES

[1]   Iijima, S.; Ichihashi, T. Single-shell carbon nanotubes of 1-nm diameter. *Nature,* **1993,** *363*(6430), 603-605.

[2]   Novoselov, K.S.; Geim, A.K.; Morozov, S.V.; Jiang, D.; Katsnelson, M.I.; Grigorieva, I.; Dubonos, S.; Firsov, A.A. Two-dimensional gas of massless Dirac fermions in graphene. *Nature, 438*(7065)**2005,** , 197-200.

[3]   Choi, W.; Lahiri, I.; Seelaboyina, R.; Kang, Y.S. Crit critical reviews in solid state and materials sciences. *Rev. Solid. State. Mater. Sci.,* **2010,** *35,* 52-71.

[4]   Castro Neto, A.H.; Guinea, F.; Peres, N.M.R.; Novoselov, K.S.; Geim, A.K. The electronic properties of graphene. *Rev. Mod. Phys.,* **2009,** *81*(1), 109-162.
        http://dx.doi.org/10.1103/RevModPhys.81.109

[5]   Novoselov, K.S.; Geim, A.K.; Morozov, S.V.; Jiang, D.; Katsnelson, M.I.; Grigorieva, I.; Dubonos, S.; Firsov, A.A. Two-dimensional gas of massless Dirac fermions in graphene. *Nature,* **2005,** *438*(7065), 197-200.

[6]   Xu, C.; Wang, X.; Yang, L.; Wu, Y. Fabrication of a graphene–cuprous oxide composite. *J. Solid State Chem.,* **2009,** *182*(9), 2486-2490.
        http://dx.doi.org/10.1016/j.jssc.2009.07.001

[7]   Bredas, J.L.; Street, G.B. Electronic properties of amorphous carbon films. *J. Phys. C Solid State Phys.,* **1985,** *18*(21), L651-L655.
        http://dx.doi.org/10.1088/0022-3719/18/21/005

[8]   Robertson, J.; O'Reilly, E.P. Electronic and atomic structure of amorphous carbon. *Phys. Rev. B Condens. Matter,* **1987,** *35*(6), 2946-2957.
        http://dx.doi.org/10.1103/PhysRevB.35.2946 PMID: 9941778

[9]   Green, D.C.; McKenzie, D.R.; Lukins, P.B. The microstructure of carbon thin films. *Mater. Sci. Forum,* **1990,** *52,* 103-124.

[10]   Jacob, W.; Möller, W. On the structure of thin hydrocarbon films. *Appl. Phys. Lett.,* **1993,** *63*(13), 1771-1773.
        http://dx.doi.org/10.1063/1.110683

[11]   Wang, Y.; Chen, H.; Hoffman, R.W.; Angus, J.C. Structural analysis of hydrogenated diamond-like carbon films from electron energy loss spectroscopy. *J. Mater. Res.,* **1990,** *5*(11), 2378-2386.
        http://dx.doi.org/10.1557/JMR.1990.2378

[12]   Koidl, P.; Wild, C.; Dischler, B.; Wagner, J.; Ramsteiner, M. *Mater. Sci. Forum,* **1990,** *52,* 41.

[13]   Liebig, J. Uber einige Stickstoff-Verbindungen. *Annalen der Pharmacie,* **1834,** *10*(1), 1-47.

[14]   Franklin, E.C. The ammono carbonic acids. *J. Am. Chem. Soc.,* **1922,** *44*(3), 486-509.

[15]   Kouvetakis, J.; Todd, M.; Wilkens, B.; Bandari, A.; Cave, N. Novel synthetic routes to carbon-nitrogen thin films. *Chem. Mater.,* **1994,** *6*(6), 811-814.
        http://dx.doi.org/10.1021/cm00042a018

[16] Gu, Y.; Chen, L.; Shi, L.; Ma, J.; Yang, Z.; Qian, Y. Synthesis of C_3N_4 and graphite by reacting cyanuric chloride with calcium cyanamide. *Carbon,* **2003**, *41*(13), 2674-2676.
http://dx.doi.org/10.1016/S0008-6223(03)00357-9

[17] Zhang, Z.; Leinenweber, K.; Bauer, M.; Garvie, L.A.J.; McMillan, P.F.; Wolf, G.H. High-pressure bulk synthesis of crystalline C(6)N(9)H(3).HCl: a novel c(3)n(4) graphitic derivative. *J. Am. Chem. Soc.,* **2001**, *123*(32), 7788-7796.
http://dx.doi.org/10.1021/ja0103849 PMID: 11493052

[18] Guo, Q.; Xie, Y.; Wang, X.; Lv, S.; Hou, T.; Liu, X. Characterization of well-crystallized graphitic carbon nitride nanocrystallites *via* a benzene-thermal route at low temperatures. *Chem. Phys. Lett.,* **2003**, *380*(1-2), 84-87.
http://dx.doi.org/10.1016/j.cplett.2003.09.009

[19] Guo, Q.; Xie, Y.; Wang, X.; Zhang, S.; Hou, T.; Lv, S. Synthesis of carbon nitride nanotubes with the C_3N_4 stoichiometry *via* a benzene-thermal process at low temperaturesElectronic Supplementary Information (ESI) available: XRD patterns. See http://www.rsc.org/suppdata/cc/b3/b311390f/. *Chem. Commun. (Camb.),* **2004**, (1), 26-27.
http://dx.doi.org/10.1039/b311390f PMID: 14737315

[20] Komatsu, T. Prototype carbon nitrides similar to the symmetric triangular form of melon. *J. Mater. Chem.,* **2001**, *11*(3), 802-803.
http://dx.doi.org/10.1039/b007165j

[21] Jürgens, B.; Irran, E.; Senker, J.; Kroll, P.; Müller, H.; Schnick, W. Melem (2,5,8-triamino-tri-s-triazine), an important intermediate during condensation of melamine rings to graphitic carbon nitride: synthesis, structure determination by X-ray powder diffractometry, solid-state NMR, and theoretical studies. *J. Am. Chem. Soc.,* **2003**, *125*(34), 10288-10300.
http://dx.doi.org/10.1021/ja0357689 PMID: 12926953

[22] Kroke, E.; Schwarz, M.; Horath-Bordon, E.; Kroll, P.; Noll, B.; Norman, A.D. Tri-s-triazine derivatives. Part I. From trichloro-tri-s-triazine to graphitic C_3N_4 structuresPart II: Alkalicyamelurates M3[C6N7O3], M = Li, Na, K, Rb, Cs, manuscript in preparation. *New J. Chem.,* **2002**, *26*(5), 508-512.
http://dx.doi.org/10.1039/b111062b

[23] Sehnert, J.; Baerwinkel, K.; Senker, J. Ab initio calculation of solid-state NMR spectra for different triazine and heptazine based structure proposals of g-C_3N_4. *J. Phys. Chem. B,* **2007**, *111*(36), 10671-10680.
http://dx.doi.org/10.1021/jp072001k PMID: 17713939

[24] Lotsch, B.V. M. Do blinger, J. Sehnert, L. Seyfarth, J. Senker, O. Oeckler and W. Schnick. *Chemistry,* **2007**, *13*, 4969-4980.
http://dx.doi.org/10.1002/chem.200601759 PMID: 17415739

[25] Zhang, X.; Xie, X.; Wang, H.; Zhang, J.; Pan, B.; Xie, Y. Enhanced photoresponsive ultrathin graphitic-phase C_3N_4 nanosheets for bioimaging. *J. Am. Chem. Soc.,* **2013**, *135*(1), 18-21.
http://dx.doi.org/10.1021/ja308249k PMID: 23244197

[26] Zhang, Y.; Pan, Q.; Chai, G.; Liang, M.; Dong, G.; Zhang, Q.; Qiu, J. Synthesis and luminescence mechanism of multicolor-emitting gC 3 N 4 nanopowders by low temperature thermal condensation of melamine. *Sci. Rep.,* **2013**, *3*(1), 1-8.
http://dx.doi.org/10.1021/ac303263n PMID: 23373468

[27]   Cheng, C.; Huang, Y.; Wang, J.; Zheng, B.; Yuan, H.; Xiao, D. Anodic electrogenerated chemiluminescence behavior of graphite-like carbon nitride and its sensing for rutin. *Analytical chemistry*, *85*(5), pp.2601-2605.

[28]   Cheng, C., Huang, Y., Wang, J., Zheng, B., Yuan, H. and Xiao, D., Anodic electrogenerated chemiluminescence behavior of graphite-like carbon nitride and its sensing for rutin. *Anal. Chem.*, **2013**, *85*(5), 2601-2605.

[29]   Liu, J.; Wang, H.; Antonietti, M. Graphitic carbon nitride "reloaded": emerging applications beyond (photo)catalysis. *Chem. Soc. Rev.*, **2016**, *45*(8), 2308-2326.
http://dx.doi.org/10.1039/C5CS00767D PMID: 26864963

[30]   Zhou, Z.; Wang, J.; Yu, J.; Shen, Y.; Li, Y.; Liu, A.; Liu, S.; Zhang, Y. Dissolution and liquid crystals phase of 2D polymeric carbon nitride. *J. Am. Chem. Soc.*, **2015**, *137*(6), 2179-2182.
http://dx.doi.org/10.1021/ja512179x PMID: 25634547

[31]   Zhu, B.; Xia, P.; Ho, W.; Yu, J. Isoelectric point and adsorption activity of porous g-C$_3$N$_4$. *Appl. Surf. Sci.*, **2015**, *344*, 188-195.
http://dx.doi.org/10.1016/j.apsusc.2015.03.086

[32]   Bai, X.; Yan, S.; Wang, J.; Wang, L.; Jiang, W.; Wu, S.; Sun, C.; Zhu, Y. A simple and efficient strategy for the synthesis of a chemically tailored g-C$_3$N$_4$ material. *J. Mater. Chem. A Mater. Energy Sustain.*, **2014**, *2*(41), 17521-17529.
http://dx.doi.org/10.1039/C4TA02781G

[33]   Look, D.C. Electrical and optical properties of p-type ZnO. *Semicond. Sci. Technol.*, **2005**, *20*(4), S55-S61.
http://dx.doi.org/10.1088/0268-1242/20/4/007

[34]   Hong, R.Y.; Li, J.H.; Chen, L.L.; Liu, D.Q.; Li, H.Z.; Zheng, Y.; Ding, J.J.P.T. Synthesis, surface modification and photocatalytic property of ZnO nanoparticles. *Powder Technology*, *189*(3), pp.426-432.

[35]   Tien, L.C., Sadik, P.W., Norton, D.P., Voss, L.F., Pearton, S.J., Wang, H.T., Kang, B.S., Ren, F., Jun, J. and Lin, J., 2005. Hydrogen sensing at room temperature with Pt-coated ZnO thin films and nanorods. *Appl. Phys. Lett.*, **2009**, *87*(22), 222106.

CHAPTER 6

Introduction to Photo-Catalysis

Abstract: The matter-energy reaction is the basis of a variety of fundamental scientific phenomena we have witnessed in nature. Photoreaction is related to the interaction of photons and the molecules of a substance. When the necessary photon in the ultra-violet or visible range of the electromagnetic spectrum is absorbed by the materials in concern, it may convert different poisonous elements into harmless substances, such as water and carbon dioxide. The study of photon physics and chemistry is fundamental for our understanding of the world we live in. The basic physiological processes through which the living species maintain their life cycles are also somehow related to the different photochemical reactions. Photo-catalysis happens to be a low-cost, versatile, and environment-friendly method that deals with a variety of harmful pollutants. Pollutants can be inorganic, organic, or even biological, and they can be found in both air and water. Photo-catalysis is a process in which the catalyst, light source, and contaminants must be in close proximity or contact. There have been numerous studies on the oxidation-induced removal of different organic pollutants as well as microorganisms, especially those found in water. In this chapter, we have covered the basics of photo-catalysis, the characteristics of various catalysts, their types, and photochemical laws, as well as the conditions and limitations of quantum yield.

Keywords: Photolysis, Photocatalysis reactions, Photochemical laws, Quantum Yield.

INTRODUCTION

The term photocatalysis may fundamentally be defined as the acceleration of a chemical reaction in the presence of light with different energy ranges, which may be visible or UV.

This method being one of the simplest, has gathered world-wide attention in the last couple of decades. The main reason behind this rapid spreading lies in the fact that this particular mechanism has two very important applications like hydrogen evolution through the splitting of water under solar power and the water purification by removing pollutants contained in small concentration. There is another practical reason also, and that is because of the fact as the mechanism, as well as application related to the phenomena, has multidirectional aspects; this has brought physicists, chemists, material scientists, pharmacists and researchers of other domains into a

Diptonil Banerjee, Amit Kumar Sharma and Nirmalya Sankar Das

single platform [1]. In the context of history and research, interest in heterogeneous photocatalysis can be traced back to many decades when Fujishima and Honda discovered in 1972 the photochemical splitting of water into hydrogen and oxygen in the presence of TiO_2. From this time, extensive research, much of it published, has been carried out to produce hydrogen from water in oxidation-reduction reactions using a variety of catalyst materials.

It is mentioned worthy that catalyst is a topic of prime interest that one might come across while studying physical chemistry, especially while learning about photochemical or chemical reactions. Some of the catalyst reactions occur rapidly or take a long time, depending upon the catalyst material and environmental media. Catalysis is of great importance as it changes our physical life. The major regions where the world economy depends upon the technology in the developing fields of nanosensors, water purification, polymer production, petroleum and energy production, chemicals, food industry and pollution control, all routes involve catalytic processes. Heterogeneous photocatalysis is explained as the acceleration of a photoreaction in the existence of a catalyst.

Generally, photocatalysis is a significant use for our environment, especially in industrial effluents and cleaning the wastewater. They help us to reduce pollution in water and air by developing such technology. This removal mode of existence would advantage everyone around the world. Today's, high-quality photocatalysts should be photoactive, competent to employ in visible or near UV light or both, photostable, inexpensive, and nontoxic for the living nature. A photocatalyst (*i.e.* semiconductor) should have positive redox potential so that the photo-generated valence band (VB) hole is photo-chemically active and can generate ˙OH radicals for oxidizing the organic pollutants. For this, the conductance band (CB) electron must be suitably negative so that they become capable to decrease absorbed oxygen to superoxide. The main benefit of photocatalysis is the reality that there is no further requisite for any secondary removal methods as the organic contaminants are changed to inorganic ions, carbon dioxide and water. Other processes, such as adsorption by activated carbon and air stripping simply concentrate the pollutants and transfer them in the adsorbent or air so that they will not transfer them into nontoxic wastes as happens in photocatalysis. Photo-catalysis is an important means of advanced oxidation processes (AOPs) and are widely used for the removal of recalcitrant organic elements from industrial and wastewater. Further, the photocatalytic systems are classified into two categories, firstly, a homogenous photocatalytic system is a promising method for wastewater treatment, but the ions remain in the solution at the end of the process. For this reason, the removal of waste in wastewater treatment becomes essential, and it will increase the cost. This drawback of homogeneous catalytic systems can be prevailing by heterogeneous photocatalysis. The current examples of semiconductor photo-catalysts are TiO_2, ZnO, CdS, ZnS, SnO_2, WO_3, WSe_2, Fe_2O_3, and so on that can be used as potential photo-catalysts in cleaning the wastewater to solve the problem of environmental pollution, globally.

The photodegradation or removal of toxic elements from environmental sources has increased remarkably using nanometer-sized semiconductor materials over the last couple of decades due to their unique characteristics like quantum efficiency effects as well as their high potential in chemical or bio-chemical reactions. The term "quantum confinement effect", was introduced to explain a wide range of electrical and optical properties of nano-sized materials in connection to the removal of toxic materials from environments. The nano-sized detection methodology, referred to as *"quantum confinement effect"*, is of significant interest from both fundamental and technological points of view for higher sensitivity, cost-effectiveness, or simple detection of environmental threats. This accounts for the removal of pesticides in the water using nanotechnology, UV source and kinetic study. Fig. (**6.1**) illustrates a fraction of the complex sequence of events that may take place in a semiconductor photocatalyst. The heterogeneous photocatalysis describes a process whereby illumination of a semiconductor particulate (TiO_2 and ZnO) with UV-visible light suitable to its bandgap energy ($\geq E_g$) ultimately generates conduction band electrons (e^-) and valence band holes (h^+) pair. Subsequent to their separation, other proposed photochemical and photophysical decay channels are poised at the particulate/solution interfaces (Fig. **6.1**).

Initially, the irradiation of the semiconductor particle generates a bound electron/hole pair (the exciton). It either recombines or dissociates to give a conduction band electron and a valence band hole. These separated charge carriers may also recombine; migrate to the surface while scanning several shallow traps (anion vacancies and Zn^{2+} for the electrons and oxygen vacancies or other defect sites for the hole). On the surface, both charge carriers scan the surface visiting several sites to reduce adsorbed electron acceptors and to oxidize adsorbed electron donors in competition with surface recombination of the surface trapped electrons and holes to produce light emission or photon emission. Oxygen is omnipresent on the particle surface and acts as an electron acceptor, whereas OH^- groups and H_2O molecules are available as electron donors to yield the strongly oxidizing $^*OH$

radicals. Trapping of electrons and holes in pristine ZnO nanoparticles collide takes place with the best efficiency.

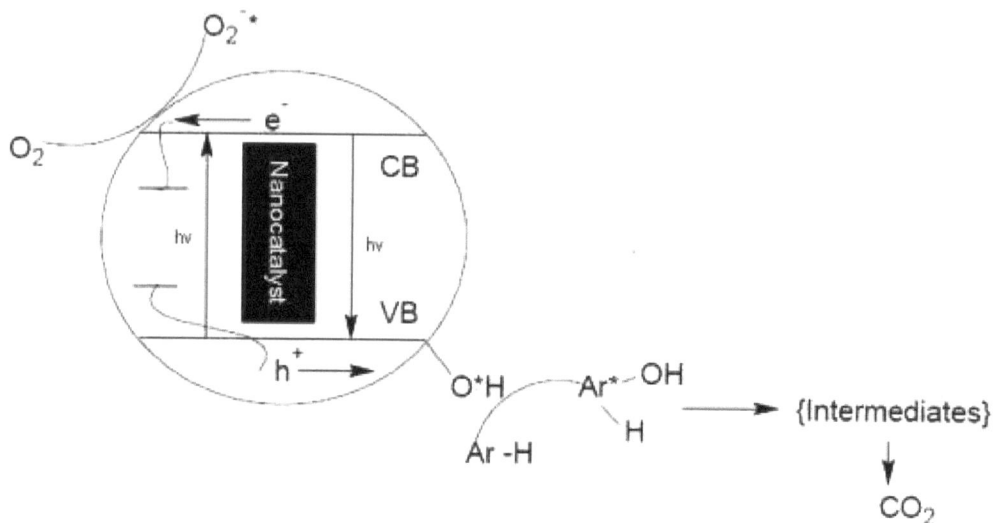

Fig. (6.1). A fraction of the complex sequence of semiconductor nanophotocatalyst.

EXPLANATION OF PHOTOCATALYSIS

Let us take a chemical reaction where A gets converted to B reversibly by some external means, as shown below:

$$A \rightleftharpoons B$$

Now to this, we add another material C_t that remained unchanged during or after the reaction, but it either initiates the reaction or speeds up the process. This is also represented in the below scheme:

$$A + C_t = B + C_t$$

This is the basic operator for a catalytic process that happens when a certain amount of a catalyst (C_t) is added. The above relations are known as equilibrium condition. After the completion of the reaction, the catalyst can be recovered exactly in the same, *i.e.*, in the initial stage. This means that the catalysts remained unaffected during the reaction process, which is the basic condition of catalysis reaction. This information is valid for acid–basic catalysis, redox catalysis and bio-catalysis and other types. The reaction rate increases if the total activation energy in the catalytic process is less than the corresponding energy in the reaction. In thermal reactions, changes in the electronic configuration of the system occur following the regrouping of nucleus/fragments without transitions to electronically excited states.

Suppose a chemical reaction occurs, as shown below, which is induced by the absorption of photons by a reactant A.

$$A + hv \longrightarrow B$$

and the corresponding photocatalytic process becomes:

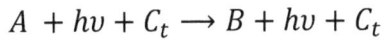

$$A + hv + C_t \longrightarrow B + hv + C_t$$

This photoreaction arises through an excited electronic state of the reagents given by the regrouping of the various nuclei/fragments. Characteristically, the photochemical reaction is irreversible. If the photon is taken as a quasi-reagent in a reaction as applied in kinetic mechanisms, the reverse reaction governs the pathway,

$$B \longrightarrow A + hv$$

Where hv indicates the photons of equal energy as those taken in the forward reaction. Evidently, such a reverse reaction is improbable, and the process must $B \longrightarrow A$ carry on by a different pathway.

TYPES AND CHARACTERISTICS OF NANOPHOTOCATALYSTS

Nanophotocatalysts exhibit an appreciably increased reactivity compared to micro-sized particles or bulk materials due to their large surface area [1, 2]. A novel nanophotocatalyst has the ability to exploit the reactivity as an application in case of water purification under controlled conditions. They are mainly classified as surface metal-organic charge-transfer-based, plasmon resonance-mediated and semiconductor-based nanophotocatalysts. They are also capable of driving various organic reactions through the same mechanism of photo-catalysis [3]. Consequently, they can be classified as graphene semiconductor composites that are composed of two semiconductors in the form of core-shell composites, first is basically a non - metal based doped semiconductor materials and another is taken from metal-modified semiconductors.

Graphitic carbon nitride nanocomposites are Metal-free catalysis, employed in photodegradation of aqueous phase organic pollutants [4], similarly polymeric graphitic carbon nitride-based Z-scheme photocatalytic systems, magnetic graphitic carbon nitride photocatalyst, and carbon quantum dot-supported graphitic carbon nitride have been respectively employed for sustainable photocatalytic water purification [5], photodegradation of 2,4-dinitrophenol [6] and degradation of Oxytetracycline antibiotic [7].

Among the various photocatalysts, metal oxide semiconductors, like TiO_2, ZnO, WO_3, and α-Fe_2O_3 are the most suitable ones, as they have photo-corrosion resistant with a wide band gap energy. At present, TiO_2 is used as the primary efficient photocatalyst and applied in wastewater treatment, because it is thermally stable, cost-effective, environment-friendly, and chemically and biologically inert, and also it can encourage the oxidation of organic compounds into water and carbon dioxide. They have various interesting structural and surface properties that include surface area, crystal composition, particle size distribution, porosity and band gap energy, thereby, they are capable of affecting the activity of the photocatalyst [8]. The following two properties usually represent nanophotocatalysts with precise encouraging gains above bulk materials [9]. Firstly, they have a large surface area to volume ratio of this; the results produce at a high particle fraction and later on, it gives a high number of active sites on the catalyst surface. Secondly, they have an energy gap between the valence band (VB) -conduction band (CB) that depends strongly on the size of the synthesized nanoparticles [10].

Furthermore, the change in the size of the nanocatalyst makes it probable to alter the absorbance wavelength. In addition, the optical and electronic properties of the nanocatalyst can be tailored by fine-tuning their sizes [11]. Due to these properties, nanophotocatalysts have found their uses in number of reactions. The basic growth of nanophotocatalysts has gained the center of attention of the researchers due to their abilities of degrading different pollutants, dyes and toxic elements from wastewater. It is because of these properties, that there is rising attention on the application of nanomaterials as catalysts for different organic reactions [12].

The efficiency rate in photocatalysis methods is mainly dependent on the size, shape, surface area and crystal structure of the material as well as on their surface properties. These surface properties can set themselves as an important factor affecting the degradation and the reaction kinetics [13]. The amount of catalyst is also directly proportional to the total rate of photocatalytic reaction. The efficiency of the nanophotocatalyst can be enhanced by increasing its quantity [13]. pH values of the solution are another key factor that influences the efficiency of the photocatalyst. This is so because pH values can drastically change the surface charge properties, thus affecting the photo-induced reaction kinetics. The pH values 6 to 8 have shown the best response. Thus more the pH deviates from the neutral value, more the system would lose its efficiency [14]. Another factor that will influence is the temperature of the system, which will, in turn, be governed by the activation energy of the reactions [15]. It is also noteworthy that one material may be very effective for a particular pollutant at a particular wavelength, but may be practically useless for another pollutant within the entire electro-magnetic spectrum, suggesting it the nature of the pollutant itself and its reaction with the catalyst that matters very crucially.

Additionally, inorganic ions present in wastewater, such as copper, zinc, phosphate, iron, chloride, sulfate, magnesium, bicarbonate and nitrate, can change the rate of degradation of the organic pollutants as they adsorbed onto the nanophotocatalyst surface [16, 17]. Table **6.1** summarizes the different photolysis processes.

Table 6.1. Different photolysis based on catalytic and non-catalytic photons [25].

Photolysis	
Catalytic Under Photon Irradiation	**Non-Catalytic Under Photon Irradiation**
Photo-generated catalysis	Catalyzed photolysis
Photon-induced catalytic reactions (Stoichiometric Photo-generated catalysis)	• Catalyzed photochemistry • Photosensitized reactions • Sensitized reactions • Photo-assisted catalysis (Stoichiometric Photo-generated catalysis) • Substance-assisted (catalyzed) Photoreactions

PHOTO-LYSIS OR PHOTO-DISSOCIATION

It is a chemical technique that was developed by the English chemists R.G.W. Norrish and G. Porter in 1949 that consists of a liquid or gas to concentrate a burst of light for an order of microseconds or mill-seconds. In this process, molecules are broken down into small units when the flashlight is absorbed by a parent molecule, for example- flash photolysis which is used in the study of intermediate short-lived molecules that occur during photochemical reactions.

The process of variation in the rate of chemical reactions or their initiation under the absorption of UV, visible or IR irradiation by a suitable material is called the photo-catalysis. The material controls the reaction kinetics by absorbing light, which remains unchanged and is called a photo-catalyst.

When one switches on light on a particular sample or molecules, they absorb a certain amount of light energy and produce the various fragments of molecules or intermediate compounds. Furthermore, the second flash on the absorbing molecule uses their identification by spectrophotometry. The technique is a helpful tool for the recognition of transient molecule or compound intermediates and, therefore, for the learning of mechanisms of speedily photochemical reactions.

Broadly speaking photolysis process, which involves the absorption of light, suitable energy and transition of an electron from lower to higher energy level, will liberate hydrogen atoms with excess kinetic energy by breaking the bonds. This simply means that when the energy of the photon is greater than the bond dissociation energy, photons break the bonds between the atoms constituting the molecule.

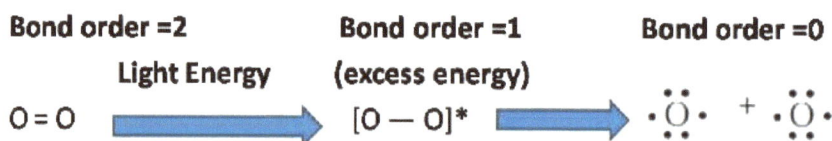

So, we need to calculate the required energy to break the bond for a hydrogen molecule (*i.e.* H_2). As we know that the formula gives the energy of light photon:

$$E = h\upsilon = hc/\lambda \ \ldots..$$

(6.1)

(Note that the frequency of light, $\upsilon = c/\lambda$)

Or, $\lambda = hc/E$....... **(6.2)**

Where h = Planck's constant = 6.63×10^{-34} J-s and the speed of light, c = 2.997×10^8 m/s

Here we have taken H_2 molecule as an example, then the bond energy for $H_2 = 4.36 \times 10^5$ J/Mol

Since, the bond energy (E) for one molecule of $H_2 = (4.36 \times 10^5$ J/Mol) / Avogadro's number

Or, $= (4.36 \times 10^5$ J/Mol) / $(6.02214076 \times 10^{23}$ molecules/mole)

$= 72.3995 \times 10^{-20}$ J

Then, from Eq. **(6.2)**, $\lambda = hc /E = (6.63 \times 10^{-34}$ J-s) $\times (2.997 \times 10^8$ m/s) / 72.3995×10^{-20} J

$= 0.2744509285 \times 10^{-6}$ m = 274.4509285 nm ~ 274 nm.

The bond dissociation energy is 941 KJ/Mol for a molecule of oxygen, and it requires photons of 240 nm to break the bond. Here, we can imagine that the bond in O=O (*i.e.*, O_2 molecule) is stronger than the H-H bond (*i.e.*, H_2 molecule) because 240 nm photons of light are very energetic or higher frequency than 274 nm.

According to Bohr's quantum atomic model and molecular orbital theory, there are various energy levels in an atom or molecule in which electrons reside. The occupied discrete electronic states are thus called orbits. Therefore, definite energy is needed for the excitation of an electron from a lower energy level to an upper energy level. The net energy of a molecule is the sum of its electronic energy (E_e), rotational energy (R_e), vibrational energy (V_e) and translational energy (T_e).

Thus, $E_{net} = E_e + R_e + V_e + T_e$

As we know, the first three energies (E_e, R_e, V_e) are quantized. It means that they exhibit state change only by a discrete energy fall or jump, although t_{ea} is the energy that is involved in the molecular movement, is not quantized. In light of the above discussion, one may conclude that the supplied external energy must be enough in order to make an electron undergoes a transition from a lower state to a higher state of energy level.

Fig. (**6.2**) represents the variation of potential energy w r t to the Inter-nuclear distance (*i.e.* The distance between two nuclei forming the bond).

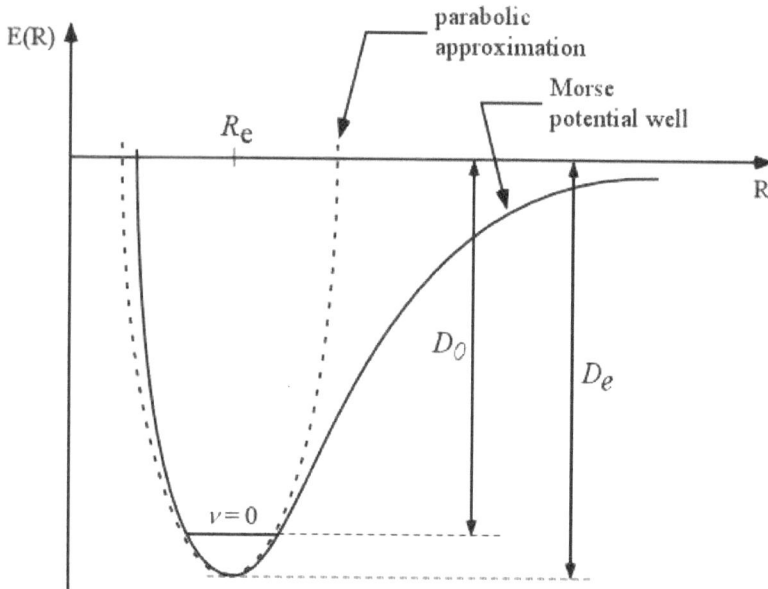

Fig. (**6.2**). Graphical representation of Morse curve.

The smallest possible potential energy is the bond length. The two nuclei will space themselves so as to have the smallest potential energy. This variation, also called Morse law, shows how potential energy varies with bond length. The smallest possible potential energy is the bond length. The two nuclei will space themselves so as to have smallest potential energy.

Now, we can imagine that when bond length is small, bond strength gets stronger, so why not push the two nuclei closer together? If that happens, the repulsion between the two nuclei will increase increasing potential energy. So, maybe we should move the two nuclei farther apart to reduce the repulsion. As the bond length increases, it will reduce repulsion, but it also reduces the strength of the interaction between the bonding electrons and the two nuclei. So considering the above facts, bonds get formed accordingly to the fact that the electrons can interact with two nuclei rather than one. So we can observe that the bond length is a cooperation between repulsion and overlap. Overlap is the idea that the electrons are interacting with two nuclei, being held in overlapping orbits on each atom [22].

Now the Question Comes why does the Morse Graph Appear as it Does?

Let's see at the lowest potential energy (*i.e.*, the bond length), as we shift to left, shifting the nuclei closer together; the mainly significant factor in calculating the energy is the repulsion of the nuclei. The force between the two nuclei can be given by Coulomb's Law:

$$F = k. \ q_1 q_2 \ / \ r^2$$

Here q_1, q_2 are charges separated by a distance r and k is the constant relating permittivity.

We see that as the distance reduces, the force increases exponentially. If the distance between the nuclei is halved, then the force becomes quadrupled in between them.

It means that the electronic transition is placed over rotational and vibrational levels. So, there are a huge number of transitions possible, which are close together that are accountable for the change in electron energy (E_e), rotational energy (R_e), and vibrational energy (V_e) levels. Basically, this is the region of the EM spectrum absorbed by the same molecule and has a wavelength of 190–780 nm, showing the change in the photochemical process. It contains 35 kcal/Mol -145 kcal/mole, which is enough energy for the electronic excitation from lower to higher molecular orbital.

Activation Energy

Generally, the transition state of a reaction has always been at a higher energy level than the reactants or a product, such that the energy associated always has a positive value which is independent of whether the reaction is endergonic or exergonic. The activation energy (E_A) shown in Fig. **(6.3)** below, is for the forward reaction, which is exergonic.

If the reaction proceeds in the reverse direction, then it is called endergonic, the transition state would remain the same, but the activation energy would be larger.

This is due to the product molecules being lower in energy and would thus need more energy to attain the transition state at the peak of the reaction.

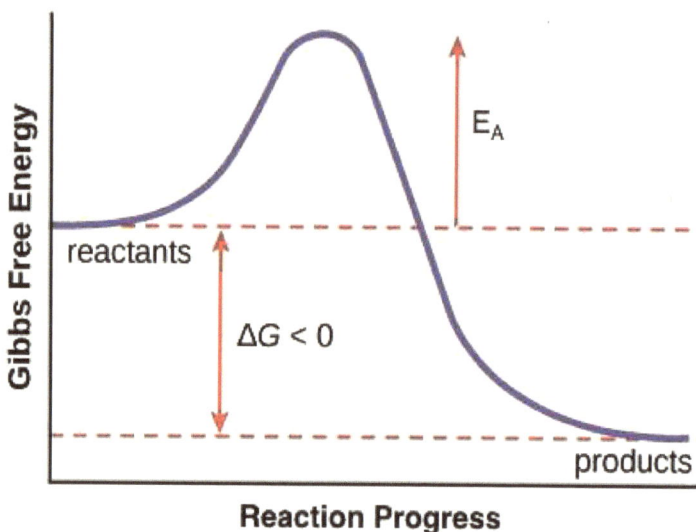

Fig. (6.3). Graph of activation energy.

The source of activation energy is normally heat, with reactant molecules absorbing thermal energy from their surroundings. This thermal energy provides the motion of the reactant molecules by increasing the frequency and force of their collisions. The activation energy also pushes the atoms and bonds nearby the unique molecules forming the surroundings, making them more favorable so that associated bonds become more vulnerable.

As a reactant molecule absorbs sufficient energy enough to make the transition, it may undergo the following two types of reactions:

1. Catalytic under Photon irradiation

2. Non-Catalytic under Photon irradiation

Catalytic under Photon Irradiation

When TiO_2 or other molecules are irradiated by photons of suitable wavelength, it exhibits catalytic properties. The activity increases with an increase in catalyst concentration up to an optimum value. After an optimum value, the activity decreases. Further, it is also classified into two categories:

Photo-Generated Catalysis

Such photo-generated activity depends on the capability of the catalyst to form electron-hole pairs that generate free radicals, for example, hydroxyl radicals ($^{\bullet}$OH), and capable to go through secondary reactions. This approach is used in the discovery of water electrolysis using titanium dioxide. This technique keeps many advantages like environmental protection, non-production of secondary pollutants and high efficient degradation of pollutants (Fig. **6.4**).

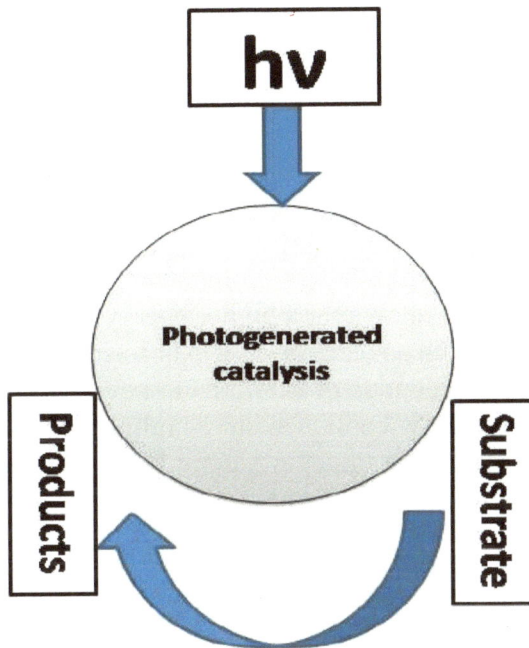

Fig. (6.4). Photogenerated Process.

A. Photo-induced catalytic reactions (Stoichiometric Photo-generated catalysis) – The recent technique have used the advantage of the visible light absorptivity of certain functional groups that enable photo-induced electron transfer (PET) [18]. PET is a single electron transfer process in which an excited electron gets transferred from donor to acceptor and thus creating a charge separation which enables a process like redox reaction in an excited state. This phenomenon is also examined in Dexter electron transfer (DET) [18]. The electrons live as an electronic band in bulk materials and live as an electron orbital in molecules. As sufficient energy of photon excites a molecule than an electron present in the ground state orbital transfers into a higher excited energy orbital and leaves a vacancy in there (*i.e.*, ground state), which is filled by a donor electron. In this process, the photoexcited molecule started to work as an oxidizing agent or a reducing agent. So, this works in two forms:

Photoinduced Oxidation Reaction

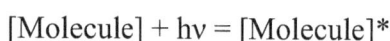

[Molecule] + hv = [Molecule]*

[Molecule]* + Donor → [Molecule]$^+$ + Donor$^-$

Photoinduced Reduction Reaction

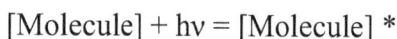

[Molecule] + hv = [Molecule] *

[Molecule] * + acceptor → [Molecule] $^+$ + acceptor$^-$

As a result, both reactions show that a higher energy electron gets delivered to an orbit which is obtained from an electron-hole pair formation. It can be more acutely observed in colloidal semiconductor hetero nano-structures [19]. In the absence of a sufficient amount of energy contained electron-hole (hole may be either a donor or an acceptor), the molecules undergo simple fluorescence emission. Therefore, the transfer of an electron is also a form of photo-quenching.

Non-Catalytic Under Photon Irradiation

It is a continuous process that works at an extensive state in an environmentally friendly way of totally keeping away from the employ of valuable catalysts and high-energy uses. The doping model recognized in this process is used to good contact for the growth of new functional substances in diverse applications [20].

Hydrogenation is an effective approach to match the property of metal oxides. It can predictably be achieved by doping hydrogen into solid substances with high-temperature and pressure annealing treatment, metal catalysis or high-energy proton, which is applied in the vacuum state. In this state, the acid solution in nature provides an affluent proton source, but it would originate corrosion to a certain extent than hydrogenation to metal oxides. In such a method, a facile approach is used to hydrogenate monoclinic vanadium dioxide (VO_2) in acid solution in an extensive form by inserting a small quantity of small work function metal (like Zn, Al, Ag, Fe, Cu) on VO_2 surface. In addition, the best selection of the metal can be achieved in two-step insulator-metal-insulator phase modulation that shows the interaction of an electron-proton co-doping mechanism [21].

For a non-catalytic process, the light energy of the radiation is absorbed by the molecules, which subsequently become activated and achieve the excited state. These undergo homolysis, heterolysis, or photoionization.

A. Catalyzed Photochemistry

Catalysis photoredox is a division of photochemistry that employs single electron transfer. Generally, these catalysts are taken from three classes of materials:

1. Transition-metal complexes,

2. Organic dyes

3. Semiconductors

While organic photoredox catalysts led during the 1990s and early 2000s now-a-day, soluble complexes of transition metal are used.

B. Photosensitised Reactions

These are molecules that are activated by light in order to generate Reactive oxygen species (ROS) that may damage cell structures from microorganisms or from diseased mammalian cells leading to cell death. So, the photosensitised reactions involve oxygen that is framed further type-I and type-II oxidation reactions having oxygen as a reagent. Such processes are used for the abstraction of an oxidizing step. For example:

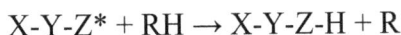

$$X\text{-}Y\text{-}Z^* + RH \rightarrow X\text{-}Y\text{-}Z\text{-}H + R$$

known as intermolecular abstraction of a hydrogen atom, the normally used sanitizer is mercury, which absorbs radiation at 184.9 nm and 253.7 nm produced in high-intensity mercury lamps and acts as sensitizers like cadmium, some noble gases (*i.e.*, xenon), *etc.*

C. Sensitized Reactions

It is defined as changes in the organism, generally, the immunochemical system, which is exposed to a substance such that further substance exposure guides the identification of the organism. These identifications lead to a reaction marked at a very low quantity level, namely hypersensitivity.

D. Photoassisted Catalysis (Stoichiometric Photogenerated Catalysis)

Stoichiometry is working on the principle of conservation of mass where the ratios of a net mass of the reactants and the products are equal to one; it leads to the relations between quantities of reactants and products normally forming a ratio of positive integers. It means that if the quantities of the reactants are known, then the quantity of the product can be calculated.

Such a process depends upon the following factors-

• The entire reagent is consumed.

• There is no deficiency of the reagent.

• There is no excess of the reagent.

This is mostly used in Green Technology.

Photoassisted reaction (catalytic), normally known as photocatalysis, explores the ground of energy and ecological uses. It is extensively famous that the breakthrough of TiO_2 -assisted photochemical reactions has gone ahead to numerous sole uses like self-cleaning glasses, hydrogen production through water splitting, the degradation of pollutants in air and water, cancer treatment, fuel conversion, antibacterial activity, and concrete. These many-sided uses of this fact may be improved and explored further if this method is prepared with additional apparatus and purposes. The word "Photoassisted" catalytic reactions evidently highlight that the photons are necessary to trigger the catalyst; this can be exceeded even into the dark if electrons are piled up in the substance for the shortly use to carry on the catalytic reactions in the absence of light. This is attained by preparing the photocatalyst with an electron pile-up substance to rise above current restrictions in the photoassisted catalytic reactions. In this situation, the appearance of such arrangements could be a consummate technology in close proximity to the future that gives also manipulates all globes of the catalytic sciences [23].

E. Substance Assisted (Catalyzed) Photoreactions

Near-infrared (NIR) photons are more useful than UV photons for biomedical uses since it breaks through the tissue and provides a smaller amount of photodamage in natural systems (since the energy of the NIR photons is lesser). The employ of NIR photons to control bio-edges has gained attention to improve interest. NIR photoreactions at bio-edges are based on up-converting nanoparticles (UCNPs) where these converts NIR to UV or visible photons that can stimulate photoreactions of photosensitive substances. This method is used as UCNP-assisted photochemistry [24-27].

PHOTOCHEMICAL LAWS

The absorption of UV- Visible light by a substance-based solution is followed by different laws that are:

a. Lambert's law

b. Beer's Law

c. Grotthus- Draper Law

d. Stark-Einstein Law

Lambert's Law

When a monochromatic beam of light (UV or Visible) is passed through a homogeneous medium, then the rate of decrease in intensity of the light with the thickness of the absorbing species-like medium is directly proportional to the intensity of the incident beam, Eq. **(6.3)**.

$$-\frac{dI}{dt} = C_a I \tag{6.3}$$

Where I is the intensity of beam after passing thickness t of medium, dI is the small decrease in the intensity of the beam on passing through dt thickness of medium, $-\frac{dI}{dt}$ is the rate of decrease in intensity of the beam through the medium and C_a is the absorption coefficient of the substance.

If I_0 is the intensity of incident beam and I is the intensity after passing through the medium, then

$$\int_{I_o}^{I} \frac{dI}{I} = -\int_{t=0}^{t=t} C_a dt$$

$$\ln \frac{I}{I_o} = C_a t$$

$$\frac{I}{I_o} = e^{-tC_a}$$

$$I = I_o\, e^{-tC_a} \tag{6.4}$$

Eq. (**6.4**) shows Lambert's Law.

Beer's Law

In spectroscopy, this is a relation that concerns the absorption of EM energy by an absorbing medium. This was introduced by German mathematician and Chemist, A. Beer in 1852. According to this law, the absorptive capacity of a dissolved substance directly depends upon its concentration in a medium (*i.e.*, solution).

As a beam of monochromatic light passes through the container containing a certain concentration of a solution, then the rate of decrease in the intensity of a monochromatic beam with the thickness of an absorbing medium is proportional to the intensity of the incident beam and the concentration of a solution, (Eq. **6.5**).

Since,　　$-\frac{dI}{dt} = C_m I C$ 　　　　　　　　　　　　　　　　　　(**6.5**)

Where C_m is the molar absorption coefficient value, which depends upon the absorbing medium, I is the intensity of a monochromatic beam, and C is the concentration of a solution in moles/unit area.

$$\int_{I_o}^{I} \frac{dI}{I} = - \int_{t=t_o}^{t=t} C_m dt C$$

$$ln\frac{I}{I_o} = C_m t C$$

$$\frac{I}{I_o} = e^{-C_m t C} \rightarrow I = I_o e^{-C_m t C} \tag{6.6}$$

So, Lambert-Beer's law directs the absorption of monochromatic light (UV or Visible) by a substance, (Eq. **6.6**).

Since, both laws have become:

$$I_a = I_0 - I_t$$

Where I_t is the intensity of transmitted light, I_o is the intensity of incident light and I_a is the intensity of the absorbed light.

From Eq. (**6.5**), $\int_{I_o}^{I} \frac{dI}{I} = - C_m C \int_a^b dt$

Or, $ln\frac{I}{I_o} = 2.303 \log\frac{I}{I_o} = - C_m C \, b$ (**6.7**)

Eq. (**6.7**) shows that the intensity of the monochromatic beam decreases exponentially with a decrease in thickness (t) and concentration (C).

$$\frac{C_m}{2.303} = \varepsilon$$

$$\log\frac{I_o}{I} = A$$

Where A is the absorbance and measured in the arbitrary unit, ε is the absorption coefficient or extinction coefficient of the substance at a precise wavelength, and depends on the temperature and the solvent. ε is measured in $dm^3 M^{-1} cm^{-1}$ (Fig. **6.7**).

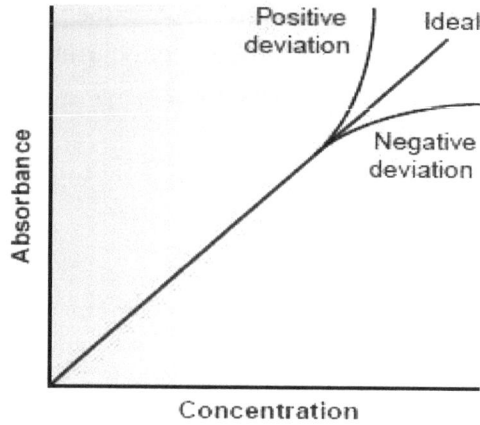

Fig. (6.5). Shows the change of absorption with concentration at positive, ideal negative deviations.

As we know, the transmittance is the ratio of the transmitted beam and incident beam. Since, $T = \frac{I}{I_o}$, and this is given by $A = -logT$

Or $T = 10^{-A} = 10^{-\varepsilon bC}$ $\hspace{4cm}$ (6.8)

This graphical representation of Eq. (6.8) has been shown in Fig. (6.6).

We see from the graph (Fig. 6.5) that A *Vs.* C is a straight line that passes through the origin, so we can write %T.

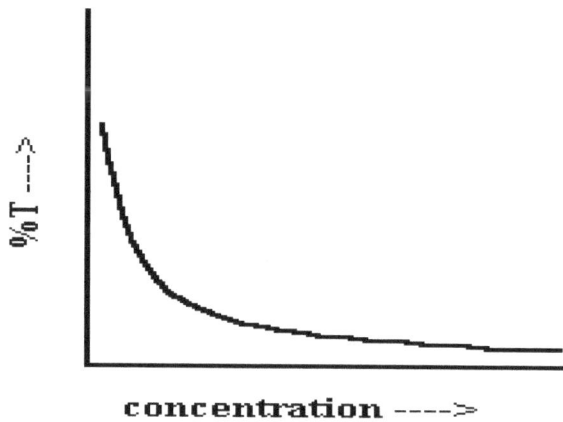

Fig. (6.6). Change of % transmittance *Vs* concentration of the solution.

Both laws (Lambert and Beer) have the following limitations:

1. It is used only in the presence of a monochromatic beam of light.
2. It is applicable for those substances that are electrolytes because inter-ionic exchanges concern the capability of the solute to absorb the beam of light, and become non-linear (*i.e.*, A *vs.* C graph).

Grotthus –Draper Law (First Law of Photochemistry)

This law is also called the principle of photochemical activation and runs as: only the light which is absorbed by a molecule can be effective in producing photochemical changes in the molecule. It means that the transmitted or reflected light does not cause any changes in the substance.

Stark-Einstein's Law (Second Law of Photochemistry)

It states that each atom or molecule of the reacting substance that takes part in a photochemical reaction absorbs one quantum of radiation (*i.e.*, photon) to undergo reaction, and the energy absorbed per mole of substance is given by:

$$E = Nh\nu = \frac{Nhc}{\lambda}$$

If ν is the frequency of absorbed radiation, then the energy of a quantum of radiation is given by $h\nu$ or $\frac{hc}{\lambda}$. This corresponds to the energy absorbed by one mole of a substance containing N molecules, where N is the number of atoms or molecules according to Stark Einstein's law. Hence, the energy absorbed by one mole is given by the Avogadro number. This energy $(\frac{Nhc}{\lambda})$ absorbed by 1 mole of reacting molecules or energy possessed by 1 mole of photons is called One Einstein. This law is also known as the principle of quantum activation.

Quantum Yield or Quantum Efficiency

There are two ways for the excitations:

1. Primary processes: In the primary process, the quantum of energy is absorbed, forming excited molecules. These processes are totally dependent on the absorption of quanta of energy from radiation. Hence the law of photochemical equivalence can be applied to these processes.

$A + h\nu \rightarrow A^*$

2. Secondary processes: In secondary processes, these excited molecules may undergo various changes leading to high or low quantum yield. These processes have nothing to do with the absorption of radiation.

The number of molecules reacting per quantum of energy absorbed is called quantum yield or quantum efficiency. It can also be defined as the number of moles reacting per reaction. It is denoted by ɸ.

Quantum Efficiency, ɸ= Einstein of energy absorbed.

$$\phi = \frac{Number\ of\ molecules\ reacting\ in\ a\ given\ time}{Number\ of\ quanta\ of\ light\ absorbed\ in\ the\ same\ time}$$

Or,
$$\phi = \frac{Number\ of\ mole\ reacting\ in\ a\ given\ time}{Number\ of\ Einstein\ absorbed\ in\ the\ same\ time}$$

So, 1 Einstein = 1 mole of quanta of beam of light or radiation

Therefore, $\phi \geq 1\ or\ \leq 1$, the quantum yield may be greater or less than 1.

For example- In the photolysis of Br_2 and H_2, Φ_{HBr} is very low and about 0.01

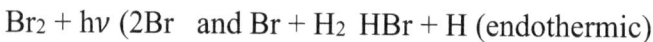

$Br_2 + h\nu$ (2Br and Br + H_2 HBr + H (endothermic)

In order to get a quantum yield of 1, the following steps are to be noted:

1. All the molecules of reacting substances should be at the same energy level, and all should be equally reacting.
2. The reactivity or the velocity of the reaction should be independent of temperature, except that the change in temperature is effective in increasing the absorbing capacity of the system.
3. The product of absorption must become unstable with respect to the original system.

REASON FOR LOW QUANTUM YIELDS

➤ The excited molecules may get deactivated before they form products. This may happen due to the collision of excited molecules with non-excited molecules or with the wall of a container which may cause them to lose their energy and get deactivated.

➤ The primary process may get reversed.

➤ The fragment obtained by the dissociation, may recombine to form the original non-excited molecule.

➤ All absorbing molecules may not receive enough energy to enable them to react because part of the absorbed energy may be lost in the form of fluorescence before they react.

CONCLUSION

This chapter deals with the basics of photochemistry and its consequence on the environmental issue. Different kinds of photo-reaction have been discussed in detail with basic mechanism advantages and disadvantages. In this consequence, photolysis has been mainly divided into two parts, *i.e.*, Catalytic under photon irradiation and Non-Catalytic under photon irradiation. The first category may again be classified into Photo-generated catalysis and Photon-induced catalytic reactions, whereas the second class may be sub-categorized into several divisions like catalysed photo-chemistry, photo-sensitized reactions, sensitized reactions, and photo-assisted catalysisted. Different laws like Beer law, Lambert law or Grotthus–Draper Law and Stark-Einstein's Law have been given and explained. The concept of quantum yield has been introduced and discussed. It is noteworthy that in this chapter basic experimental process for studying photocatalysis and its analysis has not been touched. This is due to the fact that in the next chapter, these have been taken care of rigorously.

REFERENCES

[1]   Keshavarz, M.; Tan, B.; Venkatakrishnan, K. Label-free SERS quantum semiconductor probe for molecular-level and in vitro cellular detection: a Noble-metal-free methodology. *ACS Appl. Mater. Interfaces,* **2018**, *10*(41), 34886-34904.
http://dx.doi.org/10.1021/acsami.8b10590 PMID: 30239189

[2]   Keshavarz, M.; Tan, B.; Venkatakrishnan, K. Multiplex photoluminescent silicon nanoprobe for diagnostic bioimaging and intracellular analysis. *Adv. Sci. (Weinh.),* **2018**, *5*(3), 1700548.
http://dx.doi.org/10.1002/advs.201700548 PMID: 29593957

[3]   Burwell, R.L., Jr Manual of symbols and terminology for physicochemical quantities and units— appendix II heterogeneous catalysis. *Adv. Catal.,* **1977**, *26*, 351-392.
http://dx.doi.org/10.1016/S0360-0564(08)60074-7

[4]   Sudhaik, A.; Raizada, P.; Shandilya, P.; Jeong, D.Y.; Lim, J.H.; Singh, P. Review on fabrication of graphitic carbon nitride based efficient nanocomposites for photodegradation of aqueous phase organic pollutants. *J. Ind. Eng. Chem.,* **2018**, *67*, 28-51.
http://dx.doi.org/10.1016/j.jiec.2018.07.007

[5]   Saravanan, R.; Gracia, F.; Stephen, A. Basic principles, mechanism, and challenges of photocatalysis.*Nanocomposites for visible light-induced photocatalysis*; Springer: Cham, 2017, pp. 19-40.
http://dx.doi.org/10.1007/978-3-319-62446-4_2

[6]   Hasija, V.; Sudhaik, A.; Raizada, P.; Hosseini-Bandegharaei, A.; Singh, P. Carbon quantum dots supported AgI/ZnO/phosphorus doped graphitic carbon nitride as Z-scheme photocatalyst for efficient photodegradation of 2, 4-dinitrophenol. *J. Environ. Chem. Eng.,* **2019**, *7*(4), 103272.
http://dx.doi.org/10.1016/j.jece.2019.103272

[7]   Sudhaik, A.; Raizada, P.; Shandilya, P.; Singh, P. Magnetically recoverable graphitic carbon nitride and NiFe2O4 based magnetic photocatalyst for degradation of oxytetracycline antibiotic in simulated wastewater under solar light. *J. Environ. Chem. Eng.,* **2018**, *6*(4), 3874-3883.
http://dx.doi.org/10.1016/j.jece.2018.05.039

[8]   Ahmed, S.; Rasul, M.G.; Brown, R.; Hashib, M.A. Influence of parameters on the heterogeneous photocatalytic degradation of pesticides and phenolic contaminants in wastewater: A short review. *J. Environ. Manage.,* **2011**, *92*(3), 311-330.
http://dx.doi.org/10.1016/j.jenvman.2010.08.028 PMID: 20950926

[9]   Radhika, N.P.; Selvin, R.; Kakkar, R.; Umar, A. Recent advances in nano-photocatalysts for organic synthesis. *Arab. J. Chem.,* **2019**, *12*(8), 4550-4578.
http://dx.doi.org/10.1016/j.arabjc.2016.07.007

[10]   Poole, C.P., Jr; Owens, F.J. *Introduction to nanotechnology*; John Wiley & Sons, 2003.

[11]   Kumar, P.; Singh, P.K.; Bhattacharya, B. Study of nano-CdS prepared in methanolic solution and polymer electrolyte matrix. *Ionics,* **2011**, *17*(8), 721-725.
http://dx.doi.org/10.1007/s11581-011-0570-2

[12]   Townsend, T.K.; Browning, N.D.; Osterloh, F.E. Nanoscale strontium titanate photocatalysts for overall water splitting. *ACS Nano,* **2012**, *6*(8), 7420-7426.
http://dx.doi.org/10.1021/nn302647u PMID: 22816530

[13]   Chong, M.N.; Jin, B.; Chow, C.W.K.; Saint, C. Recent developments in photocatalytic water treatment technology: A review. *Water Res.,* **2010**, *44*(10), 2997-3027.
http://dx.doi.org/10.1016/j.watres.2010.02.039 PMID: 20378145

[14]   Castillo-Ledezma, J.H.; Sánchez Salas, J.L.; López-Malo, A.; Bandala, E.R. Effect of pH, solar irradiation, and semiconductor concentration on the photocatalytic disinfection of Escherichia coli in water using nitrogen-doped TiO2. *Eur. Food Res. Technol.,* **2011**, *233*(5), 825-834.
http://dx.doi.org/10.1007/s00217-011-1579-5

[15]   Chatterjee, D.; Dasgupta, S. Visible light induced photocatalytic degradation of organic pollutants. *J. Photochem. Photobiol. Photochem. Rev.,* **2005**, *6*(2-3), 186-205.
http://dx.doi.org/10.1016/j.jphotochemrev.2005.09.001

[16]   Wang, K.; Zhang, J.; Lou, L.; Yang, S.; Chen, Y. UV or visible light induced photodegradation of AO7 on TiO2 particles: the influence of inorganic anions. *J. Photochem. Photobiol. Chem.,* **2004**, *165*(1-3), 201-207.
http://dx.doi.org/10.1016/j.jphotochem.2004.03.025

[17]   Sharma, A.K.; Tiwari, R.K.; Gaur, M.S. Nanophotocatalytic UV degradation system for organophosphorus pesticides in water samples and analysis by Kubista model. *Arab. J. Chem.,* **2016**, *9*, S1755-S1764.
http://dx.doi.org/10.1016/j.arabjc.2012.04.044

[18]   Silvi, M.; Melchiorre, P. Enhancing the potential of enantioselective organocatalysis with light. *Nature,* **2018**, *554*(7690), 41-49.
http://dx.doi.org/10.1038/nature25175 PMID: 29388950

[19]   Micheel, M.; Liu, B.; Wächtler, M. Influence of Surface Ligands on Charge-Carrier Trapping and Relaxation in Water-Soluble CdSe@CdS Nanorods. *Catalysts,* **2020**, *10*(10), 1143.
http://dx.doi.org/10.3390/catal10101143

[20]   Song, Y.; Zhang, J.; Dong, X.; Li, H. A Review and Recent Developments in Full-Spectrum Photocatalysis using ZnIn$_2$S$_4$ -Based Photocatalysts. *Energy Technol. (Weinheim),* **2021**, *9*(5), 2100033.
http://dx.doi.org/10.1002/ente.202100033

[21]   Chen, H. Perovskite catalysts for Advanced Oxidation Processes (AOPs) in wastewater treatment. **2018**.

[22]   Lindon, J.C.; Tranter, G.E.; Koppenaal, D. *Encyclopedia of spectroscopy and spectrometry*; Academic Press, 2016.

[23]   Hashimoto, K.; Irie, H.; Fujishima, A. TiO2 photocatalysis: a historical overview and future prospects. *Jpn. J. Appl. Phys.,* **2005**, *44*(12), 8269-8285.
http://dx.doi.org/10.1143/JJAP.44.8269

[24]   Sakar, M.; Nguyen, C.C.; Vu, M.H.; Do, T.O. Materials and mechanisms of photo-assisted chemical reactions under light and dark conditions: can day–night photocatalysis be achieved? *ChemSusChem,* **2018**, *11*(5), 809-820.
http://dx.doi.org/10.1002/cssc.201702238 PMID: 29316318

[25]   Serpone, N.; Salinaro, A.; Emeline, A.; Ryabchuk, V. Turnovers and photocatalysis. *J. Photochem. Photobiol. Chem.,* **2000**, *130*(2-3), 83-94.
http://dx.doi.org/10.1016/S1010-6030(99)00217-8

[26]   Murgolo, S.; De Ceglie, C.; Di Iaconi, C.; Mascolo, G. Novel TiO2-based catalysts employed in photocatalysis and photoelectrocatalysis for effective degradation of pharmaceuticals (PhACs) in water: A short review. *Curr. Opin. Green Sustain. Chem.,* **2021**, *30*, 100473.
http://dx.doi.org/10.1016/j.cogsc.2021.100473

[27]   Yashni, G.; Al-Gheethi, A.; Mohamed, R.; Hossain, M.S.; Kamil, A.F.; Abirama Shanmugan, V. Photocatalysis of xenobiotic organic compounds in greywater using zinc oxide nanoparticles: a critical review. *Water Environ. J.,* **2021**, *35*(1), 190-217.
http://dx.doi.org/10.1111/wej.12619

CHAPTER 7

Removal of Dyes by the Process of Adsorption

Abstract: Adsorption is one of the simplest ways and means to remove dyes from water. The process of adsorption simply involves the removal of water contaminants that come in contact with adsorbents, *i.e.*, the materials of interest. The material should only have sufficient surface area, porosity, and adequate numbers of adsorption sites. Besides being one of the simplest means of dye removal, the process has further advantages in that the same material may be used many times, *i.e.*, regarding the recyclability of the material. Keeping all these in mind in this chapter, a detailed discussion regarding the adsorption process has been included. The discussion not only covers the basic principle of the process but also unfolds the analysis technique regarding the performance of certain material as an efficient absorber. The quantification of removal efficiency will also be a topic of discussion. The setup for such a kind of measurement will be unveiled, and most importantly, different theoretical models for such a process will also be a topic of interest in this chapter. The different models include the Langmuir model, Freundlich model, Temkin model, and others. An effort has also been made to enlighten the readers with the different reaction kinetics like pseudo-first-order, or second-order reaction kinetics. In every subsection, a few experimental data will be shown and discussed.

Keywords: Adsorption, Absorption, Isotherm, Reaction kinetics, Regression coefficients, UV-Visible spectroscopy.

ADSORPTION: DEFINITION

Adsorption basically means adherence of a molecule or ions or functional group onto the surface of an adsorbent. It is to be noted that there is another term which is almost similar to the former, and that is absorption" [1]. The latter simply means fluid molecules get dissolved within the absorbents, and thus it becomes a bulk phenomenon. So it follows that when in absorption fluid can enter well within the inner part of the absorbers, it only loosely adheres to the surface of adsorbents in case of adsorption and thus may be considered to be a surface phenomenon.

WHAT IS ABSORPTION?

The phenomenon like the soaking of certain fluids by solid materials without any external force being applied is called adsorption. It is a bulk phenomenon, and here, external molecules to be adsorbed, which are technically called adsorbents actually

Diptonil Banerjee, Amit Kumar Sharma and Nirmalya Sankar Das

enter within the inner portion of the adsorbents. The adsorbents may get adsorbed both physically as well as chemically.

WHAT IS ADSORPTION?

Like the previous process, adsorption can also be divided into physisorption and chemisorption. In these phenomena, molecules that are supposed to be adsorbed (we call it adsorbate) get loosely adhered onto the surfaces or rather interface of particular liquids or solids (one calls it adsorbent). Table **7.1** and Fig. **(7.1)** summarizes the difference between adsorption and absorption.

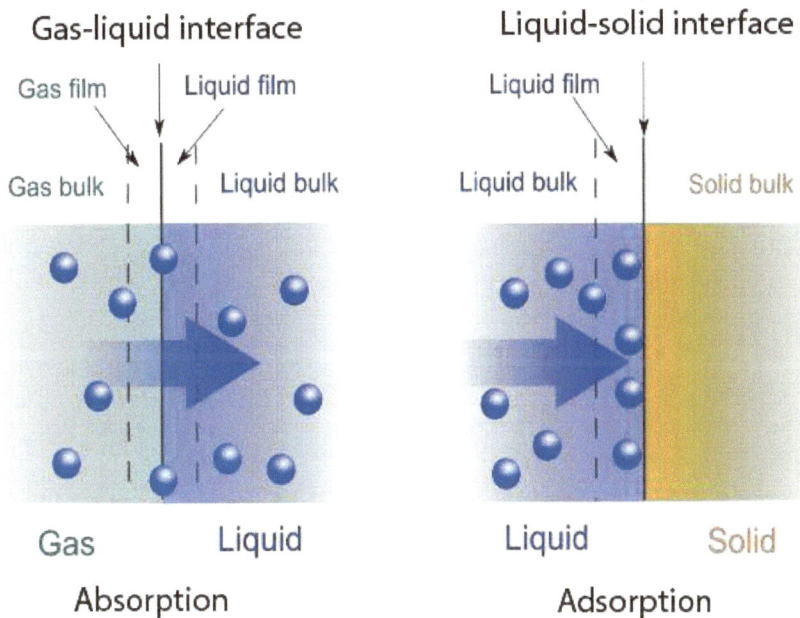

Fig. **(7.1)**. Schematics for understanding of the processes of Adsorption and absorption.

Table 7.1. Absorption and Adsorption a comparison.

Absorption	Adsorption
Here substances, *i.e.*, Molecules or atoms or ions, get inserted within the inner part of the absorbents, which are likely to be solids or liquids.	Here substances, *i.e.*, molecules or atoms or ions, get loosely adhered to the surfaces of the adsorbents, which are likely to be solids or liquids.

(Table 7.1) cont.....

It is considered to be a bulk phenomenon.	It is no doubt a surface phenomenon.
In this process, heat gets absorbed (endothermic)	In this process, heat gets released (exothermic)
It takes place with a uniform rate	Here reaction rate first shows a slow increase before ultimately attaining an equilibrium
In the medium, concentration remains the same	In the medium, concentration varies as it goes from bulk to the bottom
It is temperature-independent.	It is efficient at a lower temperature.
Few practical uses are cold storage or ice production or refrigerants, *etc.*	Few common examples are air conditioning or water puriifcatio⬚, *etc.*

TYPES OF ABSORPTION

1. Chemical absorption- As the name suggests, in this process, which is also called as a reactive absorption, a chemical reaction occurs between absorbent absorbents. Sometimes this process may also accompany physical absorption.

2. Physical absorption- In the process of physical absorption, the electronic configuration, bonding *etc.*, all get unperturbed. This process is non-reactive in nature.

TYPES OF ADSORPTION

Chemisorption: Here the adsorbates get adhered to the surface of the adsorbent by a chemical reaction forming a different chemical bond when the gas molecules are bound to the surface by a chemical bond.

Physisorption: Here no chemical reaction takes place, and adsorbate-adsorbents interactions are mainly weak attractive/repulsive forces or electrostatic forces.

BASIC OF ADSORPTION PROCESS

Adsorption study is one of the easiest processes to test the dye removal of a certain material by the process under interest. It simply requires uniform contact between the remover and the water with contaminations for a certain time. During this contact, the dyes get adsorbed onto the material, and the water slowly becomes contamination free. The process depends on time and several other parameters like the material itself, material, shape and size, material amount, initial dye concentration, types of dyes and sometimes on the pH value of water as well. The basic laboratory process for measuring absorption is rather simple and doesn't differ much from that of photo-catalysis. The process simply needs a closed container, and the contaminated water sample is taken into a beaker and placed in a magnetic stirrer. A small part of the contaminated water (may be 10 ml around) is kept before the addition of the sample, which will work as a reference sample, *i.e.*, the sample with an initial concentration of contamination. Then, as stated before, the sample is added to the contaminated water and thoroughly stirred in order to uniform dispersion of the sample into the water. After a certain interval (that depends on the response of the certain dye to the certain remover but maybe 30 minutes, for instance), a small part of the sample (may be 10 ml) is filtered, and the filtrate is kept for further use. In this way, the process gets continued. The qualitative confirmation of the removal performance can easily be estimated from the visual colour change of the water sample, just like in the process of photo-catalysis. The quantification of the removal performance may be done with the help of a UV-Visible spectrophotometer. The reason is that every dye material has specific colors in the visible range, and thus, all of them have a sharp absorption peak somewhere in the visible region. The intensity of this peak will be decreased as per the relative presence of the dyes in the water. Also, the fall of the relative intensity in the absorption spectra determines the removal efficiency mathematically; the detailed calculation has been given in the coming section. Fig. (**7.2**) explains the situation much better where two dyes, methyl orange (MO) and Rhodamine B (Rh-B), have been removed with the help of carbon nanostructures.

Another important aspect of the entire process is the initial concentration of the dye. It has been seen that the removal efficiency of any material through adsorption is very much dependent on the initial concentration of the dye and gets decreased when the initial concentration gets higher. Now if the work is a field-based work, then there is no restriction on the concentration of the contaminant, however for laboratory-based work, there is a universally accepted initial concentration, and the corresponding process involves taking raw dyes, then diluting them to the stock solution and further diluted it up to the test solution of suitable molarity and use it for actual experiment purpose. The detail process is as follows (though there are reports that many groups do not follow the same and set their own initial concentration level):

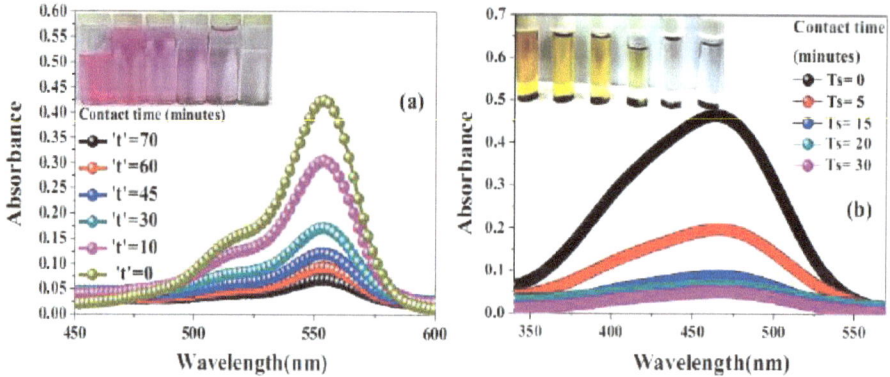

Fig. (7.2). (**a**) Effect of contact time on adsorption of RhB and (**b**) MO by a-CNTs with corresponding colour change of the solution shown inset [1].

From the raw dye, the stock solution is prepared with a concentration of 0.01 g/ml by adding a suitable amount of dyes into a particular amount of water (say 500 ml). However, this solution is further diluted to 10^{-5} M by adding a suitable amount of water into the stock solution and taken for actual laboratory test.

Laboratory-Based Experimental Setup

As discussed in the previous section, the adsorption process is nothing but cold and dark stirring of the material. It has been seen that there are several factors on which the entire process depends that include the level of dispersion of the material into the contaminated water, stirring speed, and most importantly the surface area of the sample. Thus, it should be taken into account that the materials used should have a very high surface area. The high surface area, in turn, helps increasing the amount of dangling bonds and thus the chemical activity of the material. Another factor that influences the performance of the system is the heating effect during stirring. Generally, the time taken to adsorb certain dye starts from 15 minutes and may be extended up to several hours. After stirring for a certain time, thus it is very common that the system gets sufficiently heated. The process of adsorption is sufficiently affected by this uncontrolled heating. This special arrangement should be taken to maintain the temperature of the system constant. One point that has to be remembered is adsorption does not alter.

Basic Analysis Technique

The basic experimental technique of the adsorption demands that the UV-Visible absorption peak of the dyes present as contaminants should have remained in the same position. That means there should not be any chemical reaction between the adsorbents and the dyes. If the absorption spectra show a peak shift or new peak, it is to be considered to be a chemical process associated with the adsorption, which should be avoided. The main parameter with respect to which the material is estimated in terms of its potential as dye remover is the efficiency which is measured mainly in terms of initial concentration and (C_o) and equilibrium concentration (C_e). In terms of these two parameters, η is defined as:

$$(\eta\ \%) = \frac{Co - Ce}{Co} \times 100 \tag{7.1}$$

Here one knows the initial concentration and can associate this with the UV-Visible adsorption intensity. Then it is basically the adsorption peak intensity that defines the adsorption efficiency of the material, as shown in Fig. (**7.3**) in connection to Fig. (**7.2**) [2].

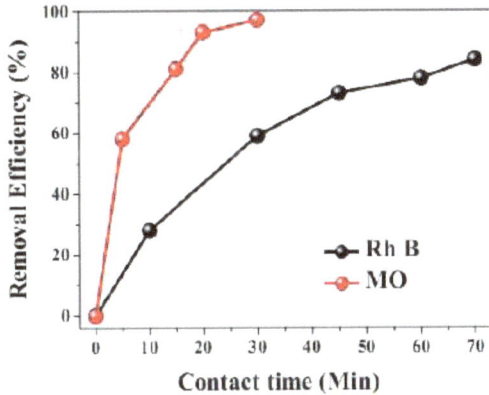

Fig. (7.3). Variation of removal efficiency with contact time for (**a**) RhB and (**b**) MO.

Another very important parameter that helps researchers studying different adsorption isotherms is adsorption coefficients Q, generally defined as:

$$Q = \frac{Co - C}{m} \times V \tag{7.2}$$

Where C is the concentration (may be instantaneous C_T or equilibrium concentration C_e), V is the volume of the solution (ml), and m is the amount of the sample working as a remover (gm). Thus:

$$Q_e = \frac{Co - Ce}{m} \times V \qquad\qquad (7.3)$$

$$Q_t = \frac{Co - Ct}{m} \times V \qquad\qquad (7.4)$$

Based on the above discussion, one can speculate the exact road map for the detailed measurement technique.

First, the initial concentration of the dye has to be set, and preferably that should be as per the internationally set reference level. The prepared sample is then taken for UV-Visible absorption spectroscopy and the corresponding spectra are recorded. Then, as stated before, from time to time sample is taken out of the parent solution after being treated with the sample under preparation. All the samples are again individually undergone absorption spectroscopy, and the relative intensities are recorded. In this process baseline correction is mandatory in order to avoid any initial offset which introduces the error.

THEORIES BEHIND ADSORPTION (ADSORPTION ISOTHERMS)

There are many theories so far developed to explain the adsorption phenomena. Depending upon the number of parameters involved, different isotherms have been developed with advantages and shortcomings. In their review article, Ayawei, Ebelegi, and Wankasi described the model in detail with their mathematical models [3].

A proper understanding and interpretation of adsorption isotherms are critical for the overall improvement of adsorption mechanism pathways and effective design of the adsorption system.

In recent times, linear regression analysis has been one of the most applied tools for defining the best fitting adsorption models because it quantifies the distribution of adsorbates, analyzes the adsorption system, and verifies the consistency of the theoretical assumptions of the adsorption isotherm model. There are many existing models, from the simplest one parameter to the complicated five-parameter models. All of these have their detailed theoretical background, which is beyond the scope of the book. However, I have followed very nice and brief articles written by Ayawei, Ebelegi, and Wankasi, as mentioned before, where the basic working formula with the detailed manifestation.

One-Parameter Isotherm

Henry's Isotherms: This model, even after being the simplest one, shows good agreement when the adsorption process has taken place at a relatively low concentration. Here the efficiency is highly dependent upon the adsorbent partial pressure [4]. In the low concentration regime in which this model holds with accuracy, each adsorbate molecule is isolated from the nearest neighbour [5], and the concentration at equilibrium follows a relation.

$$q_e = K_{HE} C_e \tag{7.5}$$

Where q_e is the amount of the adsorbate at equilibrium (mg/g), K_{HE} is Henry's adsorption constant, and C_e is the equilibrium concentration of the adsorbate on the adsorbent.

Two-Parameter Isotherm

Hill-Deboer Model

This two parameters isotherm model shows good agreement when during or even after the adsorption is taken place, there is lateral interaction between adsorbed molecules [6, 7]. The mathematical form of the model can be summarized as:

$$\ln\left[\frac{C_e\,(1-\theta)}{\theta}\right] - \frac{\theta}{1-\theta} = -\ln K_1 - \frac{K_2\theta}{RT} \tag{7.6}$$

Where K_1 is Hill-Deboer constant (Lmg^{-1}) and K_2 is the energetic constant of the interaction between adsorbed molecules (KJmol^{-1}) [8].

Fowler-Guggenheim Model: This model considers the lateral interactions of adsorbates and mathematically takes the form [9]:

$$\ln\left[\frac{C_e\,(1-\theta)}{\theta}\right] = -\ln K_{FG} + \frac{2\omega\theta}{RT} \tag{7.7}$$

Where K_{FG} is Fowler-Guggenheim equilibrium constant (Lmg^{-1}), θ is fractional coverage, R is the universal gas constant (KJmol^{-1} K^{-1}), T is temperature (k), and w is the interaction energy between adsorbed molecules (KJmol^{-1}). Here w is a crucial parameter, and it may be both positive and negative depending upon the fact that whether the interaction between adsorbate is attractive or repulsive. In these respective processes, the heat of adsorption will increase and decrease with increasing load. Thus, this model predicts here adsorption heat varies the same way as the amount of loading. W = 0 suggests no interaction, and the model gets reduced to a Langmuir equation to be discussed in the next section. It has been shown by Kumara *et al.* that the value of w is positive in the case of phenolic derivatives being adsorbed by activated carbon suggesting an attractive interaction. The limitation of the method is that it is applicable for those systems only where the surface coverage is less than 0. 6 [8].

Langmuir Isotherm: This adsorption isotherm, though, was developed to describe gas-solid phase interaction, but ultimately has shown its potential to successfully describe the adsorption capacity of many adsorbate-adsorbent combinations [10]. The model considers the dynamic equilibrium where both the adsorption and desorption process has been considered and the corresponding processes are further considered to be proportional to the fraction of exposed and covered area, respectively [11]. The mathematical form of the model looks like [12]:

$$\frac{C_e}{q_e} = \frac{1}{q_m K_e} + \frac{C_e}{q_m} \tag{7.8}$$

Where C_e is a concentration of adsorbate at equilibrium (mgg^{-1}). K_e is Langmuir constant related to adsorption capacity (mg g^{-1}).

The parameter, separation factor R_L [13] defined as:

$$R_L = \frac{1}{1 + K_L C_o} \tag{7.9}$$

is of fundamental importance in this process with K_L to be Langmuir constant (mg g^{-1}) and C_o is the initial concentration of adsorbate (mg g^{-1}).

The value of R_L defines whether the process is unfavourable ($R_L > 1$), linear ($R_L = 1$), favourable ($0 < R_L < 1$) and irreversible ($R_L = 0$). This is one of the most common analysis techniques to study the isotherm. As an example, Dabrowski reported the adsorption of green dyes (made from Uncaria Gambir extract) can be best described by the Langmuir isotherm model [12].

Freundlich Isotherm: Another very popular isotherm is *the* Freundlich isotherm which is mostly applicable to heterogeneous surface [13]. This is an effective tool to determine the surface heterogeneity, distribution of active sites with their energy [14]. The mathematical form of the linearized isotherm model takes the form as [15,16]:

$$log q_e = \log K_F + \frac{1}{n} \log C_e \tag{7.10}$$

where K_F is adsorption capacity (L/mg) and $1/n$ is adsorption intensity.

There was a report from the group of Boparai who studied the adsorption of lead (II) by coir dust and associated modified resin, and after applying many isotherm models, he has shown that the adsorption is best described by the Freundlich and Flory-Huggins isotherms and they pointed out higher correlation coefficient is the main reason for that.

Dubinin-Radushkevich Isotherm

This isotherm model is also valid for a heterogeneous surface that adequately describes the adsorption mechanism having Gaussian energy distribution [17, 18]. The limitation of this isotherm is that it produces unrealistic values when adsorbate concentration is rather high and thus, it is applicable where the absorbent concentrate is moderate [19, 20]. This semi-empirical formula follows a pore filling mechanism that considers the Van-der-Waal interaction in the multi-layer configuration and qualitatively describes well the adsorption of different vapours and gasses onto the mesoporous surface by physisorption [21].

This isotherm has the specialty that it distinguishes between physical and chemical adsorption of metal ions, and it has an acute temperature dependence. The mathematical expression of the Freundlich and Flory-Huggins isotherms runs as follows:

$$ln q_e = ln q_m - \beta E^2 \tag{7.11a}$$

$$\epsilon = RT ln \left(1 + \frac{1}{C_e}\right) \tag{7.11b}$$

$$E = \frac{1}{\sqrt{2\beta}} \tag{7.11c}$$

Where E is Polanyi potential, β is Dubinin-Radushkevich constant, R is gas constant (8.31 Jmol^{-1} k^{-1}), T is absolute temperature, and E is mean adsorption energy.

Temkin Isotherm

Temkin model, when summarized in mathematically, takes the form:

$$q_e = \frac{Rt}{b}\ln K_T + \frac{Rt}{b} + \ln C_e \tag{7.12}$$

Here b is Temkin constant related to the heat of sorption ($Jmol^{-1}$) and K_T is Temkin isotherm constant (Lg^{-1}). This model is valid for ion concentration which is neither very low nor extremely high. It considers the indirect interaction between adsorbs molecules and shows that the adsorption heat of the molecules decreases with an increment of surface coverage [22-24].

Flory-Huggins Isotherm

It gives the idea about what fraction of adsorbent surfaces get covered on the adsorbent and mathematically when it takes the linearized form.

$$\ln\left(\frac{\theta}{C_o}\right) = \ln K_{FH} + n \ln(1 - \theta) \tag{7.13}$$

where θ is degree of surface coverage, n is the number of adsorbates occupying adsorption sites, and K_{FH} is Flory-Huggins equilibrium constant ($Lmol^{-1}$). This model is alternatively a measure of the spontaneity of the process. The F-H constant helps in determination of Gibb's free energy (G) from the following relation [25]:

$$\Delta G^o = RT \ln (K_{FH}) \tag{7.14}$$

Where ΔG^o is standard free energy change, R is universal gas constant $8.314\ Jmol^{-1}$ K^{-1}, and T is absolute temperature. Hamdaoui and Nattrechoux have shown that this model is quite good or adequate when the absorption of zinc onto coir dust is explained [26].

Hill Isotherm: This model suggests a correlation between adsorbate molecules and their neighbouring sites on the adsorbent surface. It considers the homogeneous surface only and assumes that when adsorbate gets attached to the adsorbent at a particular site, it influences the other neighbouring sites and thus, the adsorption as per these isotherms becomes a cooperative phenomenon [26]. The mathematical expression of this isotherm, when linearized takes the form [26]:

$$\log \frac{q_e}{q_H - q_e} = n_H \log (C_e) - \log (K_D) \tag{7.15}$$

where K_D, n_H, and q_H are constants.

This isotherm showed remarkable success in explaining the adsorption of aniline, benzaldehyde, and benzoic acid over activated carbon as per the report published by Hamdaoui and Naffrechoux [26]. The obtained value of R^2 is to be 0.99 for various adsorbates [26].

Halsey Isotherm: This isotherm has the mathematical form as given by Eq. (**7.16**) and formulates the adsorption efficiency for multi-layer adsorption for relatively large adsorbate – adsorbent distance [27].

$$q_e = \frac{1}{n_H} I_n K_H - \frac{1}{n_H} \ln C_{qe} \tag{7.16}$$

Where K_H and n are Halsey isotherm constant and that one gets from the slope and intercept of corresponding linear plot.

There are several reports of using Halsey isotherm in describing adsorption of different materials that include textile dye like methyl orange or heavy metals like lead by different carbon atoms [27]. The analysis suggests the heterogeneous nature of the process.

Harkin-Jura Isotherm: This isotherm is applicable for multilayer adsorption and here, the surface is also heterogeneous, having substantial porosity in the structure [28]. The mathematical form of this isotherm is given below [28]:

$$\frac{1}{q_e^2} = \frac{B}{A} - \left(\frac{1}{A}\right) log C_e \tag{7.17}$$

where B and A are Harkin-Jura constants that can be obtained by plotting $1/q^2_e$ *versus* $log C_e$. When reactive clay was made adsorbed by Bentonite clay, as reported by Foo and Hameed, it is seen that this isotherm fits the most compared to Freundlich, Halsey, and Temkin isotherm [28].

Jovanovic Isotherm: The Jovanovic model is the modified form of the Langmuir model, where one considers the mechanical interaction between adsorbate and adsorbent. The isotherm may be expressed mathematically as [29]:

$$\ln q_e = \ln q_{max} - K_J C_e \tag{7.18}$$

where q_e is amount of adsorbate in the adsorbent at equilibrium (mg g^{-1}), q_{max} is maximum uptake of adsorbate obtained from the plot of $\ln q_e$ *versus* C_e, and K_J is Jovanovic constant. When adsorption of L-Lysine polymer was reported by Kiseler it was seen that the adsorption is best fitted by Jovanovic Isotherm.

Elovich Isotherm: This equation is of special importance since at first, it was successfully used to describe the chemisorption process. This also considers the multi-layer adsorption where the adsorption sites exponentially increase with adsorption [30, 31].

The mathematical forms of this Elovich model may be found in the below Eq. (**7.19**) [32]:

$$\frac{q_e}{q_m} = K_E C_e e^{\frac{q_e}{q_m}} \tag{7.19}$$

Eq. (**7.19**) when linearized takes the form of [8]:

$$\ln \frac{q_e}{C_e} = \ln K_e q_m - \frac{q_e}{q_m} \tag{7.20}$$

When copper was made adsorbed on the surface of natural sorbents as did by Rania *et al.* and the process was analysed Elovich isotherm, the value of regression coefficient came out to be rather high as compared to the other models like Langmuir isotherm.

Kiselev Isotherm. This equation also has special significance since it is the only valid equation for describing the adsorption phenomena for a surface coverage of more than 0.68, and it is valid for a single molecule layer system [33]. The linearized expression of the isotherm runs as:

$$\frac{1}{C_e(1-\theta)} = \frac{K_1}{\theta} + K_i K_n \tag{7.21}$$

Where K_i is Kiselev equilibrium constant (Lmg^{-1}) and K_n is the equilibrium constant of the formation of a complex between adsorbed molecules. Equilibrium data from adsorption processes can be modelled by plotting $1/(1 - \theta)$ *versus* $1/\theta$ [33].

Three-Parameter Isotherm

Redlich-Peterson Isotherm

This Redlich-Peterson isotherm is a combination of both the Langmuir and Freundlich isotherms. This equation is an empirical formula involving three parameters and mathematically takes the form as follows:

$$q_e = \frac{AC_e}{1+BC_e^\beta} \tag{7.22}$$

This model has parameters that come from both Langmuir and Freundlich equations and thus, it is a mixed model that does not follow ideal monolayer adsorption [34, 35].

In this Eq. (**7.22**) where A, B, C_e and β all are constant with β having between 0 and 1 and C_e equilibrium concentration of the adsorbate. At a high liquid-phase concentration of the adsorbate, (7.20) gets reduced to the Freundlich equation:

$$q_e = \frac{A}{B}C_e^{1-\beta} \tag{7.23}$$

where $A/B = K_F$ and $(1-\beta) = 1/n$ of the Freundlich isotherm model. When $\beta = 1$, (7.22) reduces to the Langmuir equation with $b = B$ (Langmuir adsorption constant (Lmg^{-1}, which is related to the energy of adsorption.

The linear form of the Redlich-Peterson isotherm can be expressed as follows:

$$\ln\frac{C_e}{q_e} = \beta \ln C_e - \ln A \tag{7.24}$$

When one plots $\ln(C_e/q_e)$ w.r.t. $\ln C_e$ the slope and intercept give the values of β and A, respectively [36]. The isotherm model has a specialty that mathematically it contains both the linear as well as exponential term, and thus, it becomes applicable to over a wide range of adsorbate concentrations. This is thus applicable in both homo as well as heterogeneous systems for its versatility [37].

Sips Isotherm. This isotherm reflects both Langmuir and Freundlich isotherms and has the following mathematical expression [38]:

$$q_e = \frac{K_s C_e^{\beta s}}{1 - \alpha_s C_e^{\beta s}} \tag{7.25}$$

Where K_s is Sips isotherm model constant (Lg^{-1}), βs is Sips isotherm exponent, and α_s is Sips isotherm model constant (Lg^{-1}). This, when linearized, gets changed to the following:

$$\beta_s \ln C_e = -\left(\frac{K_s}{q_e}\right) + \ln(a_s) \tag{7.26}$$

This model mainly differs from the Freundlich model in the sense that, unlike the later Sips model is suitable to deal with heterogeneous surfaces and thus, it is still applicable for higher adsorbate concentration. Thus, at low and high concentrations, it respectively gets reduced to Freundlich and Langmuir model. The associated parameters depend significantly on concentration, pH value, temperature and other parameters [39]. Isotherm constants differ by values when determined by linearization and non-linear regression [40].

Toth Isotherm: This isotherm is a modified form of the Langmuir isotherm, which has been developed to reduce the mismatching between theoretical prediction and experimental results [41]. The model finds its utility in describing heterogeneous surfaces satisfying both low and high adsorbate concentrations [42]. Mathematically the isotherm takes the form as:

$$\frac{q_e}{q_m} = \theta = \frac{K_e C_e}{[1 + (K_L C_e)^n]^{1/n}} \tag{7.27}$$

where K_L and n are different constants, and when $n = 1$, it is reduced to Langmuir isotherm implying the n characterizes the order of heterogeneity [41]. The linear form of the Toth equation takes the form:

$$\ln \frac{q_e^n}{q_m^n - q_e^n} = n \ln K_L + n \ln C_e \tag{7.28}$$

The associated parameters may be obtained by a simple non-linear curve fitting [43]. This has been applied to describing several multi-layer adsorption phenomena [43, 44].

Koble-Carrigan Isotherm

This three parameters isotherm model carries both Langmuir and Freundlich isotherms characteristics and represents adsorption in equilibrium [45]. Mathematically the isotherm, when expressed in linear form, looks like:

$$\frac{1}{q_e} = \left(\frac{1}{A_k C_e^p}\right) + \frac{B_k}{A_k} \tag{7.29}$$

Where A_k, B_k and p are different constants associated with Koble-Carrigan's isotherm [46]. When the concentration of the system is high, it gets reduced to Freundlich isotherm. It is also of extreme importance that the model finds its feasibility only when p is greater than or equal to unity. When p is less than unity in-spite of low error or high concentration, the model is unfit to describe the process [44].

Kahn Isotherm

The Kahn isotherm model is valid for diluted systems and mathematically expressed as [47, 48]:

$$Q_e = \frac{Q_{max} b_k C_e}{(1+b_k C_e) a_k} \tag{7.30}$$

Where a_k and b_k are constants associated with this Kahn model. Q_{max} is Khan isotherm maximum adsorption capacity (mg g^{-1}). Non-linear methods have been applied by several researchers to obtain the Khan isotherm model parameters [49, 50].

Radke-Prausniiz Isotherm: The Radke-Prausnitz isotherm model is generally preferred over others when analysing the data at low concentration. The isotherm takes the following form when expressed mathematically: is given by the following expression:

$$q_e = \frac{q_{MRP} K_{RP} C_e}{(1+K_{RP} C_e)^{MRP}} \tag{7.31}$$

Where q_{MRP} is Radke-Prausnitz maximum adsorption capacity (mg g^{-1}), K_{RP} is Radke-Prausnitz equilibrium constant, and MRP is Radke-Prausnitz model exponent.

When the adsorbate exists at a low concentration, the model takes the form of a linear isotherm. At high concentration, it gets converted to Freundlich isotherm, whereas when $M_{RP} = 0$, it becomes Langmuir isotherm. This model gives a good fit over the entire region of adsorbate concentration which may be quite wide. Here the parameter is obtained by fitting the statistical data following a non-linear approach.

Langmuir-Freundlich Isotherm: This isotherm may be considered to be a bridge between Langmuir and Freundlich isotherm. At high and low concentration of adsorbate, the isotherm equation, which mathematically looks as has been shown in Eq. (**7.32**), gets reduced to Langmuir and Freundlich, respectively.

$$q_e = \frac{q_{MLF}(K_{LF}C_e)^{MLF}}{1+(K_{LF}C_e)^{MLF}} \tag{7.32}$$

In Eq. (**7.32**) q_{MLF} is the maximum Langmuir-Freundlich adsorption capacity in the unit of mg g^{-1}, K_{LF} signifies the equilibrium constant for heterogeneous solid whereas M_{LF} has values between 0 and is nothing but heterogeneous parameter [51]. This model basically is related to heterogeneous surfaces and gives the adsorption energy distribution onto the type of surface mentioned before [44].

Jossens Isotherm: This isotherm is rather simple and is based on the interaction between adsorbate and adsorbent at the adsorbent sites [52]. The model is applicable for heterogeneous surface and mathematically runs as:

$$C_e = \frac{q_e}{H}\exp\left(Fq_e^p\right) \tag{7.33}$$

Where H, F and p are respectively the Jossens isotherm constants of which the latter carries the signature of adsorbent characteristic at all temperatures. When capacities become small, Eq. (**7.33**) gets reduced Henry's law and upon rearrangement gives [53].

$$\ln\left(\frac{C_e}{q_e}\right) = -\ln(H) + Fq_e^p \tag{7.34}$$

It is very clear that the value of F and H may be obtained by plotting $\ln(C_e/Q_e)$ *versus* q_e, which is certainly a linear plot and the slope and intercept of which gives the required values. There are reports of a good representation of equilibrium data for different systems like phenolic compounds that were adsorbed on activated carbon or on different macromolecular resins.

Four-Parameter Isotherms

Fritz-Schlunder Isotherm: This empirical equation was proposed by Fritz and Schlunder, and this can be applied to a wide range of experimental results. This is because of the fact that it is associated with a large number of coefficients belonging to the isotherm [54, 55]. Mathematically, this isotherm takes the form as follows:

$$q_e = \frac{q_{mFS5}K_{FS}C_e}{1+q_mC_e^{MFS}} \tag{7.35}$$

Here q_{mFS} is Fritz-Schlunder maximum adsorption capacity (mg g^{-1}), K_{FS} is the associated equilibrium constant (mg g^{-1}), and MFS is the model exponent. MFS = 1 implies the reduction of the Fritz - Schlunder model to the Langmuir model, however when adsorbate concentrations are high, it takes the form of the Freundlich model.

The parameters associated with Fritz-Schlunder isotherm get determined by non-linear regression analysis [56].

Baudu Isotherm: Bauder isotherm is nothing but a reduced version of Langmuir isotherm and has a mathematical form of:

$$q_e = \frac{q_m b_o C_e^{1+x+y}}{1+b_o C_e^{1+x+y}} \tag{7.36}$$

Where q_m is Bauder maximum adsorption capacity (mg g^{-1}), bo, x, Y all are constants. It is seen that when the surface coverage area is less, than Bauder isotherm gets reduced to Freundlich model. He has seen that the different Langmuir coefficients, when measured from the tangent at different equilibrium concentrations over a wide range, no longer remained constant. Thus Langmuir isotherm gets reduced to the Bauder isotherm [57]. Here the isotherm parameters are determined from non-linear regression analysis [58].

Weber-Van Vliet Isotherm: As the name suggests, this empirical formula was proposed by Weber and Van Vliet, and it shows an excellent agreement with the data obtained from a wide range of adsorption processes. The mathematical form of this isotherm goes as [59]:

$$C_e = p_1 q_e^{\left(p_2 q_e^{p_3} + p_4\right)}$$

(7.37)

In the above equation, C_e is the adsorbate equilibrium concentration (mg g^{-1}), q_e is also adsorption capacity in the same mg g^{-1} unit, p_is (I = 1, 2, 3.....) are Weber-Van Vliet isotherm parameters that can be defined by multiple non-linear curve fitting techniques. These are generally get predicted on the minimization of sum of the square of residuals.

Marczewski-Jaroniec Isotherm: The equation runs as:

$$q_e = q_{MMJ} \left(\frac{(K_{MJ} C_e)^{nMJ}}{1 + (K_{MJ} C_e)^{nMJ}} \right)^{M_{MJ}/n_{MJ}}$$

(7.38)

Where n_{MJ} and M_{MJ} are parameters describing the heterogeneity of the surface of adsorbing material. Here M_{MJ} and nMJ are respectively the distribution spreading of higher and lesser adsorption energy path. The equation takes the form of Langmuir isotherm Langmuir- Freundlich model in the respective condition of n_{MJ} and $M_{MJ} = 1$ and when $n_{MJ} = M_{MJ}$.

This equation is sometimes called a four-parameter general Langmuir equation [60].

This equation is based on the assumption of local Langmuir isotherm and the distribution of energies of adsorption on adsorbent active sites.

Five-Parameter Isotherms

A more accurate five-parameter isotherm model was also developed by which one can simulate the system more precisely and accurately for equilibrium data that may vary over a wide range. The mathematical expression of the isotherm is the following:

$$q_e = \frac{1_m FS_s K_1 C_e^{\alpha FS}}{1 + K_2 C_e^{\beta FS}}$$

(7.39)

Where K_1, K_2, α_{FS}, and β_{FS} are Fritz-Schlunder parameters.

The validity of this isotherm depends on the value of different constant and is most useful when one has $L_{FS} < 0$. At a certain condition, the isotherm approaches both Langmuir as well as a Freundlich model for the condition when constants like α_{FS} and β_{FS} equals to one and higher adsorbate concentration, respectively.

Of All these models, mainly the two-parameter variables Langmuir, Freundlich and Temkin models are most widely used. In the coming section, a real set of experimental adsorption will be taken with the corresponding analysis through all these three models for a better understanding of the readers.

ERROR ANALYSIS

In recent times linear regression analysis has been among the most pronounced and viable tools frequently applied for the analysis of experimental data obtained from the adsorption process. It has been used to define the best fitting relationship that quantifies the distribution of adsorbates and also in the verification of the consistency of adsorption models and the theoretical assumptions of adsorption models. Studies have shown that the error structure of experimental data is usually changed during the transformation of adsorption isotherms into their linearized forms. It is against this backdrop that nonlinearized regression analysis became inevitable, since it provides a mathematically rigorous method for determining adsorption parameters using the original form of isotherm equations. Unlike linear regression, non-linear regression usually involves the minimization of error distribution between the experimental data and the predicted isotherm based on its convergence criteria. This operation is no longer computationally difficult because of the availability of computer algorithms.

The Sum Square of Errors (ERRSQ)

This function may be considered to be the most widely used error analysis technique and expressed as follows:

$$\sum_{i=1}^{n}\left(q_{e,1,calc} - q_{e,i,meas}\right)^2 \tag{7.40}$$

Where $q_{e,1,calc}$ is the theoretical concentration of adsorbate on the adsorbent. These parameters may be considered from one of the standard isotherms whereas $q_{e,i,meas}$

are the measured adsorbate quantity that actually gets adsorbed on the adsorbent. It has one acute shortcoming compared to the other in the sense that at the higher value of concentration range of the liquid-phase adsorbate, the parameters related to isotherm gives a better agreement when derived from that function. One major disadvantage of this error function is that at the higher end of liquid-phase, adsorbate concentration ranges the isotherm parameters derived using this error function will provide a better fit.

Hybrid Fractional Error Function (HYBRID)

This approach was first developed by Kapoor and Yang, with the intention of improving the fit of the sum square of errors (ERRSQ) taken at lesser concentrations, and the approach consists of the division of the measured value. This function has the following expression:

$$HYBRID = \frac{100}{n-p} \sum_{i=1}^{n} \left[\frac{(q_{e,i,meas} - q_{e,i,calc})^2}{q_{e,i,meas}} \right] \tag{7.41}$$

Average Relative Error (ARE)

The method was developed by Marquardt, having an aim to minimize the fractional error distribution across the entire concentration range. It has the expression that looks like:

$$ARE = \frac{100}{n} \sum_{i=1}^{n} \left[\frac{q_{e,i,calc} - q_{e,i,meas}}{q_{e,i,meas}} \right] \tag{7.42}$$

Marquardt's Percent Standard Deviation (MPSD)

Marquardt's percent standard deviation error function is similar to a geometric mean error distribution, modified according to the degree of freedom of the system. It is given by the following expression:

$$MPSD = \sqrt{\frac{1}{n-p} \sum_{i=1}^{n} \left(\frac{(q_{e,i,exp} - q_{e,calc})}{q_{e,exp}} \right)^2} \tag{7.43}$$

Sum of Absolute Errors (EABS): This model is similar to the sum square error (ERRSQ) function. In this case, isotherm parameters determined using this error function would provide a better fit as the forward high concentration data. It is represented by the following equation:

$$EABS = \sum_{l=1}^{p}\left[q_{e,meas} - q_{e,calc}\right] \tag{7.44}$$

Sum of Normalized Errors (SNE): This particular approach has a special significance in error analysis. The reason behind this is that each error criteria are supposed to produce different parameter sets having different values. Thus, it is needed to develop a way and mean to make a comparison between the different parameter set by normalizing and combining the errors.

Below are the general steps taken by the researchers to verify the best fitting models.

Calculation procedure is as follows:

(i) Select an isotherm model and corresponding error function afterwards choose adjustable parameters that help minimize error functions.

(ii) Find the value of the other error functions taking the previously chosen parameter as a reference.

(iii) Compute the sets of other parameters that are associated with their own error functions.

(iv) Normalize and select the maximum parameter sets associated with the measurement of the largest error.

(v) Sum up each such normalized error related to the parameter set.

Coefficient of Determination (R^2) Spearman's Correlation Coefficient (Rs) and Standard Deviation of Relative Errors (SRE)

As per this approach, the expression of the determination coefficient takes the form:

$$R^2 = \frac{\Sigma(q_{ecal}-q_{mexp})^2}{\Sigma(q_{ecal}-q_{mexp})^2 + (q_{cal}-q_{mexp})^2} \tag{7.45}$$

In the above expression q_{exp} is an absorbent amount soaked by the adsorbing material (mg g^{-1}). q_{cal} is the same that is obtained from isotherm analysis (mg g^{-1}), and q_{mexp} is the average of q_{exp} (mg g^{-1}).

The coefficient here actually signifies the variance about the mean, and it is a signature of the agreement between the kinetic model and experimental data.

Non-linear Chi-Square Test (X²): The importance of this function rests in the fact that it can give the best fit of an adsorption characteristic. It comes by making the sum square difference between the calculated and adsorption system and can be defined as:

$$\sum_{i=1}^{n} \frac{(q_{ecal}-q_{emeas})^2}{q_{emeas}} \tag{7.46}$$

Coefficient of Non-Determination

This function has specific importance as it relates the data come from predicted isotherm and that experimentally transformed and thus minimizes error distribution.

Coefficient of non-determination $= 1.00 - R^2$

Where R^2 is the coefficient of determination.

EXPERIMENTAL DATA FOR ANALYSIS WITH TWO PARAMETER ISOTHERM

In connection with Fig. (**7.1**) It has been seen that when the dose of adsorbing material is kept constant and contact time is increased under constant stirring, the UV-adsorption peak intensity is decreased as expected. Depending upon the quality of the material, the fastness of the intensity reduction, *i.e.*, the fastness of the dye removal, will be determined. This, in turn, gives us the efficiency variation with time. There may be other parameters like the amount of doses, pH value of the solution and also the initial dye concentration.

When the baseline correction is done, and there are no additional peaks or peak shifting (*i.e.*, removal is completely due to adsorption and additional chemical reaction is taking place.) in the UV-Vis adsorption spectra, we can have an idea about the parameter Q_c and Q_t as shown in Fig. (**7.4**).

This data, when studied with Langmuir equation given by (**7.8**) and applicable for one-to-one interaction between adsorbate-adsorbent. If now one plots curve between C_e/q_e and C_e a linear plot will come, which in turn gives useful information regarding different adsorption parameters from the linear fitting as shown in Figs. (**7.5a-d**).

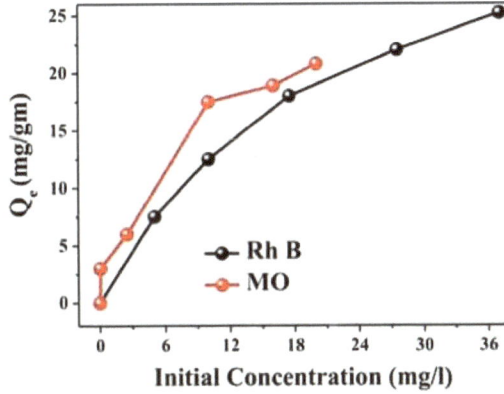

Fig. (7.4). Variation of Q_e with initial dye concentration for two types of dyes methyl orange and Rhodamine B.

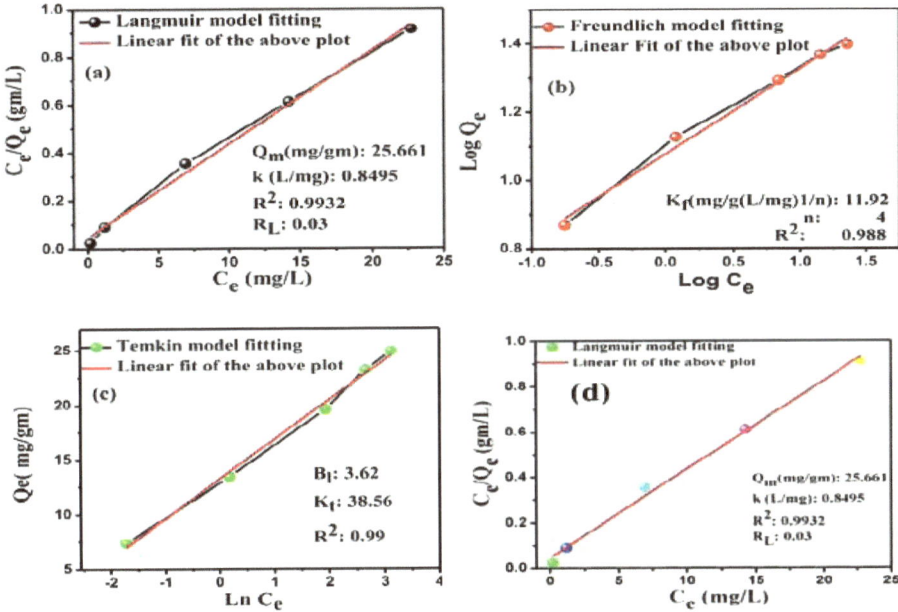

(Fig. 7.5) contd.....

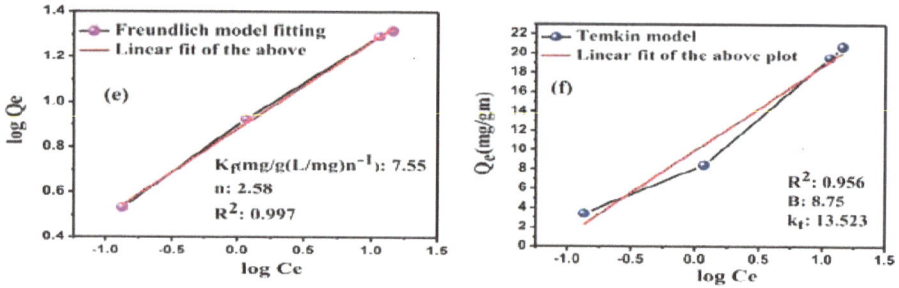

Fig. (7.5). Linear fit of adsorption isotherm data according to different models for (a-c) Rh B and (d-f) MO dye.

The same data may be analysed by the Freundlich isotherm given by Eq. (**7.10**), and in such case, the heterogeneous adsorption model should be taken into account. In these cases, the presence of different functional groups as well as complex adsorbate-adsorbent reactions, cause heterogeneity. Following Eq. (**7.10a**) plot between q_e and c_e ideally should give a linear nature of and the slope and intercept of the corresponding linear fit give the value of K_f and $1/n$ as shown in Figs. (**7.5b, e**).

The Temkin isotherm model assumes that the adsorption heat of all molecules decreases linearly with the increase in coverage of the adsorbent surface, and that adsorption is characterized by a uniform distribution of binding energies, up to a maximum binding energy. The mathematical expression of this model has already been given in Eq. (**7.12**). Here also, the plot of Q_e with respect to $\ln C_e$ is supposed to give a straight line, the linear fit of which provides us the required information regarding Temkin's isotherm constant and the whole state of affair has been shown in Fig. (**7.5c, f**).

Table **7.2** summarizes the different parameters obtained from all these linear fits for the same.

2 dyes methyl orange and Rhodamine B.

Table 7.2. Constants of Langmuir, Freundlich and Temkin models for adsorption of two dyes onto CNTs.

Isotherm Models	Parameter	Value	
		Rh B	**MO**
	q_m (mg/g)	25.661	21.514

(Table 7.2) cont.....

Langmuir	K (L/mg)	0.849	1.364
	R^2	0.993	0.974
	R_L	0.03	0.036
Freundlich	K_f (mg/g(L/mg)1/n)	11.92	7.55
Temking	n	4	2.58
	R^2	0.988	0.988
	B	3.623	8.75
	k_t	38.56	13.52
	R^2	0.993	0.956

REACTION KINETICS

Pseudo-First-Order and Second Order Kinetic Model

For any reaction, one of the most important parameters is the reaction kinetics which determines the particular applications of the reaction and how favourable a reaction is in terms of its occurrence. Adsorption is one of the most popular tools to produce pure water *i.e.*, free from certain industry effluents specially in the area of paper and textile industry. Thus, after the performance of the sample as absorber is witnessed it is generally customary to study the reaction kinetics. One can see that the most popular reaction kinetics that is available in the literature are the pseudo 1st order and second order reaction kinetics. Of which the pseudo-first-order reaction kinetics (afterward it would be denoted as K_1) was first proposed by Lagergren [8] at the end of 19th century, whereas the second order kinetics was reported by the middle of the 80's [9, 10]. However, it should be noted that even after the evolution of the concept of reaction kinetics, it took a long time to be popularized especially until 1999, when the group of Ho and McKay did substantial work by taking adsorption data from literature and analysing them by both the reaction kinetics [54]. They reached a conclusion that "for all of the systems studied, The pseudo-second-order reaction kinetics provides the best correlation of the experimental data". Since the publication of the work reported by them, the concept of the K2 analysing technique gets the world wide popularity [55]. The reason behind this is the fact that most of the studies made a comparison between K1 and K2, and it was seen that K2 gives that remains much closer to reality.

Theoretical Background

Pseudo-First Order Rate Law, K1

The equation for pseudo-first-order kinetics was introduced initially by Lagergren [56]. It is generally used in the form proposed by Ho and McKay [54],

$$\ln [q_e - q(t)] = \ln q_e - k_1 t \tag{7.47}$$

with q the amount of adsorbed solute, qe its value at equilibrium, k1 the pseudo-first-order rate constant and t the time. This equation may also be written in the following alternative way,

$$q(t) = q_e[1 - \exp(-k_1 t)] \tag{7.48}$$

If q_e is determined from the experiment, the fractional uptake (with respect to equilibrium),

$$F(t) \equiv q(t)/q_e \tag{7.49}$$

may be computed. Then one would have, in the case of K1,

$$F(t) = 1 - \exp(-\mathbf{k_1 t}) \tag{7.50}$$

Pseudo-Second Order Rate Law, K2

The formula for pseudo-second order kinetics [57] is generally employed (*e.g.*, in refs. [58] in the form proposed by Ho and McKay [54] as,

$$\frac{t}{q(t)} = \frac{t}{q_e} + \frac{1}{k_2 q_e^2} \tag{7.51}$$

in which k2 is the pseudo-second-order kinetic rate constant. Eq. (**7.5**) may be rewritten as,

$$q(t) = q_e \frac{k_2^* t}{1 + k_2^* t} \tag{7.52}$$

with k * 2 ≡ k2qe, or,

$$F(t) = \frac{k_2^* t}{1 + k_2^* t} \tag{7.53}$$

Rate law of Arbitrary Order

If we suppose that the rate law is of arbitrary order n, then q obeys the equation,

$$\frac{dq(t)}{dt} = k_n [q_e - q(t)]^n \tag{7.54}$$

For n ≠ 1, the solution to this equation with q(t = 0) = 0 is,

$$q(t) = q_e - [q_e^{n-1} + (n-1)k_n t]^{\frac{1}{1-n}} \tag{7.55}$$

or

$$F(t) = 1 - (1 + k_n^* t)^{\frac{1}{1-n}} \tag{7.56}$$

$$k_n^* = (n-1)q_e^{n-1}k_n \tag{7.57}$$

Assessment of Data Correlation Quality

Eqs. (**7.47** and **7.51**) gives the linear form of K1 and K2 that are used extensively in studying the adsorption phenomena and are of the general form $\hat{y}(t) = at + b$ where a and b are respectively the slope and intercept.

The use of this fits transformed experimental data y, obtained at instant t_i for i = 1, ..., N (N is the number of data). The above Eqs. (**7.47** and **7.51**) give $y^{(1)}(t) = \ln[q_e - q(t)]$ and $\hat{y}^1(t) = \ln(q_e - k_1 t)$ for K1 and $y^{(2)}(t) = t/q(t)$ and $\hat{y}^2(t) = \frac{t}{q_e} + 1/k_2 q_e^2$ for K2 respectively.

The average absolute relative deviation (AARD) for the function y is defined by,

$$AARD_y \equiv \frac{1}{N}\sum_{i=1}^{N} \left| \frac{\hat{y}_i - y_i}{y_i} \right| \tag{7.58}$$

where $y_i \equiv y(t = t_i)$ and $\hat{y}_i \equiv \hat{y}(t = t_i)$ are the values of y and $\hat{y}$, respectively, at time t_i . For any kind of models, both linear and non-linear, the value of R may be defined as [59],

$$R^2 = 1 - \frac{\sum_{i=1}^{N}(y_i - \hat{y}_i)^2}{\sum_{i=1}^{N}(y_i - \langle y \rangle)^2} \qquad\qquad (7.59)$$

where $\langle y \rangle$ is the average value of the y_i 's ($i = 1, ..., N$), which is defined as $\langle y \rangle \equiv (1/N) \sum_{i=1}^{N} y_i$. It is to be noted that R^2 may sometimes take negative values < 0 in case of a very bad fitting using non-linear regression.

Issues in Data Analysis

In most cases of adsorption kinetic studies, the experimental values for $y^{(1)} = \ln[q_e - q]$ or $\log_{10}[q_e - q]$, and $y^{(2)} = t/q$, are plotted at times $t_1, t_2,...,t_N$ by using the experimental values of q_e generally denoted by $q_{e,exp}$. The subsequent step is to perform a linear fit with double parameters for the associated plot mentioned before in the case of K1 and K2. The slope and especially the y axis (*i.e.*, t axis) intercept of this linear fit gives a calculated q_e value denoted by q_{ecal} and R^2 for both the K1 and K2. The accuracy of the fit basically determines the validity of the model. The differences between $q_{e,cal}$ and q_{eexp}, values of R^2 all are the measures of accuracies as well as the deviation and reliabilities of the fitting in the case of both K1 and K2.

So far, it has been seen, on the basis of the data reported by previous workers, that, in most of the cases, K2 gives significantly better results as compared to the K1. However, as pointed out by different workers like Jean-Pierre Simonina, the reason for better results obtained in the case of K2 may be hidden in the fact that most of the experimental data used in this case lay in or close to the equilibrium. As a consequence, the method has an inherent pitfall that in the K2 plots of $t/q(t)$ *vs.* t, the points close to the equilibrium are already aligned due to the fact that $t/q \simeq t/q_e$ when $q \simeq q_e$. Thus, with this, when one tries to fit the data, he/she is expected to get a value of R^2 closed to unity.

On the contrary, in the case of K1, when the value of q is getting close to q_e the value of ($q_e - q$) becomes almost zero and the quantity $\ln(q_e - q)$ takes indefinite large values, thus reducing the accuracy in the K1 analysis approach. As a result, the methodology itself favours K2 over K1.

Another issue lies in the homogeneity in the condition of the approach. It is related to the fact that when two statistical models are needed to be compared, then they should be done under the original scale, not on the transformed scale.

There is another issue in this usual treatment, which is connected with the statistical method itself. As pointed out by statisticians, two fitting models must be compared to the original scale, not on the transformed scale. Thus, in the present context, one cannot compare R^2 for K1 and K2, where two functions are completely different. Rather, it is advised to use Eq. **(7.59)** to get a reliable comparison between K1 and K2 in terms of R^2.

Data represented in the absorption spectra when analysed by both the kinetic theories gives the linear plots as shown in Fig. **(7.6)**, and the corresponding parameter values are summarized in Table **7.3** for both the two dyes.

Table 7.3. Kinetic parameters for the adsorption of dyes onto a-CNTs

Isotherm Models	Parameter	Value	
		Rh B	**MO**
Pseudo-First Order	q_e, exp(mg/g)	6.42	11.67
	q_e, graph(mg/g)	6.62	10.46
	K_1(min^{-1})	0.044	0.154
	R^2	0.998	0.990
Pseudo-Second Order	K_2(g/mg min)	0.011	0.007
	q_e, graph(mg/g)	7.165	12.06
	R^2	0.988	0.988

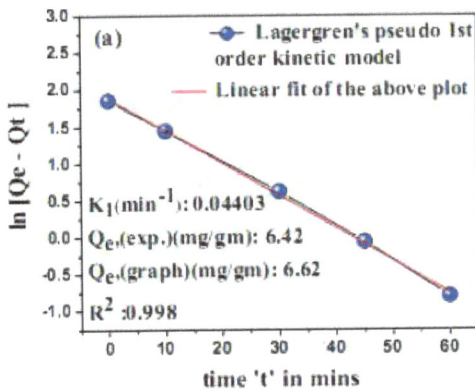

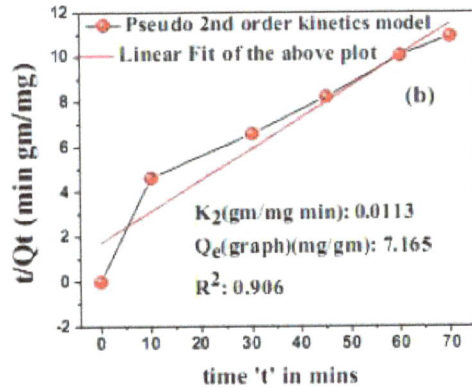

(Fig. 7.6) contd.....

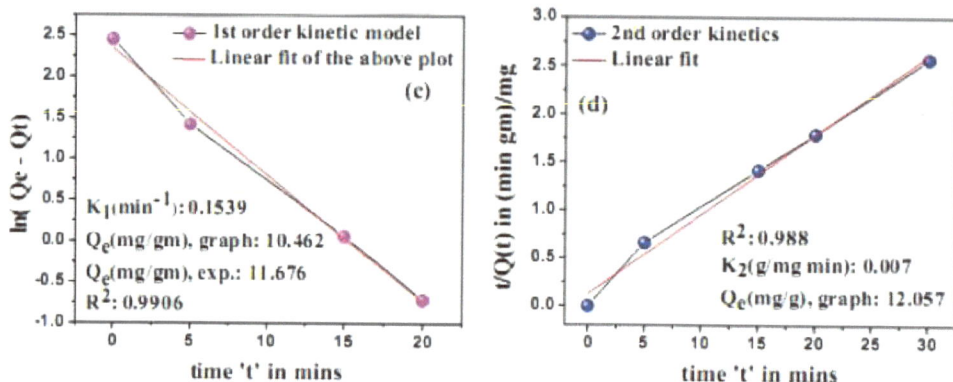

Fig. (7.6). Linear fit of the adsorption isotherm for Rh B (**a, b**) and MO (**c, d**) according to (**a, c**) 1st order and (**b, d**) pseudo 2nd order kinetics model.

CONCLUSION

In this chapter, we have discussed the process of adsorption in much detail with a clear distinction between the two terms **"adsorption"** and **"absorption"**. The emphasis has been given to explaining how this process helps purify wastewater from different textile effluents. In this consequence, the basic procedure for the performance of the experiments, preparation of the dyes, and basic analysis processes have all been described in detail. Efforts have been made to make the reader get acquainted with the basic isotherms that are commonly used in analysing adsorption data. In these consequences, Langmuir, Freundlich and Temkin models have been discussed in more detail with actual experimental data taken for two dyes Rhodamine B as well as methyl orange. Detail error analysis process and its importance have also been a part of this particular **chapter 7**.

The basic theoretical part of the 1st and 2nd order chemical kinetics model has been given here, and the same data has been analysed with these two kinetic models. Beginners will benefit from knowing the actual analysis of adsorption data and the parameters with respect to which a good adsorbing material is described.

REFERENCES

[1]   https://www.toppr.com/guides/biology/difference-between/absorption-and-adsorption/ **2021**.

[2]   Banerjee, D.; Bhowmick, P.; Pahari, D.; Santra, S.; Sarkar, S.; Das, B.; Chattopadhyay, K.K. Pseudo first ordered adsorption of noxious textile dyes by low-temperature synthesized amorphous carbon nanotubes. *Physica E,* **2017**, *87*, 68-76.

http://dx.doi.org/10.1016/j.physe.2016.11.024

[3]   Ayawei, N.; Ebelegi, A.N.; Wankasi, D. Modelling and interpretation of adsorption isotherms. *J. Chem.,* **2017**, 2017.

[4]   Faust, S.D.; Aly, O.M. *Adsorption processes for water treatment*; Elsevier, 2013.

[5]   Ruthven, D.M. *Principles of adsorption and adsorption processes*; John Wiley & Sons, 1984.

[6]   Hill, T.L. Statistical mechanics of multimolecular adsorption II. Localized and mobile adsorption and absorption. *J. Chem. Phys.,* **1946**, *14*(7), 441-453.

http://dx.doi.org/10.1063/1.1724166

[7]   Deboer, J.H. The dynamical character of adsorption. *LWW,* **1953**, *76*(2), 166.

[8]   Kumar, P.S.; Ramalingam, S.; Kirupha, S.D.; Murugesan, A.; Vidhyadevi, T.; Sivanesan, S. Adsorption behavior of nickel(II) onto cashew nut shell: Equilibrium, thermodynamics, kinetics, mechanism and process design. *Chem. Eng. J.,* **2011**, *167*(1), 122-131.

http://dx.doi.org/10.1016/j.cej.2010.12.010

[9]   Sampranpiboon, P.; Charnkeitkong, P.; Feng, X. Equilibrium isotherm models for adsorption of zinc (II) ion from aqueous solution on pulp waste. *WSEAS Trans. Environ. Dev.,* **2014**, *10*, 35-47.

[10]   Elmorsi, T.M. Equilibrium isotherms and kinetic studies of removal of methylene blue dye by adsorption onto miswak leaves as a natural adsorbent. *J. Environ. Prot. (Irvine Calif.),* **2011**, *2*(6), 817-827.

http://dx.doi.org/10.4236/jep.2011.26093

[11]   Günay, A.; Arslankaya, E.; Tosun, İ. Lead removal from aqueous solution by natural and pretreated clinoptilolite: Adsorption equilibrium and kinetics. *J. Hazard. Mater.,* **2007**, *146*(1-2), 362-371.

http://dx.doi.org/10.1016/j.jhazmat.2006.12.034 PMID: 17261347

[12]   Dąbrowski, A. Adsorption — from theory to practice. *Adv. Colloid Interface Sci.,* **2001**, *93*(1-3), 135-224.

http://dx.doi.org/10.1016/S0001-8686(00)00082-8 PMID: 11591108

[13]   Ayawei, N.; Angaye, S.S.; Wankasi, D.; Dikio, E.D. Synthesis, characterization and application of Mg/Al layered double hydroxide for the degradation of congo red in aqueous solution. *Open J. Phys. Chem.,* **2015**, *5*(3), 56-70.

http://dx.doi.org/10.4236/ojpc.2015.53007

[14]   Ayawei, N.; Ekubo, A.T.; Wankasi, D.; Dikio, E.D. Adsorption of congo red by Ni/Al-CO₃: equilibrium, thermodynamic and kinetic studies. *Orient. J. Chem.,* **2015**, *31*(3), 1307-1318.

http://dx.doi.org/10.13005/ojc/310307

[15]   Boparai, H.K.; Joseph, M.; O'Carroll, D.M. Kinetics and thermodynamics of cadmium ion removal by adsorption onto nano zerovalent iron particles. *J. Hazard. Mater.,* **2011**, *186*(1), 458-465.

http://dx.doi.org/10.1016/j.jhazmat.2010.11.029 PMID: 21130566

[16]   Hutson, N.D.; Yang, R.T. Theoretical basis for the Dubinin-Radushkevitch (D-R) adsorption isotherm equation. *Adsorption,* **1997**, *3*(3), 189-195.

http://dx.doi.org/10.1007/BF01650130

[17]   Travis, C.C.; Etnier, E.L. A survey of sorption relationships for reactive solutes in soil. *J. Environ. Qual.*, **1981**, *10*(1), 8-17.
http://dx.doi.org/10.2134/jeq1981.00472425001000010002x

[18]   Çelebi, O.; Üzüm, Ç.; Shahwan, T.; Erten, H.N. A radiotracer study of the adsorption behavior of aqueous Ba2+ ions on nanoparticles of zero-valent iron. *J. Hazard. Mater.*, **2007**, *148*(3), 761-767.
http://dx.doi.org/10.1016/j.jhazmat.2007.06.122 PMID: 17686578

[19]   Theivarasu, C.; Mylsamy, S. Removal of malachite green from aqueous solution by activated carbon developed from cocoa (Theobroma Cacao) shell-A kinetic and equilibrium studies. *E-J. Chem.*, **2011**, *8*(s1), S363-S371.
http://dx.doi.org/10.1155/2011/714808

[20]   Vijayaraghavan, K.; Padmesh, T.; Palanivelu, K.; Velan, M. Biosorption of nickel(II) ions onto Sargassum wightii: Application of two-parameter and three-parameter isotherm models. *J. Hazard. Mater.*, **2006**, *133*(1-3), 304-308.
http://dx.doi.org/10.1016/j.jhazmat.2005.10.016 PMID: 16297540

[21]   Israel, U.; Eduok, U.M. Biosorption of zinc from aqueous solution using coconut (Cocos nucifera L) coir dust. *Arch. Appl. Sci. Res.*, **2012**, *4*(2), 809-819.

[22]   Ringot, D.; Lerzy, B.; Chaplain, K.; Bonhoure, J.; Auclair, E.; Larondelle, Y. In vitro biosorption of ochratoxin A on the yeast industry by-products: Comparison of isotherm models. *Bioresour. Technol.*, **2007**, *98*(9), 1812-1821.
http://dx.doi.org/10.1016/j.biortech.2006.06.015 PMID: 16919938

[23]   Shahbeig, H.; Bagheri, N.; Ghorbanian, S.A.; Hallajisani, A.; Poorkarimi, S. A new adsorption isotherm model of aqueous solutions on granular activated carbon. *World Journal of Modelling and Simulation*, **2013**, *9*(4), 243-254.

[24]   Samarghandi, M.R.; Hadi, M.; Moayedi, S.; Barjesteh, A.F. Two-parameter isotherms of methyl orange sorption by pinecone derived activated carbon. **2009**.

[25]   Amin, M.; Alazba, A.; Shafiq, M. Adsorptive removal of reactive black 5 from wastewater using bentonite clay: isotherms, kinetics and thermodynamics. *Sustainability (Basel)*, **2015**, *7*(11), 15302-15318.
http://dx.doi.org/10.3390/su71115302

[26]   Hamdaoui, O.; Naffrechoux, E. Modeling of adsorption isotherms of phenol and chlorophenols onto granular activated carbonPart I. Two-parameter models and equations allowing determination of thermodynamic parameters. *J. Hazard. Mater.*, **2007**, *147*(1-2), 381-394.
http://dx.doi.org/10.1016/j.jhazmat.2007.01.021 PMID: 17276594

[27]   Fowler, R.H. *Statistical thermodynamics*; CUP Archive, 1939.

[28]   Foo, K.Y.; Hameed, B.H. Insights into the modeling of adsorption isotherm systems. *Chem. Eng. J.*, **2010**, *156*(1), 2-10.
http://dx.doi.org/10.1016/j.cej.2009.09.013

[29]   Kiselev, A.V. Vapor adsorption in the formation of adsorbate molecule complexes on the surface. *Kolloid Zhur*, **1958**, *20*, 338-348.

[30]   Gubernak, M.; Zapala, W.; Kaczmarski, K. Analysis of amylbenzene adsorption equilibria on an RP-18e chromatographic column. *Acta Chromatogr.*, **2003**, •••, 38-59.

[31]   Hamdaoui, O.; Naffrechoux, E. Modeling of adsorption isotherms of phenol and chlorophenols onto granular activated carbonPart II. Models with more than two parameters. *J. Hazard. Mater.,* **2007**, *147*(1-2), 401-411.
http://dx.doi.org/10.1016/j.jhazmat.2007.01.023 PMID: 17289259

[32]   Achmad, A.; Kassim, J.; Suan, T.K.; Amat, R.C.; Seey, T.L. Equilibrium, kinetic and thermodynamic studies on the adsorption of direct dye onto a novel green adsorbent developed from Uncaria gambir extract. *J. Physiol. Sci.,* **2012**, *23*(1), 1-13.

[33]   Weber, T.W.; Chakravorti, R.K. Pore and solid diffusion models for fixed-bed adsorbers. *AIChE J.,* **1974**, *20*(2), 228-238.
http://dx.doi.org/10.1002/aic.690200204

[34]   Davoudinejad, M.; Ghorbanian, S.A. Modeling of adsorption isotherm of benzoic compounds onto GAC and introducing three new isotherm models using new concept of Adsorption Effective Surface (AES). *Sci. Res. Essays,* **2013**, *8*(46), 2263-2275.

[35]   Brouers, F.; Al-Musawi, T.J. On the optimal use of isotherm models for the characterization of biosorption of lead onto algae. *J. Mol. Liq.,* **2015**, *212*, 46-51.
http://dx.doi.org/10.1016/j.molliq.2015.08.054

[36]   Ng, J.C.Y.; Cheung, W.H.; McKay, G. Equilibrium studies of the sorption of Cu(II) ions onto chitosan. *J. Colloid Interface Sci.,* **2002**, *255*(1), 64-74.
http://dx.doi.org/10.1006/jcis.2002.8664 PMID: 12702369

[37]   Sips, R. On the structure of a catalyst surface. *J. Chem. Phys.,* **1948**, *16*(5), 490-495.
http://dx.doi.org/10.1063/1.1746922

[38]   Jeppu, G.P.; Clement, T.P. A modified Langmuir-Freundlich isotherm model for simulating pH-dependent adsorption effects. *J. Contam. Hydrol.,* **2012**, *129-130*, 46-53.
http://dx.doi.org/10.1016/j.jconhyd.2011.12.001 PMID: 22261349

[39]   Chen, C. Evaluation of equilibrium sorption isotherm equations. *Open Chem. Eng. J.,* **2013**, *7*(1), 24-44.
http://dx.doi.org/10.2174/1874123101307010024

[40]   Toth, J. State equation of the solid-gas interface layers. *Acta Chir. Hung.,* **1971**, *69*, 311-328.

[41]   Behbahani, T.J.; Behbahani, Z.J. A new study on asphaltene adsorption in porous media. *Petrol. Coal,* **2014**, *56*(5), 459-466.

[42]   Padder, M.S.; Majunder, C.B.C. Studies on Removal of As (II) and S (V) onto GAC. *MnFe, 804 Composite: Isotherm Studies and Error Analysis.,* **2012**.

[43]   Benzaoui, T.; Selatnia, A.; Djabali, D. Adsorption of copper (II) ions from aqueous solution using bottom ash of expired drugs incineration. *Adsorpt. Sci. Technol.,* **2018**, *36*(1-2), 114-129.
http://dx.doi.org/10.1177/0263617416685099

[44]   Koble, R.A.; Corrigan, T.E. Adsorption isotherms for pure hydrocarbons. *Ind. Eng. Chem.,* **1952**, *44*(2), 383-387.
http://dx.doi.org/10.1021/ie50506a049

[45]   Alahmadi, S.; Mohamad, S.; Maah, M. Comparative Study of Tributyltin Adsorption onto Mesoporous Silica Functionalized with Calix[4]arene, p-tert-Butylcalix[4]arene and p-Sulfonatocalix[4]arene. *Molecules,* **2014**, *19*(4), 4524-4547.

http://dx.doi.org/10.3390/molecules19044524 PMID: 24727422

[46] Khan, A.R.; Ataullah, R.; Al-Haddad, A. Equilibrium adsorption studies of some aromatic pollutants from dilute aqueous solutions on activated carbon at different temperatures. *J. Colloid Interface Sci.,* **1997**, *194*(1), 154-165.
http://dx.doi.org/10.1006/jcis.1997.5041 PMID: 9367594

[47] Sathishkumar, M.; Binupriya, A.R.; Vijayaraghavan, K.; Yun, S.I. Two and three-parameter isothermal modeling for liquid-phase sorption of Procion Blue H-B by inactive mycelial biomass of Panus fulvus. *Journal of Chemical Technology & Biotechnology: International Research in Process. Environmental & Clean Technology,* **2007**, *82*(4), 389-398.

[48] Amrhar, O.; Nassali, H.; Elyoubi, M.S. Application of non-linear regression analysis to select the optimum absorption isotherm for Methylene Blue adsorption onto Natural Illitic Clay. *Bull. Soc. R. Sci. Liege,* **2015**.

[49] Varank, G.; Demir, A.; Yetilmezsoy, K.; Top, S.; Sekman, E.; Bilgili, M.S. Removal of 4-nitrophenol from aqueous solution by natural low-cost adsorbents. **2012**.

[50] Aarden, F.B. *Adsorption onto heterogeneous porous materials: equilibria and kinetics*; Technische Universiteit Eindhoven, 2001, pp. 129-129.

[51] Jossens, L.; Prausnitz, J.M.; Fritz, W.; Schlünder, E.U.; Myers, A.L. Thermodynamics of multi-solute adsorption from dilute aqueous solutions. *Chem. Eng. Sci.,* **1978**, *33*(8), 1097-1106.
http://dx.doi.org/10.1016/0009-2509(78)85015-5

[52] Dilekoglu, M.F. Use of Genetic Algorithm Optimization Technique in the Adsorption of Phenol on Banana and Grapefruit Peels. *J. Chem. Soc. Pak.,* **2016**, *38*(6)

[53] Juang, R.S.; Wu, F.C.; Tseng, R.L. Adsorption isotherms of phenolic compounds from aqueous solutions onto activated carbon fibers. *J. Chem. Eng. Data,* **1996**, *41*(3), 487-492.
http://dx.doi.org/10.1021/je950238g

[54] Ho, Y.S.; McKay, G. Pseudo-second order model for sorption processes. *Process Biochem.,* **1999**, *34*(5), 451-465.
http://dx.doi.org/10.1016/S0032-9592(98)00112-5

[55] Ho, Y. Review of second-order models for adsorption systems. *J. Hazard. Mater.,* **2006**, *136*(3), 681-689.
http://dx.doi.org/10.1016/j.jhazmat.2005.12.043 PMID: 16460877

[56] Lagergren, Š. Zur theorie der sogenannten adsorption geloster stoffe. **1898**.

[57] Gosset, T.; Trancart, J.L.; Thévenot, D.R. Batch metal removal by peat. Kinetics and thermodynamics. *Water Res.,* **1986**, *20*(1), 21-26.
http://dx.doi.org/10.1016/0043-1354(86)90209-5

[58] Chiou, M.S.; Li, H.Y. Equilibrium and kinetic modeling of adsorption of reactive dye on cross-linked chitosan beads. *J. Hazard. Mater.,* **2002**, *93*(2), 233-248.
http://dx.doi.org/10.1016/S0304-3894(02)00030-4 PMID: 12117469

[59] Kvålseth, T.O. Cautionary note about R^2. *Am. Stat.,* **1985**, *39*(4), 279-285.

[60] Simonin, J.P. On the comparison of pseudo-first order and pseudo-second order rate laws in the modeling of adsorption kinetics. *Chem. Eng. J.,* **2016**, *300*, 254-263.
http://dx.doi.org/10.1016/j.cej.2016.04.079

CHAPTER 8

The Efficiency of a Few Potential Nanosystems as Dye Removers

Abstract: In the preceding section, we have discussed the basic characteristics of nanomaterials, their properties, as well as the basic science behind the fact that nanomaterials behave differently from their bulk form. The discussion has also been done regarding the basic features of different microscopes and their working principle. We now have an idea regarding the features of different textile dyes, their toxicity, classifications, and properties. Not only that, the two main ways, like catalysis and adsorption, that help remove dyes from water have also been discussed. We are also now familiar with the basic structures and properties of a few effective dye removers like carbon nanostructure, oxides, and others. As a continuation of all the previous discussions, this chapter deals with the efficiencies/performances of a few potential materials in removing different textile dyes from water. This chapter considers mainly the performance of materials in removing dyes through catalysis and adsorption. We have taken only those materials whose structures and properties we have discussed in the preceding section. The results related to the performance of materials in removing different dyes from water have been taken from outside literature as well as from our own experimental results. It has been shown that the dye removing efficiency of the materials depends upon the material itself, its structure, surface area, porosity as well as numbers of effective adsorption sites.

Keywords: Adsorption, Carbon, Carbon Nitride, Nanostructures, Photo-catalysis, Zinc Oxide.

GRAPHITIC CARBON NITRIDE (NITRIDE SYSTEM)

The introduction to the materials has been given in Chapter 5, along with its basic crystal structures, synthesis processes, and optical and other basic properties. This chapter will deal with the use of pure and doped GCN as the remover of dyes from wastewater.

Application of GCN as a Catalyst Material for Dye Degradation

The presence of the amine group in the GCN structure plays an important role in determining its property [1, 2]. As has been shown by Zhu *et al.*, amine-induced defects help delocalization and re-localization of electrons, introducing interesting

Diptonil Banerjee, Amit Kumar Sharma and Nirmalya Sankar Das

surface properties like Lewis base functionalities as well as electron-rich features. All these, together with the good thermal and chemical stability of GCN, have made them efficient catalysts.

Photocatalysis

Due to different properties mentioned before, like suitable optical gap (that can make the material suitable to utilize solar energy in the visible range), or perfect VB and CB redox edge potential GCN may be considered to be a good photo-catalyst. However, the bulk form of this material suffers the shortcomings like rapid electron-hole recombination, less surface area and lesser quantum efficiency, all collectively affect the catalytic activity of the material in a negative way [3]. This discrepancy has been overcome by structural and morphological modifications as reported by many researchers. However, here the focus has been given mainly on the pristine GCN, which has not been modified by any other foreign material.

Among the numbers of reports available, only a few have been taken in order to make the reader aware of the efficiency of pure GCN as a photo-catalyst. In this regard, the work done by Paul *et al.* is worth mentioning. In separate work, they have shown how GCN can effectively remove both the cationic (RhB and MB) and as well as anionic (MO) dyes [4, 5] with a rate constant value of 0.0029, 0.0076 and 0.008 min^{-1}, respectively. Porous GCN with a surface area of 109.3 m^2/g synthesized from melamine also efficiently removes MO with a rate constant, as has been reported by Gu *et al.* [6]. Mesoporous GCN developed from guanidine hydrochloride by a template-assisted method, as reported by Erdogan *et al.*, has shown excellent UV and visible light-induced photo-degradation of the RhB, MO and MB [7].

There are other separate reports also, for example, those that came from Cui *et al.* or Shi *et al.* All showing GCN with different morphology may be used to remove dye like RhB or MO or MB rapidly with better efficiency as well as with good recyclability [8, 9]. In separate work, Gao developed mesoporous GCN with cashew morphology, which when used to remove MO under the visible Xenon lamp with 500 W power. He observed a 100 % degradation of the dye with a rate constant value of 0.0935 min^{-1}. He concluded, as stated above that the increased surface area, number of effective sites, and favourable transport from bulk to surface make the recombination phenomenon so easily happen, giving good photo-catalytic activity to the sample [10]. Wang *et al.* have shown that porous nanosheet of GCN synthesized from melamine can give rate constant 8 times greater than that of the bulk GCN in case of the removal of RhB [11]. Liu *et al.* reported an interesting phenomenon. In their urea-modified GCN, it was surprisingly seen that over 63 % of MB got removed in simple dark conditions (*i.e.*, the process may be considered as adsorption) before switching on the photon source. After that, 100 % removal of dye took place within 100 minutes [12]. Apart from them, Chang *et al.*, Ibad *et al.*, and Li *et al.* separately reported the efficient catalytic activity of the different GCN nanostructures synthesized from melamine or urea [13-15]. They all have reported a different rate constant in their respective studies. The noteworthy conclusion was taken by Chang *et al.*, who inferred that the main species that take part in photocatalysis are the superoxide anion and the photon-induced generated hole. They supported their conclusion from data obtained from the ESR study [13]. Though different workers have synthesized different structures of GCN and used them to degrade different dyes under photo irradiation with different energy ranges, the reasons they have given are almost the same, like enhanced surface area and a number of active sites reduce recombination rate. The work reported by Pwaer *et al.* for their porous GCN micro rods is not an exception [16]. Reports from Zhang *et al.*, Dong *et al.*, Cui *et al.* or Huang *et al.* followed the same trend [17-20]. Of all these, Huang *et al.* observed a 100 % removal of dyes within just 1 hour and calculated a rate constant 0.051 min^{-1}, which is much better compared to the bulk GCN [20]. Relatively inferior performance was achieved by Yan *et al.* for their ultrathin GCN sheets for the removal of RhB, where the degradation time is near about 100 minutes [21]. In this discussion, the work done by Sundaram and co-workers is mentioned as worthy [22]. They developed GCN isotype hetero-junction by separately taking the combination of melamine, urea, melamine, thiourea, thiourea and urea, all with a 1:1 ratio. They have shown that in-spite of having the greatest surface area in case GCN synthesized out of melamine-thiourea combination that synthesized from melamine, urea combination gives the best catalytic activity. If we go beyond the pristine GCN, there are efforts of doping of GCN by both metals as well as non-metals. In this regard, it should be mentioned that Liu and co-workers incorporated an oxygen functional group into GCN to get an enhanced removal activity for MB [23]. Photo-catalytic activity of doped GCN as well as $CsPbBrCl_2/g-C_3N_4$ type II hetero-junction system has been reported by author's group also where the prolonged separation of –electron-hole pairs were considered to be the key reason for enhanced photo-catalytic activity [24, 25]. The phenomenon has been schematically shown in Fig. (**8.1**).

In two separate works, Ye *et al.* and Tonda *et al.* reported excellent photo-catalytic activities of AgVO₃/GCN and Fe/GCN system, respectively, both under visible light [26, 27]. The reasons behind that, as pointed out, were strong couplings, suitable ban bending or enhanced charge separation due to trapping.

Thus it is no doubt that GCN may be regarded as an efficient photo-catalyst whose performance may still be made better by suitable mix and matching with other systems.

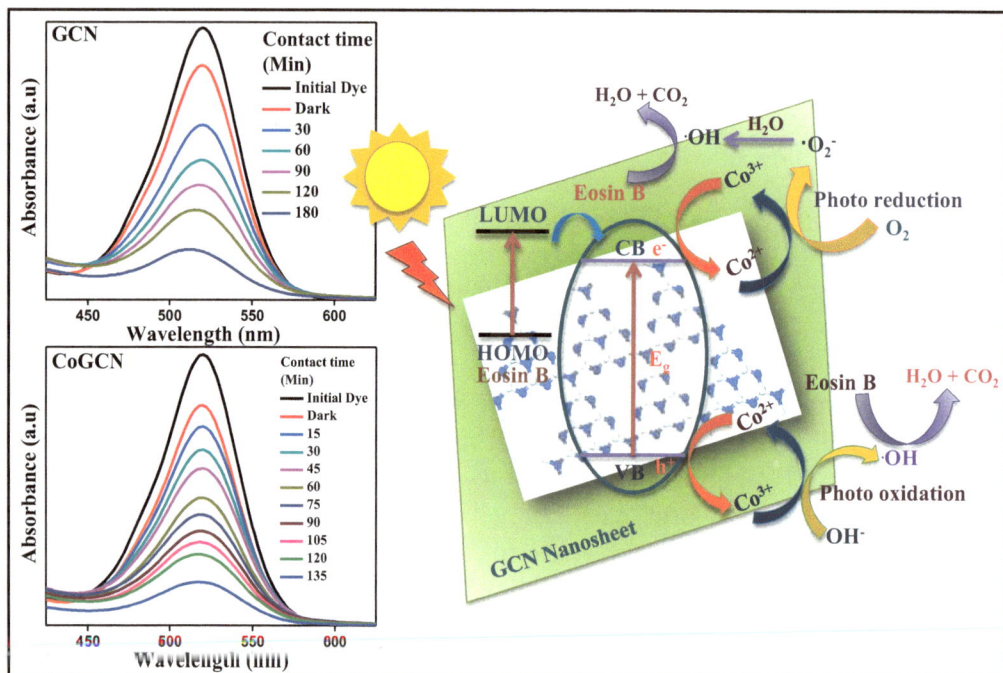

Fig. (8.1). Schematic of removal of Eosin B dye by GCN doped with cobalt [24].

Chemical Catalysis

Chemical catalysis is basically a sodium borohydride-induced catalysis process that can be seen in the GCN-based system but not in pure GCN. Thus, one needs transition metal or noble metal doped GCN for chemical catalysis to occur. The reason is that electron transfer between sodium borohydride and the dye complex through pure GCN is very poor or impossible where the metal nanoparticles play an important role. Pure metallic nanoparticle suffers various constraints, including high cost, less stability, poor availability, non-reusability, agglomeration tendency and other which hinder them from being used in real applications. Thus, it becomes necessary to incorporate, immobilize and anchor them in some carbon-based matrix to improve their stability and usability [28-30].

From the author's group, the reports on chemical catalysis of nickel-doped GCN were published [31], and it was observed that when pure GCN could not remove MO even after 45 minutes through chemical catalysis, Ni-doped GCN took just 4 minutes to degrade the dye with a rate constant 1.42193 min^{-1} for the sample showing the best result.

In Fig. (**8.2**), the entire schematics of chemical catalysis induced removal of MO by nickel doped GCN has been shown along with the actual result [31].

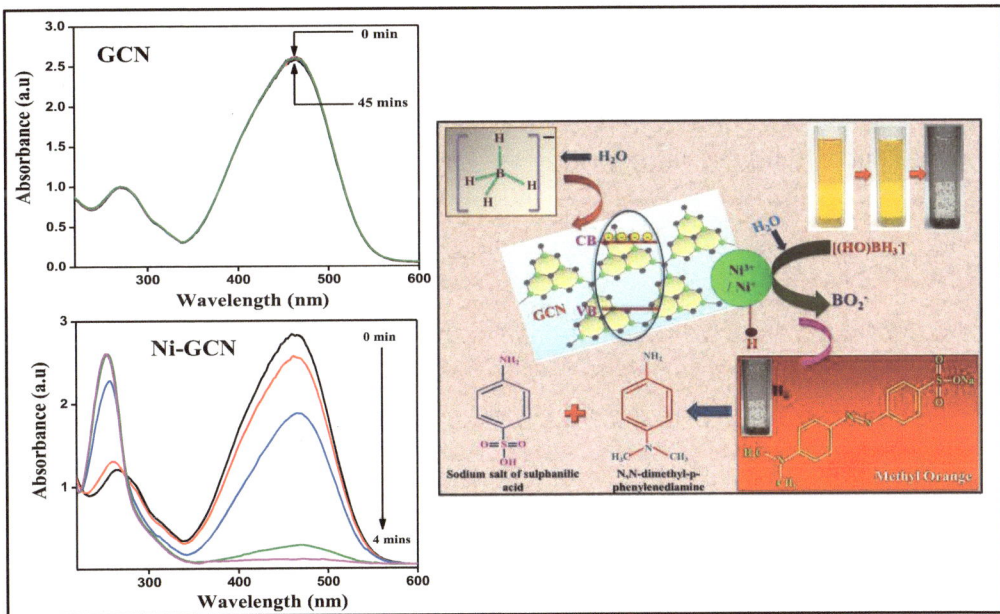

Fig. (8.2). Schematic of NaBH$_4$ induced degradation of MO by Ni doped GCN through chemical catalysis [31].

Mitra *et al.* reported the catalytic activities of a Cu_2O/GCN system that degrade MO even lesser than 5 minutes with a rate constant of 1.3 min^{-1} with simultaneous generation of hydrogen and other non-toxic materials [32].

Within just 70 and 100 seconds, complete degradation MO and RhB, respectively, were reported by Mohammadi *et al.* with their silver decorated GCN through chemical catalysis [33]. The new information they observed is that reduction is not possible when the catalyst and $NaBH_4$ solution are present separately, and only the simultaneous presence of both can enable the system to remove the dye. Apart from these, Sarangapany *et al.*, Fu *et al.* Kumar *et al.* separately reported Ag/GCN or Pd/GCN system, induced removal of different dyes through chemical catalysis within a few minutes only (in maximum cases, it is within a few seconds) [34-36]. For the last one, Kumar *et al.* inferred that the porous structure of GCN helped the adsorption of dye molecules as well as $NaBH_4$ on the catalyst surface, giving much better results.

Murugan *et al.* also used Ag/GCN system to remove another toxic textile pollutant crystal violet, by the same chemical catalysis [37] with success. In all the cases of the kinetics, the system followed the pseudo-first-order reaction kinetics because here, the concentration of $NaBH_4$ is always kept much higher than that of pollutants [31]. Thus, here the reaction rates are independent of $NaBH_4$ concentration.

Application of GCN as an Adsorbent for Dye Removal

Due to almost the same reason mentioned in the previous section, *i.e.*, High surface area, more surface active sites porous GCN in nano-regime also acts as an excellent adsorbent capable of removing different dyes through the process of adsorption.

Mohanty *et al.* reported 99 % adsorption efficiency of GCN when applied on MB [38] synthesized GCN a dose of 0.02 gm GCN at a pH value of 9 could able to remove MB having initial concentration 100 ppm to that extend in 120 minutes. The corresponding isotherm followed a monolayer, *i.e.*, Langmuir isotherm. The corresponding reaction kinetics obey pseudo-second-order reaction kinetics.

A comparison was made by Zhu *et al.* between the adsorption ability of porous GCN when applied for cationic dyes like MB and anionic dye MO. A GCN sample with a surface area 40 m^2/g and isoelectric point 5.1 can adsorb MB much more efficiently than MO [39]. This is because of the negative surface charge of GCN as confirmed by the zeta potential. The system showed the highest adsorption capacity of 2.51 mg/g. Here also, the reaction kinetics followed a pseudo-second-order reaction kinetics.

The possibility of Na-doped GCN as the remover of MB through adsorption was first reported by Fronczak and co-workers [40]. In their case, the sample was best active at normal pH, and concentration and equilibrium were achieved within 5 minutes. The enhanced electrostatic interaction for the Na-GCN system was believed to be the reason behind the better adsorption performance of the system. Other workers like Yousefi *et al.*, Bhowmik *et al.*, Zou *et al.* or Zhang *et al.* have established nano-porous pure or modified GCN to be an excellent adsorbent, but in most cases, the studies got limited to RhB, MB or MO [41-44]. Bhowmik *et al.* attributed negative surface charge in the case of Au decorated GCN for its better adsorption performance compared to the pure one [42]. Zou *et al.* developed a rather less studied system of GCN/β-cyclodextrin hybrid showing efficient multilayer adsorption of MO and thus best described by Freundlich isotherm [43]. ZnO-GCN hybrid, as developed by Zhang *et al.*, was employed to remove MB and anionic Orange II dye. Though it could remove the MB within 20 minutes still, the performance was much inferior compared to that was shown against orange II and the reason, as pointed out by the author, was a positive surface charge of the sample, which made a much higher impact in case of anionic orange II dye [44]. Ren *et al.* Adsorbed MB with their as synthesized carbon doped sample [45]. The adsorption was monolayer adsorption and best described by Langmuir isotherm. However, it took almost 5 hrs. to degrade the dye. In their case, π-conjugation for the doped sample was an additional reason for the observed phenomenon. Apart from them, Xie *et al.* made MB adsorbed onto the surface of one-dimensional GCN nanostructure [46] with surface areas of 74.25 m^2/g and 60.16 m^2/g for nanowire and nanofiber-like morphology, respectively. In a separate work, Fronczak *et al.* did an extensive study of adsorption kinetics of GCN by doping it with 's' block materials like Li, Na, K, Mg, Ca, Sr and Ba for removing MB [47]. It has been seen that even after having a moderate surface area, Mg-doped GCN has a remarkable adsorption efficiency of MB with a corresponding capacity 7500 mg/g whereas the pristine sample has the same as 11 mg/g. The reason is the same as that discussed before and nothing but the favourable adsorption anionic MB over the surface of positive charge induced doped GCN *via* electrostatic interaction. The equilibrium, in that case, was achieved within 60-90 minutes. A GCN- polyoxoniobate-based nanocomposite was developed by Gan *et al.* to remove MB [48] but took almost 4 hours to set the equilibrium and thus is of little significance. A new hybrid based on strontium carbonate-GCN was developed by Lu *et al.* for the effective removal of crystal violet [49]. This system, however, remained unresponsive towards the pH values. There is a possibility of chemisorption as pseudo-second-order kinetic model fits the characteristics much better. Non-metal doping into GCN has been done by Chegeni and co-workers who developed phosphorous-doped GCN [50]. They have studied the adsorption of MB as a function of many parameters like time, dose, temperature, pH, *etc.* The adsorption was supposed to be monolayer adsorption as suggested from best fitting with Langmuir isotherm. Xu *et al.* and Yang *et al.* also reported the adsorption of MB or RhB by the different GCN-based systems [51-53]. Of which the work reported by Yang *et al.* may be mentioned separately as they have developed GCN from melamine and mushroom waste and studied its efficacy on MB removal [52]. A 1D-2D system consisting of 1 D $W_{18}O_{49}$ planer GCN sheet shows a removal efficiency of over 99 % in just 30 minutes when applied on MB [51]. GCN-RGO (reduced graphene oxide) based hybrid has been shown to be less efficient to remove RhB as the system shows removal efficiency of around 89 % but with a much higher time of 240 minutes [53]. Considering all, it is no doubt that pure and doped GCN, as well as GCN-based hybrids, are really efficient to remove different contaminants from water through different processes like photocatalysis, adsorption or chemical catalysis.

CARBON NANOTUBES (CARBON SYSTEM)

CNTs are one of the most effective materials for the removal of different textile dyes in different ways and means that mainly include photocatalysis and adsorption. This part will mainly focus on the literature and a few results related to this particular material.

Adsorption Properties of CNTs

The adsorption-related applications of CNTs to solve environmental pollution problems have received considerable attention in recent years. CNTs are relatively new adsorbents and hold interesting positions in carbon-based absorptive materials for many reasons. On the one hand, they provide chemically inert surfaces for physical adsorption, and their high specific surface areas stand in comparison with those of activated carbons (ACs). On the other hand, CNTs are essentially different from ACs in the aspect that their structure at the atomic scale is far more well-defined and uniform. Parameters such as the pore diameter distribution and adsorption energy distribution are needed to quantify adsorption on ACs, while, as for CNTs, one can deal directly with various well-defined adsorption sites available to the adsorbed molecules. From the viewpoint of CNT structures, the relationship between CNTs and other carbonaceous adsorptive materials is almost the same as that between single crystals and polycrystalline materials [54].

Carbon Nanotubes for Adsorption of Organic Chemicals

The possibility of multilayer adsorption of organic molecules on the surface of CNT has been reported by Pan *et al.* [55], where it has been reported that the first couple of layers only interact with the CNT surface and the rest layer interact with each other only. As expected, the energy associated with this process is a strong function of the distances between the surface and the molecular layers. For an open-end CNT or even the interstitial area within the CNT bundle, the capillary condensation is supposed to take place if the pore size varies as per the Kelvin-Laplace equation. This, in turn, results in adsorption energy depending upon the load and heterogeneous adsorption. It has been speculated that the organic molecule first sits on the sites with higher energy and then gets distributed to the region of less energy suggesting a concentration-dependent thermodynamics. The later has been verified for organic vapours, but yet to be studied with the aqueous phase. There are also substantial efforts from the worker to explain the long adsorption-desorption kinetics in view of heterogeneous adsorption sites. Hydrophobic interaction has been employed to explain the adsorption of several molecules like protein, naphthalene and other organic molecules. This assumption, however, has few drawbacks because of the fact that if hydrophobic contacts are the reason behind the interaction with adsorbate (organic molecules) and adsorbent (CNTs), the adsorption may be predicted from associated parameters like KOW or KHW, and the modelling of the environmental behaviour of all these chemicals would be rather straightforward. Unfortunately, the situation is not so simple, and it has been shown by Chen *et al.* [57] that there actually exists poor or no correlation between adsorption inclination and hydrophobic nature of several aromatic derivatives.

Other mechanisms include π-π interactions (between bulk π system on CNT surfaces and organic molecules with C=C double bonds or benzene rings), hydrogen bonds (because of the functional groups on CNT surfaces), and electrostatic interactions (because of the charged CNT surface) [58]. For example, the importance of π-π interactions was demonstrated by comparing the adsorption of several carefully selected organic chemicals on CNTs [59].

Carbon Nanotubes for Adsorption of Textile Dyes

As everyone knows, CNTs are nothing but rolled sp^2 hybridized graphite sheets having a typical sidewall curvature. It is the number of parallel graphite sheets that get rolled up, which is defined in the as-developed CNTs are single-walled, double-walled or multi-walled. The π conjugated structure introduces very high hydrophobicity into the CNT surface. Different parameters like morphology, defects as well as active sites all play a very important role in explaining the adsorption of different dyes into the CNT surface. The unique properties of the CNT surfaces make them unable to interact with the foreign material through π–π electronic as well as hydrophobic interactions, making them a material of potential for dye adsorption into CNT surfaces [60]. The adsorption of CNTs comes out to be even more acute compared to the same properties exhibited by conventional porous activated carbon [61]. Dye adsorption properties of CNT play an important role in the environmental issues of nanotechnology/nanomaterial. In this consequence, there are a number of studies that include studies done by Yao *et al.* [62] or Shahryari *et al.* [63] or others, showing the adsorption capacity of CNT is a function of other physical properties of the material.

The adsorption capacity also depends on the experimental conditions and the nature and type of the adsorbent. In the above-mentioned study, Shahryari *et al.* [63] used MWCNTs as adsorbents, while Yao *et al.* [62] reported the use of CNTs. The comparative adsorption of cationic methylene blue (MB) and anionic orange II (OII), from an aqueous solution by using MWNTs and carbon nanofibers (CNF) as adsorbents were studied in batch experiments by Rodríguez and coworkers [64]. Yan *et al.* and Chao *et al.* reported synthesis, characterization and application of amino-functionalized multi-walled carbon nanotubes for effective, fast removal of methyl orange from aqueous solution [65]. Wang *et al.* reported the adsorption mechanism of activated carbon fibre/carbon nanotube composites for Rhodamine b from an aqueous solution [66]. Thus, from the past review works, it can be obviously acclaimed that CNTs undoubtedly act as all-rounder scavengers encompassing almost all genres of environmental pollutants. In our laboratory, we have got a few interesting results regarding the adsorption-induced removal of organic molecules, textile dyes or even heavy metals like arsenic.

Figs. (**8.3** and **8.4a-d**) shows the response of different textile dyes like methyl orange (MO) or Rhodamine B (Rh-B) towards carbon nanotubes exposed to different times or adsorbent dosage [67].

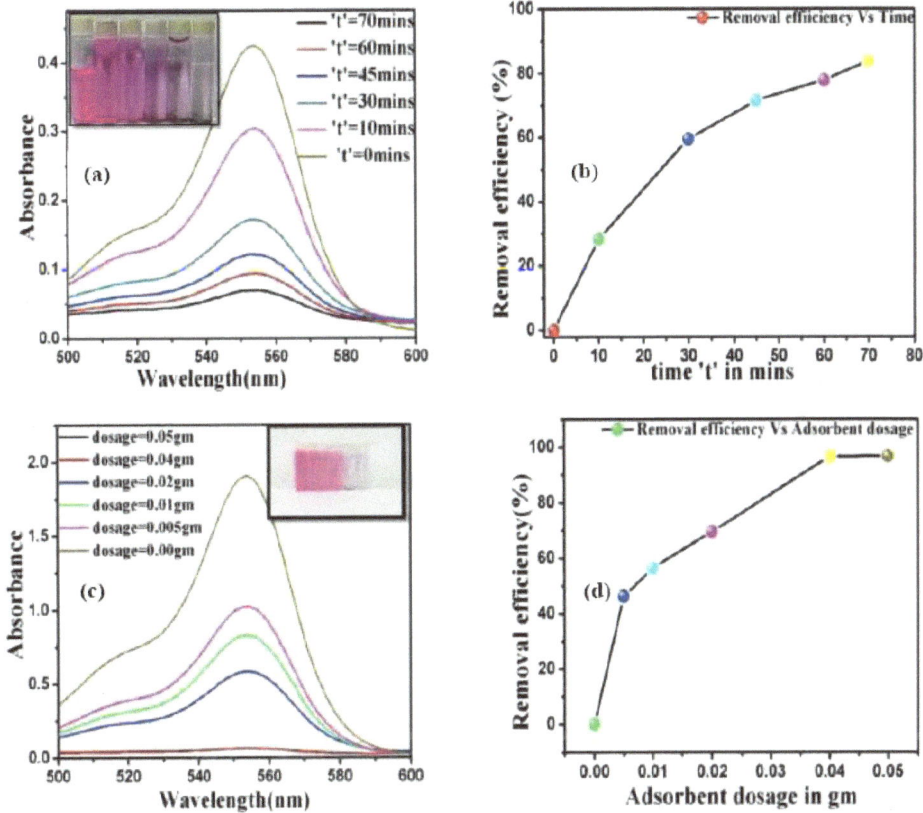

Fig. (8.3). Removal of Rh-B by a-CNTs exposed for different times (**a**) and with different dosage (**c**) with corresponding variation of efficiency (**b** and **d**) [67].

However, instead of crystalline CNTs, here we have used amorphous CNTs (a-CNTs) that have huge numbers of inherent defects favouring the process of adsorption. Here, a-CNTs have further advantages of cost-effective, high yield gram scale synthesis process. Also, the as-synthesized a-CNTs do not have to be fictionalized externally in order to overcome the Steric – hindrance force.

We have fitted the obtained data with all the existing models as well as with the existing kinetics. The results for both the dyes are summarized in Tables (**8.1-8.4**) [67].

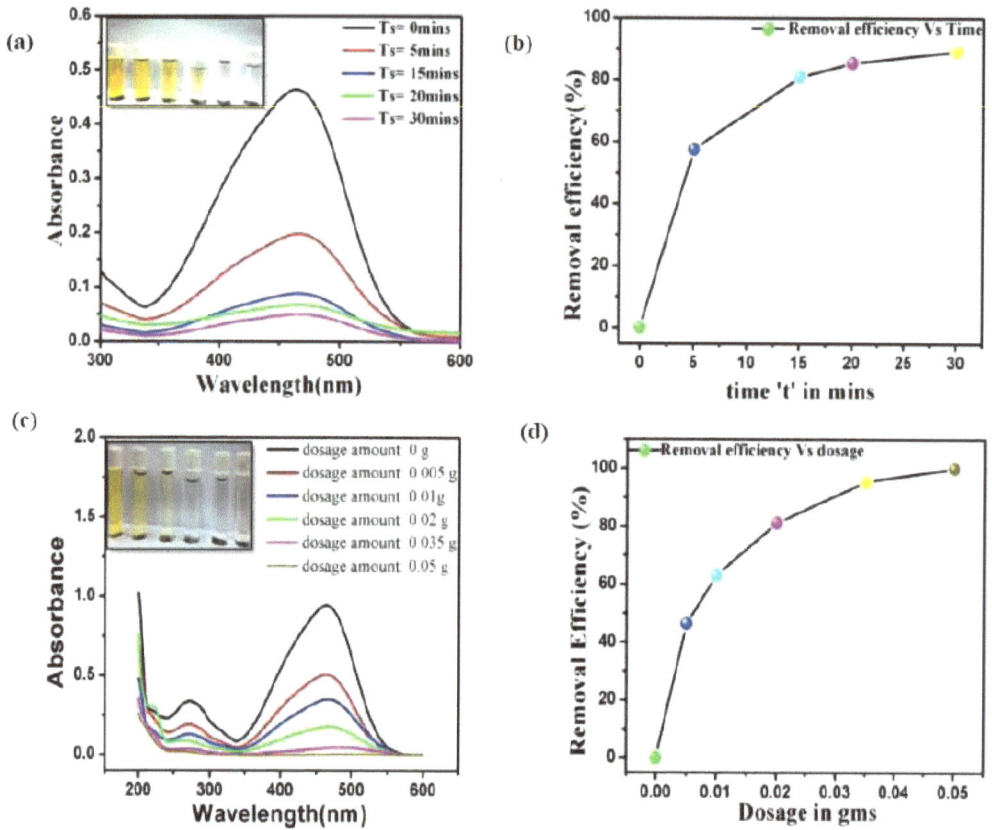

Fig. (8.4). Removal of MO by a-CNTs exposed for different times (**a**) and with different dosage (**c**) with corresponding variation of efficiency (**b** and **d**) [67].

Table 8.1. Constants of Langmuir, Freundlich and Temkin models for adsorption of RhB on to a-CNTs.

Isotherm Models	Parameters	Value
Langmuir	q_m (mg/g)	25.661
	k(L/mg)	0.849
	R^2	0.993
	R_L	0.03

(Table 8.1) cont.....

Freundlich	k_f(mg/g(L/mg)1/n)	11.92
	n	4
	R^2	0.988
Temkin	B	3.623
	k_t	38.56
	R^2	0.993

Table 8.2. Kinetic parameters for the adsorption of RhB onto a-CNTs.

Kinetic Models	Parameters	Value
Pseudo-first order	q_e, exp(mg/g)	6.42
	q_e, graph(mg/g)	6.62
	K_1(min^{-1})	0.044
	R^2	0.998
Pseudo-second order	K_2(g/mg min)	0.011
	q_e, graph(mg/g)	7.165
	R^2	0.906

Table 8.3. Constants of Langmuir, Freundlich and Temkin models for adsorption of MO on to a-CNTs.

Isotherm Models	Parameters	Value
Langmuir	q_m (mg/g)	21.514
	k(L/mg)	1.364
	R^2	0.974
	R_L	0.036
Freundlich	k_f(mg/g(L/mg)1/n)	7.55
	n	2.58
	R^2	0.988

(Table 8.3) cont.....

Temkin	B	8.75
	k_t	13.52
	R^2	0.956

Table 8.4. Kinetic parameters for the adsorption of MO onto a-CNTs.

Kinetic Models	Parameters	Value
Pseudo-first order	q_e, exp(mg/g)	11.67
	q_e, graph(mg/g)	10.46
	$K_1(min^{-1})$	0.154
	R^2	0.990
Pseudo-second order	K_2(g/mg min)	0.007
	q_e, graph(mg/g)	12.06
	R^2	0.988

It is also shown that the adsorption when done in the presence of UV light, improvement is not very considerable as compared to the pure adsorption process suggesting later is the key factor for textile dye removals. The result is shown in Fig. (**8.5**), where sample A means pure a-CNTs and sample B means a-CNTs with iron impurities [68].

Apart from that, the other unconventional carbon nanostructures like porous carbon spheres have shown considerable potential towards photo-induced removal of the same dyes Rh-B or MO. The performance gets significant enhancement when the as-synthesized carbon spheres are coupled with TiO_2 nanoparticles. The responses of both the dyes towards all the samples have been shown in Fig. (**8.6**). Here, C is the pure carbon sample and CT-0, 1, 2 are hybridized with TiO_2 particle with increasing TiO_2 concentration [69].

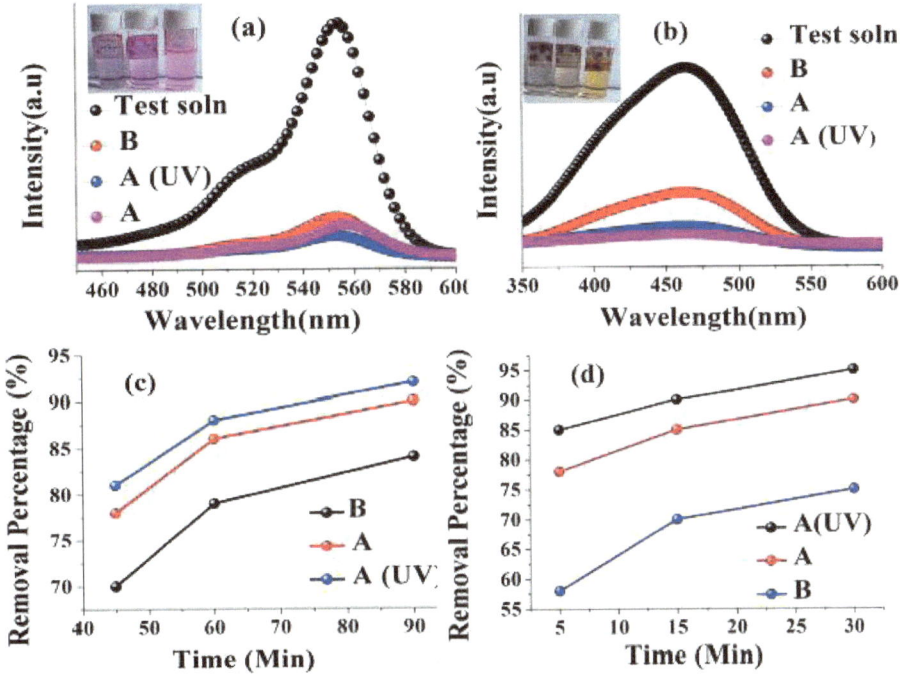

Fig. (8.5). The sample response towards different dyes (**a**) Rh-B; and (**b**) MO under different conditions (inset corresponding digital picture) and the associated removal efficiency (**c**) Rh-B; and (**d**) MO [68].

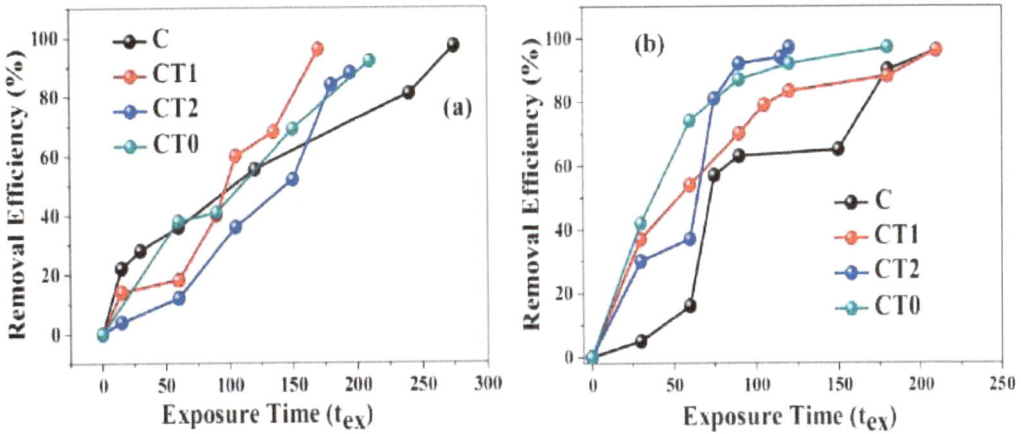

Fig. (8.6). Variation of different photo-catalytic efficiency with time for (**a**) MO and (**b**) RhB solution [69].

Oxide Nanostructure

Different oxide nanostructures have shown tremendous promise regarding the photo-induced removal of textile dyes from water. Oxide materials include TiO_2, ZnO, Fe_2O_3 CuO or others. Among which, the TiO_2 nanostructures are one of the most promising regarding the removal of textile dyes. ZnO also shows prominent potential in the same application. Thus this section will though discuss the responses of dyes to oxide structures, but the emphasis would be given on TiO_2 and ZnO as the remover of dyes from water. The basic structure and property information on both materials have been given in **chapter 5**. In recent years, interest in photocatalysis has focused on the use of semiconductor materials as photocatalysts for the removal of ambient concentrations of organic and inorganic species from aqueous or gas phase systems in environmental clean-up, drinking, water treatment, industrial and health applications. This is because of the ability of TiO_2 to oxidize organic and inorganic substrates in air and water through redox processes. In this context, TiO_2 has not only emerged as one of the most fascinating materials in both homogeneous and heterogeneous catalysis, but has also succeeded in engaging the attention of physical chemists, physicists, material scientists and engineers in exploring distinctive semiconducting and catalytic properties.

TiO_2 generally exists mainly in two forms, retail and anatase, both having a high band gap of 3 and 3.2 eV, respectively, and it is their ability to UV light that has made them extremely important in environment-related research [70] both in terms of photocatalysis induced contaminant removal as well as solar fuel production. Scheme **8.1** shows how the band gap excitation of TiO_2 results in charge separation and subsequent scavenging of electrons and holes by species adsorbed by the surface [71].

Doping of TiO_2 with low band gap material thus makes it suitable for being used as a photocatalyst and even the property may be fine-tuned to a particular application of suitable surface treatment. The application may be effective mineralization of different contaminants present in the air or others. However, the situation is not so easy when it is used for the removal of water contaminants mainly because of two reasons. First, it is a challenging task to make the material effectively in the visible region, and secondly, there may be issues like catalyst poisoning or deactivation. To make TiO_2 active and perform the degradation process, it has to be excited by an energy greater than its band gap, *i.e.*, 3.2 eV for anatase TiO_2, to be specific, meaning the excitation wavelength should be lesser than 387 nm. However, the natural sunlight contains only 5–8 % UV component and thus, it is not possible to achieve significant photocatalysis until a dedicated external UV source is used, which raises the cost concern. Thus the main goal of the current researchers is to develop a system where natural sunlight may be effectively used in the catalytic action. At present, significant enhancement has been achieved regarding the development of different metal oxide nanostructure-based catalysts with different shapes and sizes and thus with enhanced surface area. Not only that, tuning the optical gap by suitable doping now has become an effective tool that, in turn, selects the portion of the spectrum material is supposed to utilize. In the present discussion, doping of TiO_2 with foreign elements like V, Cr, Mn, Fe and Ni as well as Ag, Au and Ru causes redshifts of the band gap from UV to the visible region, increasing efficiency in the visible range solar cell. It has been observed that when TiO_2 gets exposed to a photon of a suitable wavelength, it acquires a strong oxidizing ability that in turn, decomposes organic contaminants present in water. TiO_2 has established itself to be one of the most efficient photocatalyst, and thus, other potential oxide materials like ZnO are lesser studied though they also are beneficial in some particular applications. To activate TiO_2, UVB has been found to be more effective than UVA, and sometimes activation is modified by adding other materials like nitrogen, carbon sulphur or other semiconductors. To avoid free movement of catalyst particles, sometimes catalyst in thin film form is also used [72].

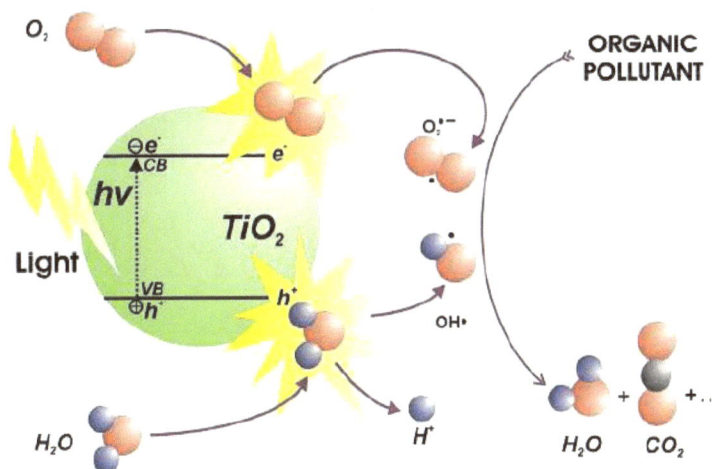

Scheme 8.1. Schematic of semiconductor excitation by band gap illumination leading to the creation of "electrons" in the conduction band and "holes" in the valance band [71].

As the Scheme **8.2** has shown, the activation of the catalyst results formation of electron hole pair.

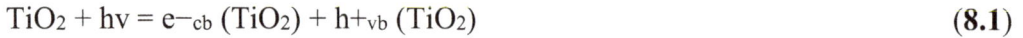

$$TiO_2 + hv = e^-_{cb} (TiO_2) + h^+_{vb} (TiO_2) \tag{8.1}$$

Here cb and vb are, respectively, the conduction and valence bands. This suggests that the catalyst may now be acted as an electron acceptor or donor for the molecules that come in contact with the catalyst. The generated electron-hole pair may undergo non-radiative recombination, releasing absorbed light energy as heat, or they can show a strong redox reaction as holes in VB and electrons in CB are respectively very strong oxidizing and reducing agents. These, when participate in the redox reaction with the adsorbed organic molecules along with water and hydroxyl group mineralization of the pollutant gets, occurred.

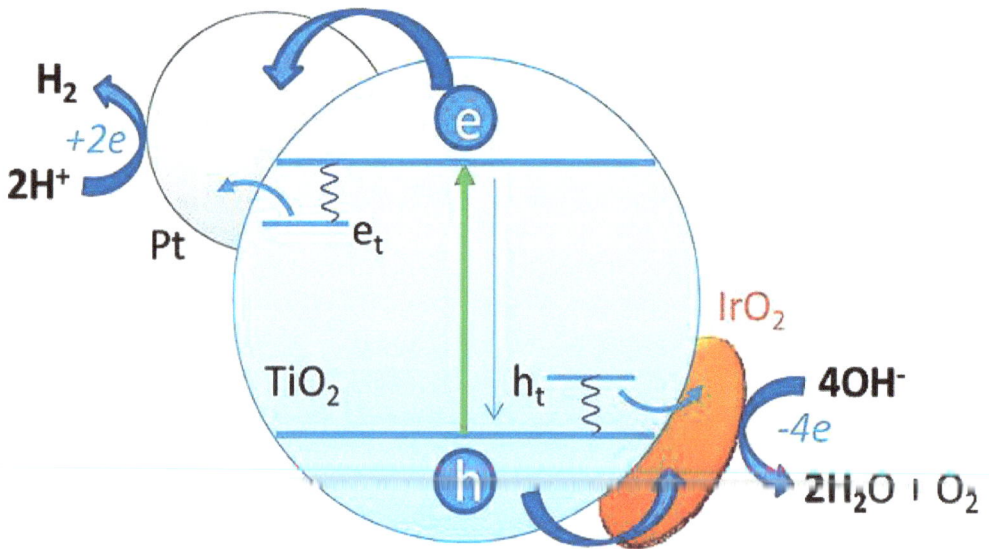

Scheme 8.2. of the charge transfer across semiconductor interface.

After a systematic study, it has been seen that any kinds of charges are able to react with the present contaminant, however as the relative abundance of water is much higher compared to the contaminates, the reaction with water is much expected to form hydroxyl radical, which is one of the most powerful oxidants. Apart from the as mentioned electron-hole pair, other useful water disinfectants are O_3, H_2O_2, HOCl. Cl, *etc.* As mentioned, OH radical is a very powerful decontaminants and is one of the main species formed in TiO_2 photocatalysis. Here conduction band electrons attack adsorbed oxygen, converting them to oxygen radicals, which in turn prevents immediate electron-hole recombination. The process ultimately increases the amount of oxygen radicals that, in turn, take part in contaminant degradation.

TiO_2 overall satisfy all the qualities of a good photo-catalyst, *i.e.*, photo-activity, inertness, photo-corrosion resistance, ability of energy harnessing in visible and UV region, cost-effective and environment-friendly [73] and thus, this semiconductor has remarkable popularity to be used as a water contaminant both in the form of suspension as well thin film. P-25 is one of the most effective and popular photo-catalyst that consist of 25% retail and 75% anatase phase of TiO_2 as it has been seen that Anatase phase is more stable as well as more efficient as photo-catalyst compared to retail phase. The conversion from anatase to rutile needs heating up the material at a temperature as high as 700 °C or above.

In current days, many semiconductors like TiO_2, WO_3, ZnO, *etc.* are being used as efficient photo-catalyst owing to the same properties mentioned before [74]. However, it is a proven fact that metal oxides are less efficient catalysts than noble metals but still are mostly used because of their strong resistance towards poisoning and deactivation.

If we compare the potential of a few popular materials as photo-catalyst, we will see that TiO_2 is no doubt the most potential materials due to the various reasons mentioned above; but a few other materials also show catalysis with some inherent disadvantages. For instance, binary metal sulfides like CdS and PbS are not very stable and more importantly, toxic. ZnO is also lacking stability. WO_3, though satisfies the other entire criterion, but it has a poor catalytic performance [75, 76]. Considering all, the combination of different materials and simultaneous application of two or more catalysts can offer more efficiency to the photo-catalytic activity over a wide spectrum range.

Combined studies of Teoh, Kamat and other on ZnO-based photo-catalyst has shown that though energetically ZnO shows almost the same characteristics as that of TiO_2, it suffers photo-induced corrosion in solution [77-79]. Another low-cost, highly available oxide material, *i.e.*, Hematite (α-Fe_2O_3), suffers from the drawbacks of very rapid electron-hole recombination and short carrier diffusion length. Thus, there has been an effort to overcome the drawback by modifying with suitable doping, mainly Cr, Mo or Si in order to improve the transport properties of the carrier [80].

There are considerable works on WO_3 having an optical band gap 2.7 eV but still suffering the problem of low electron concentration in the conduction band, however, this can be overcome by combining this with Pt nanoparticle. The compound system is now under the establishment of its position as an efficient single-phase photo-catalyst with very high visible light activity.

Silicon Nanowire

It is reported that the SiNWs prepared by 20% H_2O_2 etching solution exhibit the best activity in the decomposition of the target organic pollutant, Rhodamine B (RhB), under Xenon (Xe) arc lamp irradiation [81]. To enhance the photo-catalytic activity of the SiNWs, its surfaces were doped with Hydrogen fluoride [82], Platinum, Copper, Palladium, Gold [82, 83], Silver, Rhodium [82-84], *etc.* When SiNWs were doped with HF, hydrogen-terminated SiNWs (H-SiNWs) were obtained. The H-SiNWs enhance the performance of SiNW arrays by increasing their photo-catalytic activity. This was mainly due to the hydrogen atoms on the H-SiNWs surface, which produce much more •OH radicals and have high reductive activity. Also, they have high stability and catalytic activity with at least 10 recycle times. H-SiNWs were ideal photocatalysts for the selective hydroxylation of benzene into phenol and for methyl red decomposition [85]. When the SiNWs were doped with the platinum nanoparticle, they were used as effective photocatalysts for photo-catalytic degradation of organic dyes and toxic pollutants for organic waste treatment and environmental remediation [86]. When the SiNWs were decorated with copper nanoparticles (SiNWs–Cu), they exhibited excellent and stable activity for the catalytic reduction of 4-nitrophenol to 4-aminophenol by sodium borohydride in an aqueous solution. This novel catalyst also shows excellent catalytic performance for the degradation of organic dyes, such as methylene blue (MB) and rhodamine B (RhB). The degree of degradation for SiNWs–Cu was about 64% in comparison to 30% for H–SiNWs after 20 min [87]. Thus, SiNWs–Cu acted as an attractive catalyst, because it was highly efficient, cost-effective, eco-friendly and could replace noble metals for certain catalytic applications. When SiNWs were modified with Platinum and Gold, it degraded 95% of the methylene blue dye under UV light irradiation, whereas for H-SiNWs, it was only 91% degradation [82]. Both SiNWs and SiNWs doped with silver

(SiNWs–Ag) photocatalysts displayed a comparable activity, but remained less efficient than the H–SiNWs substrate.

In our laboratory, we are presently working on the performance of SiNW (both pure and doped) regarding the removal of different textile dyes, mainly rhodamine B. The SiNWs were synthesized *via* the conventional metal-assisted chemical etching method described in previous **Chapter 5.** The etching times were varied between 5 to almost 30 minutes, which in turn varied the aspect ratios of the sample. The samples were further modified by different materials like hydrofluoric acid (to make the sample hydrogen terminated) or with metals like silver and copper. The FESEM images of the samples are shown in Figs. (**8.7a-c** and **8.8a-c**). For an open-end CNT or even the interstitial area within the CNT bundle, the capillary condensation is supposed to take place if the pore size varies as per the Kelvin-Laplace equation [56].

Fig. (8.7). (a-c) FESEM Images of SiNWs 27mins modified with Cu with different magnification.

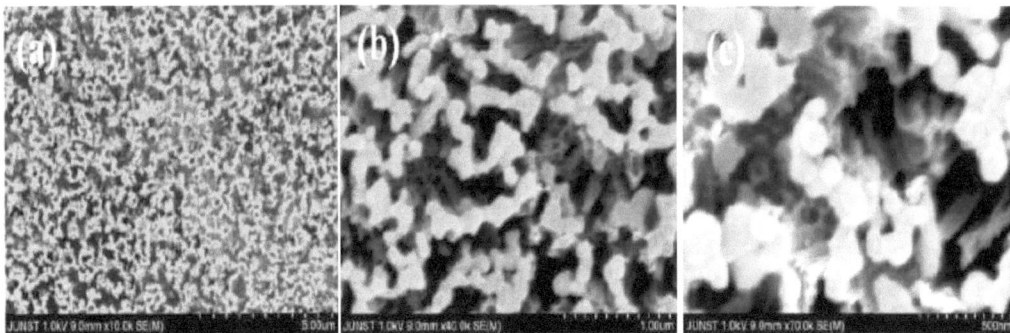

Fig. (8.8). (a-c) FESEM Images of SiNWs 27 mins modified with silver with different magnification.

The photo-catalytic performance was measured by the decay of the absorption of the dye as a function of irradiation time. Fig. (**8.9a-d**) shows the absorbance *versus* wavelength of oxygen terminated SiNWs etched for different times. From the graphs, it is clearly visible that the characteristic absorption band of RhB at 554 nm decreased significantly with increasing irradiation time.

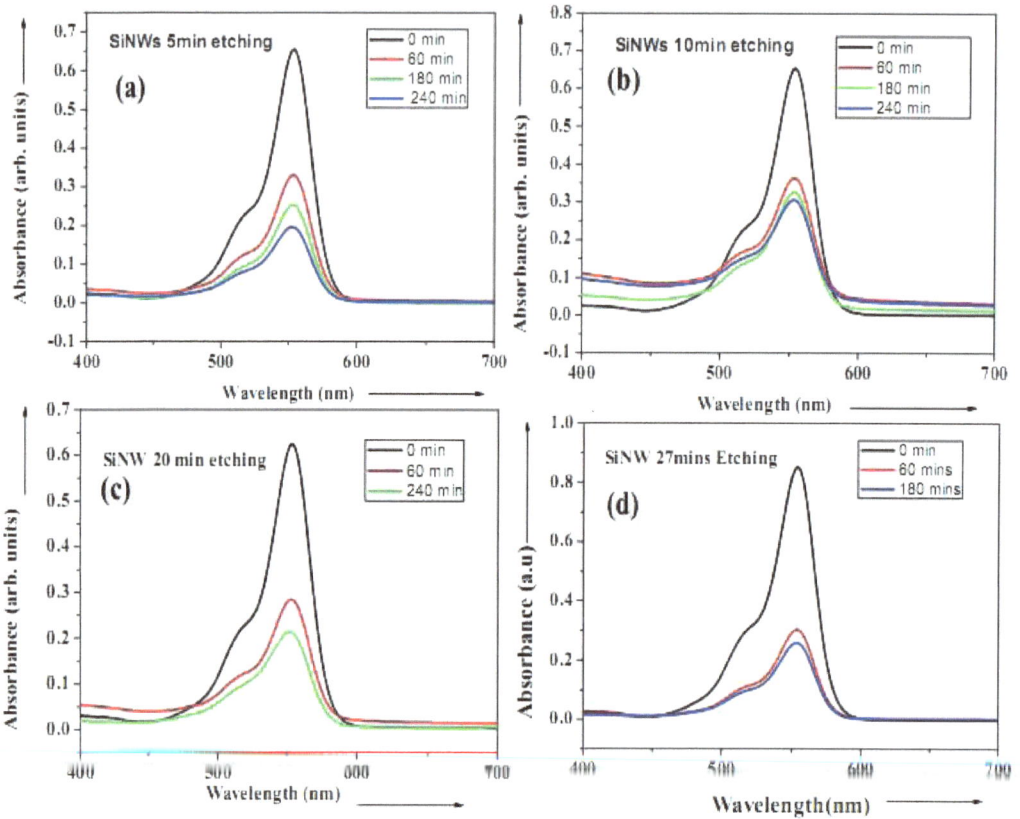

Fig. (8.9). Absorbance *vs* wavelength graph of SiNWs etched for (**a**) 5 mins, (**b**) 10 mins (**c**) 20 mins (**d**) 27 mins.

We have further examined the photo-catalytic activity of HF-treated samples as well as metal-modified samples, for which the result has been summarized in Fig. (**8.10a-f**).

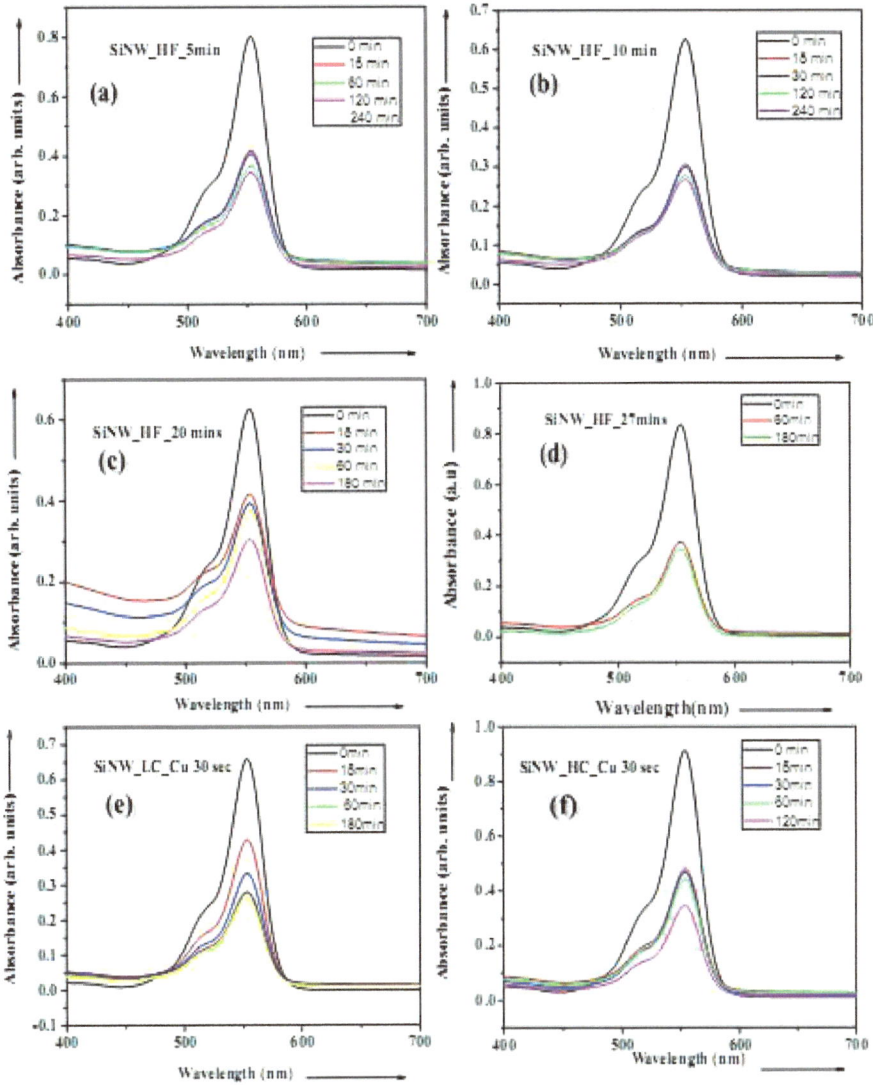

Fig. (8.10). Absorbance *vs* Wavelength graph of HF treated SiNWs etched for 5 mins (**b**) 10 mins (**c**) 20 mins and (**d**) 27 mins. (**e**) SiNWs modified with less (**e**) (LC_Cu) and high (**f**) (HC_Cu) concentration of copper treated for 30secs.

From the above figures, we observe that the absorption band of Rhodamine B (RhB) at 554 nm decreased significantly with increasing irradiation time. The results indicate that the photodegradation yield of the oxygen terminated samples of SiNWs substrate is much lower than that of the hydrogenated sample. Loading the SiNWs substrate with silver nanoparticles, the photodegradation was less when compared to the hydrogen-terminated samples. Modification of the SiNWs substrate with LC_Cu nanoparticles (SiNWs LC_Cu) did not cause any enhancement in the photo-catalytic efficiency, as compared to hydrogen-terminated SiNWs (H-SiNWs). However, when the concentration of the Cu nanoparticles was significantly increased, a significant increase in the photoactivity was obtained in the sample of SiNWs HC_Cu. The degree of degradation for SiNWs HC_Cu is about 65.2% in comparison to 56.1% for SiNWs LC_Cu. However, there are many reports showing much better results than this.

There are many reports where researchers have adopted a two-way approach where an initial theoretical study presumes the effectiveness of certain material as a remover of dyes and then experimentally verifies the fact. Or on the contrary, certain experimental facts have well been explained with the help of theoretical approaches. In this regards, few particular works are needed to be mentioned separately. For instance, Akber Raza and a co-worker took a machine-learning approach in order to predict the defluorination of Per- and Polyfluoroalkyl substance (PFAS) [88]. Through this approach, they very accurately predict carbon-fluorine bond dissociation energy. In the current year also, Zhang *et al.* in-depth discussed the utility of machine – learning approach to store and analyse environmental science and technology [89].

Regti *et al.* have shown an excellent agreement between their theoretical assumption and the experimental fact regarding the efficiency of activated carbon in removing different cationic dyes [90]. On the contrary, a combined theoretical and experimental study on removing anionic dyes by $ZnCl_2$ activated carbon was reported by Paredes-Laverde *et al.* [91]. Few other combined theoretical and experimental studies include the reports given by Kassimi *et al.*, Gonzalez *et al.* or Mohammed *et al.* [92-94]. In the year 2019, Elias and co-workers studied the photo-catalytic behaviour of Ce doped ZnO/CNT system, both theoretically as well as experimentally [95]. A relatively new system of non-calcined Mg-Al layered double hydroxide has been used in the adsorption-induced removal of both different anionic and cationic dyes by Aguiar *et al.* [96]. They have also found a good agreement between theoretical and experimental analysis. In the year 2018, de-Souza reported the adsorption of basic dyes by activated carbon [97] and speculated the possible reasons theoretically. Shen *et al.* reported a systematic combined study on the adsorption of basic dyes by organ-vermiculite [98]. The author's previous studies also combined theoretical and experimental studies have been done explaining chemical catalysis-induced removal of dyes by nickel-doped graphitic carbon nitride [31].

Here, the DFT study of the system has been done to get ideas regarding the bonding gap density of states, as well as the probable sites where the dopant may be accommodated.

The result is given in Fig. (**8.11**), where it is seen for g-CN nitrogen 2p 2s as well as carbon 2s from the valence band edge, whereas carbon 2p and nitrogen 2s are the main contributors towards the development of valence band edges. When the sample further gets doped with nickel, as shown in Figs. (**8.11b and c**), a number of states have been formed, which manifests in the corresponding changes in DOS. The DOS spectra show a considerable number of states formed within the forbidden gap.

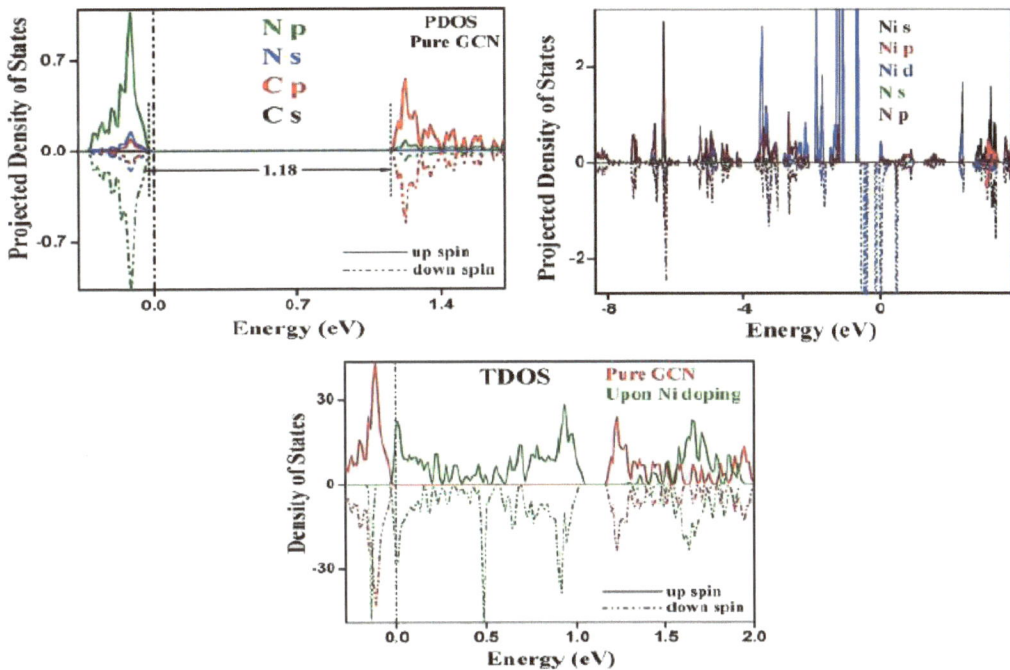

Fig. (8.11). Projected density of states plot for (**a**) Pure GCN and (**b**) magnified PDOS plot for doped g-CN (**c**) total density of states plot for pristine and doped g-CN.

CONCLUSION

In this chapter, we have discussed the photo-catalytic and adsorption properties of a few established materials like graphitic carbon nitride, carbon nanotubes, and TiO$_2$, as well as a few emerging materials like carbon microspheres, silicon nanowires, *etc.* The chapter shows a few interesting results as having been reported by the other workers as well as done from the author's own lab. This chapter shows the numerical values regarding the time constant, removal time, efficiency, *etc.* of all the materials discussed.

It has also been shown that TiO$_2$ is probably the most efficient photo-catalyst in the oxide family, followed by ZnO and others. Carbon-based material is also a promising candidate in this application point of view. The carbon-based materials that show potential for the removal of dyes include carbon nanotubes, graphene, and graphitic carbon nitride. It has been shown that all these materials show considerable potential regarding the removal efficiencies in the purest form and the performances become considerably enhanced when doped or made hybrid with other elements.

REFERENCES

[1]   Zhu, J.; Xiao, P.; Li, H.; Carabineiro, S.A.C. Graphitic carbon nitride: synthesis, properties, and applications in catalysis. *ACS Appl. Mater. Interfaces,* **2014**, *6*(19), 16449-16465.
      http://dx.doi.org/10.1021/am502925j PMID: 25215903

[2]   Das, D.; Banerjee, D.; Pahari, D.; Ghorai, U.K.; Sarkar, S.; Das, N.S.; Chattopadhyay, K.K. Defect induced tuning of photoluminescence property in graphitic carbon nitride nanosheets through synthesis conditions. *J. Lumin.,* **2017**, *185*, 155-165.
      http://dx.doi.org/10.1016/j.jlumin.2017.01.007

[3]   Sun, S.; Liang, S. Recent advances in functional mesoporous graphitic carbon nitride (mpg C$_3$N$_4$) polymers. *Nanoscale,* **2017**, *9*(30), 10544-10578.
      http://dx.doi.org/10.1039/C7NR03656F PMID: 28726962

[4]   Paul, D.R.; Sharma, R.; Nehra, S.P.; Sharma, A. Effect of calcination temperature, pH and catalyst loading on photodegradation efficiency of urea derived graphitic carbon nitride towards methylene blue dye solution. *RSC Advances,* **2019**, *9*(27), 15381-15391.
      http://dx.doi.org/10.1039/C9RA02201E PMID: 35514817

[5]   Paul, D.R. Nehra, S. P. Graphitic carbon nitride: a sustainable photocatalyst for organic pollutant degradation and antibacterial applications. *Environ. Sci. Pollut. Res. Int.,* **2020**.

[6]   Gu, S.; Xie, J.; Li, C.M. Hierarchically porous graphitic carbon nitride: large-scale facile synthesis and its application toward photocatalytic dye degradation. *RSC Advances,* **2014**, *4*(103), 59436-59439.
      http://dx.doi.org/10.1039/C4RA10958A

[7]   Erdogan, D.A.; Sevim, M.; Kısa, E.; Emiroglu, D.B.; Karatok, M.; Vovk, E.I.; Bjerring, M.; Akbey, Ü.; Metin, Ö.; Ozensoy, E. Photocatalytic activity of mesoporous graphitic carbon nitride (mpg-C_3N_4) towards organic chromophores under UV and VIS light illumination. *Top. Catal.,* **2016**, *59*(15-16), 1305-1318.
      http://dx.doi.org/10.1007/s11244-016-0654-3

[8]   Cui, Y.; Tang, Y.; Wang, X. Template-free synthesis of graphitic carbon nitride hollow spheres for photocatalytic degradation of organic pollutants. *Mater. Lett.,* **2015**, *161*, 197-200.
      http://dx.doi.org/10.1016/j.matlet.2015.08.106

[9]   Shi, L.; Liang, L.; Wang, F.; Liu, M.; Zhong, S.; Sun, J. Tetraethylorthosilicate induced preparation of mesoporous graphitic carbon nitride with improved visible light photocatalytic activity. *Catal. Commun.,* **2015**, *59*, 131-135.
      http://dx.doi.org/10.1016/j.catcom.2014.10.014

[10] Gao, X.; Jiao, X.; Zhang, L.; Zhu, W.; Xu, X.; Ma, H.; Chen, T. Cosolvent-free nanocasting synthesis of ordered mesoporous g-C_3N_4 and its remarkable photocatalytic activity for methyl orange degradation. *RSC Advances,* **2015**, *5*(94), 76963-76972.
      http://dx.doi.org/10.1039/C5RA13438B

[11] Wang, C.; Fan, H.; Ren, X.; Fang, J.; Ma, J.; Zhao, N. Porous graphitic carbon nitride nanosheets by pre-polymerization for enhanced photocatalysis. *Mater. Charact.,* **2018**, *139*, 89-99.
      http://dx.doi.org/10.1016/j.matchar.2018.02.036

[12] Liu, J.; Zhang, T.; Wang, Z.; Dawson, G.; Chen, W. Simple pyrolysis of urea into graphitic carbon nitride with recyclable adsorption and photocatalytic activity. *J. Mater. Chem.,* **2011**, *21*(38), 14398-14401.
      http://dx.doi.org/10.1039/c1jm12620b

[13] Chang, F.; Li, C.; Luo, J.; Xie, Y.; Deng, B.; Hu, X. Enhanced visible-light-driven photocatalytic performance of porous graphitic carbon nitride. *Appl. Surf. Sci.,* **2015**, *358*, 270-277.
      http://dx.doi.org/10.1016/j.apsusc.2015.08.124

[14] Ibad, M.F.; Kosslick, H.; Tomm, J.W.; Frank, M.; Schulz, A. Impact of the crystallinity of mesoporous polymeric graphitic carbon nitride on the photocatalytic performance under UV and visible light. *Microporous Mesoporous Mater.,* **2017**, *254*, 136-145.
      http://dx.doi.org/10.1016/j.micromeso.2017.04.052

[15] Li, X.; Zhang, H.; Huang, J.; Luo, J.; Feng, Z.; Wang, X. Folded nano-porous graphene-like carbon nitride with significantly improved visible-light photocatalytic activity for dye degradation. *Ceram. Int.,* **2017**, *43*(17), 15785-15792.
      http://dx.doi.org/10.1016/j.ceramint.2017.08.144

[16]   Pawar, R.C.; Kang, S.; Park, J.II.; Kim, J.; Ahn, S.; Lee, C.S. Room-temperature synthesis of nanoporous 1D microrods of graphitic carbon nitride (g-C$_3$N$_4$) with highly enhanced photocatalytic activity and stability. *Sci. Rep.,* **2016**, *6*(1), 31147.
http://dx.doi.org/10.1038/srep31147 PMID: 27498979

[17]   Zhang, X.S.; Hu, J.Y.; Jiang, H. Facile modification of a graphitic carbon nitride catalyst to improve its photoreactivity under visible light irradiation. *Chem. Eng. J.,* **2014**, *256*, 230-237.
http://dx.doi.org/10.1016/j.cej.2014.07.012

[18]   Dong, F.; Sun, Y.; Wu, L.; Fu, M.; Wu, Z. Facile transformation of low cost thiourea into nitrogen-rich graphitic carbon nitride nanocatalyst with high visible light photocatalytic performance. *Catal. Sci. Technol.,* **2012**, *2*(7), 1332-1335.
http://dx.doi.org/10.1039/c2cy20049j

[19]   Cui, Y.; Huang, J.; Fu, X.; Wang, X. Metal-free photocatalytic degradation of 4-chlorophenol in water by mesoporous carbon nitride semiconductors. *Catal. Sci. Technol.,* **2012**, *2*(7), 1396-1402.
http://dx.doi.org/10.1039/c2cy20036h

[20]   Huang, Z.; Li, F.; Chen, B.; Yuan, G. Nanosheets of graphitic carbon nitride as metal-free environmental photocatalysts. *Catal. Sci. Technol.,* **2014**, *4*(12), 4258-4264.
http://dx.doi.org/10.1039/C4CY00832D

[21]   Yan, J.; Zhou, C.; Li, P.; Chen, B.; Zhang, S.; Dong, X.; Xi, F.; Liu, J. Nitrogen-rich graphitic carbon nitride: Controllable nanosheet-like morphology, enhanced visible light absorption and superior photocatalytic performance. *Colloids Surf. A Physicochem. Eng. Asp.,* **2016**, *508*, 257-264.
http://dx.doi.org/10.1016/j.colsurfa.2016.08.067

[22]   Sundaram, I.M.; Kalimuthu, S.; Gomathi priya, P. Metal-free heterojunction of graphitic carbon nitride composite with superior and stable visible-light active photocatalysis. *Mater. Chem. Phys.,* **2018**, *204*, 243-250.
http://dx.doi.org/10.1016/j.matchemphys.2017.10.041

[23]   Liu, S.; Li, D.; Sun, H.; Ang, H.M.; Tadé, M.O.; Wang, S. Oxygen functional groups in graphitic carbon nitride for enhanced photocatalysis. *J. Colloid Interface Sci.,* **2016**, *468*, 176-182.
http://dx.doi.org/10.1016/j.jcis.2016.01.051 PMID: 26845029

[24]   Das, D.; Banerjee, D.; Das, B.; Das, N.S.; Chattopadhyay, K.K. Effect of cobalt doping into graphitic carbon nitride on photo induced removal of dye from water. *Mater. Res. Bull.,* **2017**, *89*, 170-179.
http://dx.doi.org/10.1016/j.materresbull.2017.01.034

[25]   Paul, T.; Das, D.; Das, B.K.; Sarkar, S.; Maiti, S.; Chattopadhyay, K.K. CsPbBrCl$_2$/g-C$_3$N$_4$ type II heterojunction as efficient visible range photocatalyst. *J. Hazard. Mater.,* **2019**, *380*120855
http://dx.doi.org/10.1016/j.jhazmat.2019.120855 PMID: 31325693

[26]   Ye, M.Y.; Zhao, Z.H.; Hu, Z.F.; Liu, L.Q.; Ji, H.M.; Shen, Z.R.; Ma, T.Y. 0D/2D heterojunctions of vanadate quantum dots/graphitic carbon nitride nanosheets for enhanced visible light driven photocatalysis. *Angew. Chem. Int. Ed.,* **2017**, *56*(29), 8407-8411.
        http://dx.doi.org/10.1002/anie.201611127 PMID: 28052568

[27]   Tonda, S.; Kumar, S.; Kandula, S.; Shanker, V. Fe-doped and -mediated graphitic carbon nitride nanosheets for enhanced photocatalytic performance under natural sunlight. *J. Mater. Chem. A Mater. Energy Sustain.,* **2014**, *2*(19), 6772-6780.
        http://dx.doi.org/10.1039/c3ta15358d

[28]   Veisi, H.; Kazemi, S.; Mohammadi, P.; Safarimehr, P.; Hemmati, S. Catalytic reduction of 4-nitrophenol over Ag nanoparticles immobilized on Stachys lavandulifolia extract-modified multi walled carbon nanotubes. *Polyhedron,* **2019**, *157*, 232-240.
        http://dx.doi.org/10.1016/j.poly.2018.10.014

[29]   Li, J.; Liu, C.; Liu, Y. Au/graphene hydrogel: synthesis, characterization and its use for catalytic reduction of 4-nitrophenol. *J. Mater. Chem.,* **2012**, *22*(17), 8426-8430.
        http://dx.doi.org/10.1039/c2jm16386a

[30]   Zhang, P.; Shao, C.; Zhang, Z.; Zhang, M.; Mu, J.; Guo, Z.; Liu, Y. *In situ* assembly of well-dispersed Ag nanoparticles (AgNPs) on electrospun carbon nanofibers (CNFs) for catalytic reduction of 4-nitrophenol. *Nanoscale,* **2011**, *3*(8), 3357-3363.
        http://dx.doi.org/10.1039/c1nr10405e PMID: 21761072

[31]   Das, D.; Banerjee, D.; Mondal, M.; Shett, A.; Das, B.; Das, N.S.; Ghorai, U.K.; Chattopadhyay, K.K. Nickel doped graphitic carbon nitride nanosheets and its application for dye degradation by chemical catalysis. *Mater. Res. Bull.,* **2018**, *101*, 291-304.
        http://dx.doi.org/10.1016/j.materresbull.2018.02.004

[32]   Mitra, A.; Howli, P.; Sen, D.; Das, B.; Chattopadhyay, K.K. Cu_2O/g-C_3N_4 nanocomposites: an insight into the band structure tuning and catalytic efficiencies. *Nanoscale,* **2016**, *8*(45), 19099-19109.
        http://dx.doi.org/10.1039/C6NR06837E PMID: 27824200

[33]   Mohammadi, P.; Heravi, M.M.; Sadjadi, S. Green synthesis of Ag NPs on magnetic polyallylamine decorated g-C_3N_4 by Heracleum persicum extract: efficient catalyst for reduction of dyes. *Sci. Rep.,* **2020**, *10*(1), 6579.
        http://dx.doi.org/10.1038/s41598-020-63756-4 PMID: 31913322

[34]   Sarangapany, S.; Mohanty, K. Facile Green Synthesis of Ag@gC_3N_4 for Enhanced Photocatalytic and Catalytic Degradation of Organic Pollutant. *J. Cluster Sci.,* **2020**.

[35]   Fu, Y.; Huang, T.; Zhang, L.; Zhu, J.; Wang, X. Ag/g-C_3N_4 catalyst with superior catalytic performance for the degradation of dyes: a borohydride-generated superoxide radical approach. *Nanoscale,* **2015**, *7*(32), 13723-13733.
        http://dx.doi.org/10.1039/C5NR03260A PMID: 26220662

[36]   Kumar, Y.; Rani, S.; Shabir, J.; Kumar, L.S. Nitrogen-Rich and Porous Graphitic Carbon Nitride Nanosheet-Immobilized Palladium Nanoparticles as Highly Active and Recyclable Catalysts for the Reduction of Nitro Compounds and Degradation of Organic Dyes. *ACS Omega,* **2020**, *5*(22), 13250-13258.

http://dx.doi.org/10.1021/acsomega.0c01280

[37]   Murugan, E.; Santhosh Kumar, S.; Reshna, K.M.; Govindaraju, S. Highly sensitive, stable g-CN decorated with AgNPs for SERS sensing of toluidine blue and catalytic reduction of crystal violet. *J. Mater. Sci.,* **2019,** *54*(7), 5294-5310.

http://dx.doi.org/10.1007/s10853-018-3184-5

[38]   Mohanty, L.; Dash, S.K. Adsorptive Removal Of MB Dye By Graphitic-C_3N_4 From Industrial Effluents. *Int. J. Sci. Technol. Res.,* **2020,** *9*(03), 2029-2034.

[39]   Zhu, B.; Xia, P.; Ho, W.; Yu, J. Isoelectric point and adsorption activity of porous g-C_3N_4. *Appl. Surf. Sci.,* **2015,** *344*, 188-195.

http://dx.doi.org/10.1016/j.apsusc.2015.03.086

[40]   Fronczak, M.; Krajewska, M.; Demby, K.; Bystrzejewski, M. Extraordinary adsorption of methyl blue onto sodium-doped graphitic carbon nitride. *J. Phys. Chem. C,* **2017,** *121*(29), 15756-15766.

http://dx.doi.org/10.1021/acs.jpcc.7b03674

[41]   Yousefi, M.; Villar-Rodil, S.; Paredes, J.I.; Moshfegh, A.Z. Oxidized graphitic carbon nitride nanosheets as an effective adsorbent for organic dyes and tetracycline for water remediation. *J. Alloys Compd.,* **2019,** *809*151783

http://dx.doi.org/10.1016/j.jallcom.2019.151783

[42]   Bhowmik, T.; Kundu, M.K.; Barman, S. Ultra small gold nanoparticles–graphitic carbon nitride composite: an efficient catalyst for ultrafast reduction of 4-nitrophenol and removal of organic dyes from water. *RSC Advances,* **2015,** *5*(48), 38760-38773.

http://dx.doi.org/10.1039/C5RA04913J

[43]   Zou, Y.; Wang, X.; Ai, Y.; Liu, Y.; Ji, Y.; Wang, H.; Hayat, T.; Alsaedi, A.; Hu, W.; Wang, X. β-Cyclodextrin modified graphitic carbon nitride for the removal of pollutants from aqueous solution: experimental and theoretical calculation study. *J. Mater. Chem. A Mater. Energy Sustain.,* **2016,** *4*(37), 14170-14179.

http://dx.doi.org/10.1039/C6TA05958A

[44]   Zhang, L.; Li, L.; Sun, X.; Liu, P.; Yang, D.; Zhao, X. ZnO-layered double hydroxide@ graphitic carbon nitride composite for consecutive adsorption and photodegradation of dyes under UV and visible lights. *Materials (Basel),* **2016,** *9*(11), 927.

http://dx.doi.org/10.3390/ma9110927 PMID: 28774047

[45]   Ren, B.; Xu, Y.; Zhang, L.; Liu, Z. Carbon-doped graphitic carbon nitride as environment-benign adsorbent for methylene blue adsorption: Kinetics, isotherm and thermodynamics study. *J. Taiwan Inst. Chem. Eng.,* **2018,** *88*, 114-120.

http://dx.doi.org/10.1016/j.jtice.2018.03.041

[46]   Xie, M.; Wei, W.; Jiang, Z.; Xu, Y.; Xie, J. Carbon nitride nanowires/nanofibers: A novel template-free synthesis from a cyanuric chloride–melamine precursor towards enhanced adsorption and visible-light photocatalytic performance. *Ceram. Int.,* **2016,** *42*(3), 4158-4170.

http://dx.doi.org/10.1016/j.ceramint.2015.11.089

[47]   Fronczak, M.; Demby, K.; Strachowski, P.; Strawski, M.; Bystrzejewski, M. Graphitic carbon nitride doped with the s-block metals: adsorbent for the removal of methyl blue and copper (II) ions. *Langmuir,* **2018**, *34*(25), 7272-7283.
http://dx.doi.org/10.1021/acs.langmuir.8b01041 PMID: 29856628

[48]   Gan, Q.; Shi, W.; Xing, Y.; Hou, Y. A polyoxoniobate/g-C$_3$N$_4$ nanoporous material with high adsorption capacity of methylene blue from aqueous solution. *Front Chem.,* **2018**, *6*, 7.
http://dx.doi.org/10.3389/fchem.2018.00007 PMID: 29445725

[49]   Lu, P.; Hu, X.; Li, Y.; Zhang, M.; Liu, X.; He, Y.; Dong, F.; Fu, M.; Zhang, Z. One-step preparation of a novel SrCO$_3$/g-C$_3$N$_4$ nano-composite and its application in selective adsorption of crystal violet. *RSC Advances,* **2018**, *8*(12), 6315-6325.
http://dx.doi.org/10.1039/C7RA11565B PMID: 35540413

[50]   Chegeni, M.; Dehghan, N. Preparation of Phosphorus Doped Graphitic Carbon Nitride Using a Simple Method and Its Application for Removing Methylene Blue. *Phys. Chem. Res.,* **2020**, *8*(1), 31-44.

[51]   Xu, L.; Gu, D.; Chang, X.; Chai, L.; Li, Z.; Jin, X.; Sun, S. Adsorption and photocatalytic study of dye degradation over the g-C$_3$N$_4$/W$_{18}$O$_{49}$ nanocomposite. *Micro & Nano Lett.,* **2018**, *13*(4), 541-545
http://dx.doi.org/10.1049/mnl.2017.0719

[52]   Yang, H.C.; Chao, M.W.; Chou, C.J.; Wang, K.H.; Hu, C. Mushroom waste-derived g-C$_3$N$_4$ for methyl blue adsorption and cytotoxic test for Chinese hamster ovary cells. *Mater. Chem. Phys.,* **2020**, *244*122715
http://dx.doi.org/10.1016/j.matchemphys.2020.122715

[53]   Xu, C.; Wang, J.; Gao, B.; Dou, M.; Chen, R. Synergistic adsorption and visible-light catalytic degradation of RhB from recyclable 3D mesoporous graphitic carbon nitride/reduced graphene oxide aerogels. *J. Mater. Sci.,* **2019**, *54*(12), 8892-8906.
http://dx.doi.org/10.1007/s10853-019-03531-7

[54]   Grujicic, M.; Cao, G.; Gersten, B. Reactor length-scale modeling of chemical vapor deposition of carbon nanotubes. *J. Mater. Sci.,* **2003**, *38*(8), 1819-1830.
http://dx.doi.org/10.1023/A:1023252432202

[55]   Pan, B.; Xing, B. Adsorption mechanisms of organic chemicals on carbon nanotubes. *Environ. Sci. Technol.,* **2008**, *42*(24), 9005-9013.
http://dx.doi.org/10.1021/es801777n PMID: 19174865

[56]   Yudasaka, M.; Fan, J.; Miyawaki, J.; Iijima, S. Studies on the adsorption of organic materials inside thick carbon nanotubes. *J. Phys. Chem. B,* **2005**, *109*(18), 8909-8913.
http://dx.doi.org/10.1021/jp050980f PMID: 16852059

[57]   Chen, W.; Duan, L.; Zhu, D. Adsorption of polar and nonpolar organic chemicals to carbon nanotubes. *Environ. Sci. Technol.,* **2007**, *41*(24), 8295-8300.
http://dx.doi.org/10.1021/es071230h PMID: 18200854

[58] Lin, D.; Xing, B. Adsorption of phenolic compounds by carbon nanotubes: role of aromaticity and substitution of hydroxyl groups. *Environ. Sci. Technol.,* **2008**, *42*(19), 7254-7259.

http://dx.doi.org/10.1021/es801297u PMID: 18939555

[59] Sumanasekera, G.U.; Pradhan, B.K.; Romero, H.E.; Adu, K.W.; Eklund, P.C. Giant thermopower effects from molecular physisorption on carbon nanotubes. *Phys. Rev. Lett.,* **2002**, *89*(16)166801

http://dx.doi.org/10.1103/PhysRevLett.89.166801 PMID: 12398745

[60] Stafiej, A.; Pyrzynska, K. Adsorption of heavy metal ions with carbon nanotubes. *Separ. Purif. Tech.,* **2007**, *58*(1), 49-52.

http://dx.doi.org/10.1016/j.seppur.2007.07.008

[61] Simonyan, V.V.; Johnson, J.K.; Kuznetsova, A.; Yates, J.T., Jr Molecular simulation of xenon adsorption on single-walled carbon nanotubes. *J. Chem. Phys.,* **2001**, *114*(9), 4180-4185.

http://dx.doi.org/10.1063/1.1344234

[62] Yao, Y.; Xu, F.; Chen, M.; Xu, Z.; Zhu, Z. Adsorption behavior of methylene blue on carbon nanotubes. *Bioresour. Technol.,* **2010**, *101*(9), 3040-3046.

http://dx.doi.org/10.1016/j.biortech.2009.12.042 PMID: 20060712

[63] Shahryari, Z.; Goharrizi, A.S.; Azadi, M. Experimental study of methylene blue adsorption from aqueous solutions onto carbon nano tubes. *International Journal of Water Resources and Environmental Engineering,* **2010**, *2147483647*(2), 016-028.

[64] Rodríguez, A.; Ovejero, G.; Sotelo, J.L.; Mestanza, M.; García, J. Adsorption of dyes on carbon nanomaterials from aqueous solutions. *J. Environ. Sci. Health Part A Tox. Hazard. Subst. Environ. Eng.,* **2010**, *45*(12), 1642-1653.

http://dx.doi.org/10.1080/10934529.2010.506137 PMID: 20730657

[65] Liu, Y.; Cui, G.; Luo, C.; Zhang, L.; Guo, Y.; Yan, S. Synthesis, characterization and application of amino-functionalized multi-walled carbon nanotubes for effective fast removal of methyl orange from aqueous solution. *RSC Advances,* **2014**, *4*(98), 55162-55172.

http://dx.doi.org/10.1039/C4RA10047F

[66] Wang, L.; Zhang, M.; Huang, Z.; Zhao, C.; Pei, X. Adsorption Mechanism of Activated Carbon Fibre/Carbon Nanotube Composites for Rhodamine B from Aqueous Solution. *Asian J. Chem.,* **2013**, *25*(18), 10509-10514.

http://dx.doi.org/10.14233/ajchem.2013.15769

[67] Banerjee, D.; Bhowmick, P.; Pahari, D.; Santra, S.; Sarkar, S.; Das, B.; Chattopadhyay, K.K. Pseudo first ordered adsorption of noxious textile dyes by low-temperature synthesized amorphous carbon nanotubes. *Physica E,* **2017**, *87*, 68-76.

http://dx.doi.org/10.1016/j.physe.2016.11.024

[68] Dutta, A.K.; Ghorai, U.K.; Chattopadhyay, K.K.; Banerjee, D. Removal of textile dyes by carbon nanotubes: A comparison between adsorption and UV assisted photocatalysis. *Physica E,* **2018**, *99*, 6-15.

http://dx.doi.org/10.1016/j.physe.2018.01.008

[69] Kumar, D.; Banerjee, D.; Sarkar, S.; Das, N.S.; Chattopadhyay, K.K. Easy synthesis of porous carbon mesospheres and its functionalization with titania nanoparticles for enhanced field emission and photocatalytic activity. *Mater. Chem. Phys.,* **2016,** *175,* 22-32.

http://dx.doi.org/10.1016/j.matchemphys.2016.02.002

[70] Cho, M.; Chung, H.; Choi, W.; Yoon, J. Linear correlation between inactivation of E. coli and OH radical concentration in TiO2 photocatalytic disinfection. *Water Res.,* **2004,** *38*(4), 1069-1077.

http://dx.doi.org/10.1016/j.watres.2003.10.029 PMID: 14769428

[71] Bhadon, A.O.; Fitzpatrick, P. Heterogeneous photocatalysis: recent advances and applications. *Catalysts,* **2013,** *3*(1), 189-218.

[72] Sunada, K.; Watanabe, T.; Hashimoto, K. Studies on photokilling of bacteria on TiO2 thin film. *J. Photochem. Photobiol. Chem.,* **2003,** *156*(1-3), 227-233.

http://dx.doi.org/10.1016/S1010-6030(02)00434-3

[73] Al-Rasheed, R.A. Water treatment by heterogeneous photocatalysis: an review. *Proceedings of the 4th SWCC Acquired Experience Symposium,* Jeddah, Saudi Arabia**2005.**

[74] Genscher, H. Electrochemical behavior of semiconductors under illumination. *J. Electrochem. Soc.,* **1966,** *113*(11), 1174.

http://dx.doi.org/10.1149/1.2423779

[75] Minero, C.; Pelizzetti, E.; Sega, M.; Friberg, S.E.; Sjö blom, J. The role of humic substances in the photo-catalytic degradation of water contaminants. *J. Dispers. Sci. Technol.,* **1999,** *20*(1-2), 643-661.

http://dx.doi.org/10.1080/01932699908943812

[76] Eggleston, C.M. Geochemistry. Toward new uses for hematite. *Science,* **2008,** *320*(5873), 184-185.

http://dx.doi.org/10.1126/science.1157189 PMID: 18403697

[77] Teoh, W.Y.; Denny, F.; Amal, R.; Friedmann, D.; Mädler, L.; Pratsinis, S.E. Photocatalytic mineralisation of organic compounds: a comparison of flame-made TiO2 catalysts. *Top. Catal.,* **2007,** *44*(4), 489-497.

http://dx.doi.org/10.1007/s11244-006-0096-4

[78] Teoh, W.; Mädler, L.; Amal, R. Inter-relationship between Pt oxidation states on TiO2 and the photocatalytic mineralisation of organic matters. *J. Catal.,* **2007,** *251*(2), 271-280.

http://dx.doi.org/10.1016/j.jcat.2007.08.008

[79] Teoh, W.Y.; Mädler, L.; Beydoun, D.; Pratsinis, S.E.; Amal, R. Direct (one-step) synthesis of TiO2 and Pt/TiO2 nanoparticles for photo-catalytic mineralisation of sucrose. *Chem. Eng. Sci.,* **2005,** *60*(21), 5852-5861.

http://dx.doi.org/10.1016/j.ces.2005.05.037

[80] Tran, H.; Chiang, K.; Scott, J.; Amal, R. Understanding selective enhancement by silver during photocatalytic oxidation. *Photochem. Photobiol. Sci.,* **2005,** *4*(8), 565-567.

http://dx.doi.org/10.1039/b506320e PMID: 16052260

[81] Lian, S.; Tsang, C.H.A.; Kang, Z.; Liu, Y.; Wong, N.; Lee, S.T. Hydrogen-terminated silicon nanowire photocatalysis: Benzene oxidation and methyl red decomposition. *Mater. Res. Bull.,* **2011**, *46*(12), 2441-2444.
http://dx.doi.org/10.1016/j.materresbull.2011.08.027

[82] Qu, Y.; Zhong, X.; Li, Y.; Liao, L.; Huang, Y.; Duan, X. Photocatalytic properties of porous silicon nanowires. *J. Mater. Chem.,* **2010**, *20*(18), 3590-3594.
http://dx.doi.org/10.1039/c0jm00493f PMID: 22190767

[83] Megouda, N.; Cofininier, Y.; Szunerits, S.; Hadjersi, T.; ElKechai, O.; Boukherroub, R. Photocatalytic activity of silicon nanowires under UV and visible light irradiation. *Chem. Commun. (Camb.),* **2011**, *47*(3), 991-993.
http://dx.doi.org/10.1039/C0CC04250A PMID: 21113518

[84] Yang, X.; Zhong, H.; Zhu, Y.; Jiang, H.; Shen, J.; Huang, J.; Li, C. Highly efficient reusable catalyst based on silicon nanowire arrays decorated with copper nanoparticles. *J. Mater. Chem. A Mater. Energy Sustain.,* **2014**, *2*(24), 9040-9047.
http://dx.doi.org/10.1039/c4ta00119b

[85] Brahiti, N.; Hadjersi, T.; Menari, H. 2014. Photo-catalytic degradation of Methylene blue by modified porous silicon nanowires. *Journal of New Technology and Materials,* **1747**, *277*, 1-4.

[86] Liu, Y.; Ji, G.; Wang, J.; Liang, X.; Zuo, Z.; Shi, Y. Fabrication and photocatalytic properties of silicon nanowires by metal-assisted chemical etching: effect of H2O2 concentration. *Nanoscale Res. Lett.,* **2012**, *7*(1), 663.
http://dx.doi.org/10.1186/1556-276X-7-663 PMID: 22214494

[87] Dawood, M.K.; Zheng, H.; Liew, T.H.; Leong, K.C.; Foo, Y.L.; Rajagopalan, R.; Khan, S.A.; Choi, W.K. Mimicking both petal and lotus effects on a single silicon substrate by tuning the wettability of nanostructured surfaces. *Langmuir,* **2011**, *27*(7), 4126-4133.
http://dx.doi.org/10.1021/la1050783 PMID: 21355585

[88] Raza, A.; Bardhan, S.; Xu, L.; Yamijala, S.S.R.K.C.; Lian, C.; Kwon, H.; Wong, B.M. A machine learning approach for predicting defluorination of per-and polyfluoroalkyl substances (PFAS) for their efficient treatment and removal. *Environ. Sci. Technol. Lett.,* **2019**, *6*(10), 624-629.
http://dx.doi.org/10.1021/acs.estlett.9b00476

[89] Zhong, S.; Zhang, K.; Bagheri, M.; Burken, J.G.; Gu, A.; Li, B.; Ma, X.; Marrone, B.L.; Ren, Z.J.; Schrier, J.; Shi, W.; Tan, H.; Wang, T.; Wang, X.; Wong, B.M.; Xiao, X.; Yu, X.; Zhu, J.J.; Zhang, H. Machine Learning: New Ideas and Tools in Environmental Science and Engineering. *Environ. Sci. Technol.,* **2021**, *55*(19)acs.est.1c01339
http://dx.doi.org/10.1021/acs.est.1c01339 PMID: 34403250

[90] Regti, A.; Ayouchia, H.B.E.; Laamari, M.R.; Stiriba, S.E.; Anane, H.; Haddad, M.E. Experimental and theoretical study using DFT method for the competitive adsorption of two cationic dyes from wastewaters. *Appl. Surf. Sci.,* **2016**, *390*, 311-319.
http://dx.doi.org/10.1016/j.apsusc.2016.08.059

[91]   Paredes-Laverde, M.; Salamanca, M.; Diaz-Corrales, J.D.; Flórez, E.; Silva-Agredo, J.; Torres-Palma, R.A. Understanding the removal of an anionic dye in textile wastewaters by adsorption on ZnCl2 activated carbons from rice and coffee husk wastes: A combined experimental and theoretical study. *J. Environ. Chem. Eng.,* **2021**, *9*(4)105685
        http://dx.doi.org/10.1016/j.jece.2021.105685

[92]   El Kassimi, A.; Boutouil, A.; El Himri, M.; Rachid Laamari, M.; El Haddad, M. Selective and competitive removal of three basic dyes from single, binary and ternary systems in aqueous solutions: A combined experimental and theoretical study. *J. Saudi Chem. Soc.,* **2020**, *24*(7), 527-544.
        http://dx.doi.org/10.1016/j.jscs.2020.05.005

[93]   García-González, A.; Zavala-Arce, R.E.; Avila-Pérez, P.; Rangel-Vazquez, N.A.; Salazar-Rábago, J.J.; García-Rivas, J.L.; García-Gaitán, B. Experimental and theoretical study of dyes adsorption process on chitosan-based cryogel. *Int. J. Biol. Macromol.,* **2021**, *169*, 75-84.
        http://dx.doi.org/10.1016/j.ijbiomac.2020.12.100 PMID: 33338526

[94]   Ba Mohammed, B.; Lgaz, H.; Alrashdi, A.A.; Yamni, K.; Tijani, N.; Dehmani, Y.; El Hamdani, H.; Chung, I.M. Insights into methyl orange adsorption behavior on a cadmium zeolitic-imidazolate framework Cd-ZIF-8: A joint experimental and theoretical study. *Arab. J. Chem.,* **2021**, *14*(1)102897
        http://dx.doi.org/10.1016/j.arabjc.2020.11.003

[95]   Elias, M.; Uddin, M.N.; Hossain, M.A.; Saha, J.K.; Siddiquey, I.A.; Sarker, D.R.; Diba, Z.R.; Uddin, J.; Rashid Choudhury, M.H.; Firoz, S.H. An experimental and theoretical study of the effect of Ce doping in ZnO/CNT composite thin film with enhanced visible light photo-catalysis. *Int. J. Hydrogen Energy,* **2019**, *44*(36), 20068-20078.
        http://dx.doi.org/10.1016/j.ijhydene.2019.06.056

[96]   Aguiar, J.E.; Bezerra, B.T.C.; Braga, B.M.; Lima, P.D.S.; Nogueira, R.E.F.Q.; de Lucena, S.M.P.; José da Silva, I., Jr Adsorption of anionic and cationic dyes from aqueous solution on non-calcined Mg-Al layered double hydroxide: experimental and theoretical study. *Sep. Sci. Technol.,* **2013**, *48*(15), 2307-2316.
        http://dx.doi.org/10.1080/01496395.2013.804837

[97]   de Souza, T.N.V.; de Carvalho, S.M.L.; Vieira, M.G.A.; da Silva, M.G.C.; Brasil, D.S.B. Adsorption of basic dyes onto activated carbon: Experimental and theoretical investigation of chemical reactivity of basic dyes using DFT-based descriptors. *Appl. Surf. Sci.,* **2018**, *448*, 662-670.
        http://dx.doi.org/10.1016/j.apsusc.2018.04.087

[98]   Shen, T.; Wang, L.; Zhao, Q.; Guo, S.; Gao, M. Single and simultaneous adsorption of basic dyes by novel organo-vermiculite: A combined experimental and theoretical study. *Colloids Surf. A Physicochem. Eng. Asp.,* **2020**, *601*125059
        http://dx.doi.org/10.1016/j.colsurfa.2020.125059

SUBJECT INDEX

A

Ability 2, 3, 17, 79, 123, 127, 173, 187, 201, 211, 212, 282, 285
  light harvesting 187
  sensing 173
Absorbers 232, 257
  efficient 232
Absorption 25, 26, 127, 184, 211, 214, 223, 224, 225, 226, 227, 228, 232, 233, 234, 235, 238
  coefficient 224, 225
  measuring 235
  photon 184, 211
  spectroscopy 238
Acid 22, 31, 145, 185, 211, 243, 287
  acetic 145
  benzoic 243
  corrosion 185
  hydrofluoric 287
  stearic 22
  sulphuric 31
Activated carbons (ACs) 99, 274
Activity 11, 31, 128, 183, 189, 202, 203, 212, 218, 219, 222, 287
  antibacterial 222
  anti-bacterial 189
  photocatalytic 202
  photo-generated 219
Adsorbate equilibrium concentration 250
Adsorbents 232, 233, 234, 237, 239, 242, 243, 244, 248, 250, 251, 252, 275, 276
Adsorption 161, 233, 237, 238, 240, 241, 243, 245, 251, 262, 272, 273, 275
  and absorption 233
  catalysis 161
  efficiency 237, 243, 272, 273
  energy 241, 245, 275
  isotherms 237, 238, 240, 251, 262
  kinetics 273
Advanced oxidation processes (AOPs) 208
Air pollution 156

Amorphous carbon 161, 162, 174, 175, 176, 177
Analysis 238, 249, 251
  linear regression 238, 249, 251
  nonlinear regression 249
Applications 10, 21, 32, 33, 49, 93, 135, 161, 176, 186, 187, 193, 194, 195, 196, 200, 201, 274, 282
  adsorption-related 274
  high-performance engineering 187
Armchair nanotubes 166
Atomic 5, 76, 116, 121, 122, 123, 192
  force microscopy (AFM) 76, 116, 121, 122, 123
  layer deposition (ALD) 5, 192
  resolution 122
Avogadro number 227

B

Ballistic conductor 68
Band gap energy 212
Beer's law 223, 224
Biomedical sensors 202
Bohr radius and quantum confinement 60
Boltzmann 43, 72
  constant 43, 72
  relation 72
Bond dissociation energy 214, 215
Bonding force 16
Borohydride-induced catalysis process 270
Bravais lattices 38
Broglie relation 48

C

Capacities 224, 228
  absorbing 228
  absorptive 224

www.ingramcontent.com/pod-product-compliance
Lightning Source LLC
Chambersburg PA
CBHW050809220326
41598CB00006B/160